Results and Problems in Cell Differentiation

Series Editors:
W. Hennig, L. Nover, U. Scheer

32

Springer
Berlin
Heidelberg
New York
Barcelona
Hong Kong
London
Milan
Paris
Singapore
Tokyo

C. G. dos Remedios · D. D. Thomas (Eds.)

Molecular Interactions of Actin

Actin Structure and Actin-Binding Proteins

With 63 Figures

Springer

Dr. CRISTOBAL G. DOS REMEDIOS
University of Sydney
Department of Anatomy and Histology
Institute for Biomedical Research
Sydney 2006
Australia

Dr. DAVID D. THOMAS
University of Minnesota
Department of Biochemistry
Minneapolis, MN 55455
USA

ISSN 0080-1844
ISBN 978-3-642-53675-5

Library of Congress Cataloging-in-Publication Data

Molecular interactions of actin: actin structure and actin-binding proteins / Cristobal G. dos Remedios, David D. Thomas, (ed).
p. cm. – (Results and problems in cell differentiation; 32)
Includes bibliographical references and index.
ISBN 978-3-642-53675-5 ISBN 978-3-540-46560-7 (eBook)
DOI 10.1007/978-3-540-46560-7

1. Actin. I. Dos Remedios, Cristobal G. II. Thomas, David D.M.D. III. Series.

QP552.A27 M65 2000
572'.66 – dc21

Springer-Verlag Berlin Heidelberg New York
a member of BertelsmannSpringer Science+Business Media GmbH

Softcover reprint of the hardcover 1st edition 2001

Cover design: Meta Design, Berlin
Typesetting: Best-set Typesetter Ltd., Hong Kong
SPIN 10674819 39/3130 – 5 4 3 2 1 0 – Printed on acid-free paper

Contents

Two Conformations of G-Actin Related to Two Conformations of F-Actin
Edward H. Egelman and Albina Orlova

Actin Structure Function Relationships Revealed by Yeast Molecular Genetics
Lisa D. Belmont and David G. Drubin

Actin-Binding Proteins: An Overview
Enrique M. De La Cruz

An Overview of Actin Structure and Actin-Binding Proteins

Cristobal G. dos Remedios and Dave D. Thomas[1]

The Origins of this Book

The origins of this book and its companion volume go back more than 15 years to the first International Meeting devoted to the structure and function of actin. That conference, held in Sydney in 1982 as a satellite of the International Union of Biochemistry in Perth, Australia, resulted in the first Volume in this Series (dos Remedios and Barden 1983). Every 5 years since, an Actin-centric meeting has been held. In 1987, Roberto Colombo (1987) conducted the second meeting in Monza, Italy, and then in 1993 Jim Estes and his colleagues ran a large and successful third meeting in Rochester, New York (Estes et al. 1994).

The stimulus for this book and its companion volume (Molecular Interactions of Actin: Actomyosin, Motility Assays and Actin-Based Regulation) came from the Fourth Pentennial Actin Conference held at the new Sheraton Maui in Hawaii in 1998. Naturally, the venue was chosen purely because of its geographic convenience to the USA, Japan and Australia, but many actin researchers managed to sacrifice valuable research time to spend 6 days in beautiful Maui. Although our multinational Maui meeting formed the basis for the contents of these books, others who could not attend have made important contributions.

This first Volume covers the structure of actin, particularly the structure of filamentous actin for which there is a well-defined model but no categorical proof. This Volume also contains an up-to-the-minute account of how celllular actin microfilaments are regulated by actin-binding proteins.

A Historical Perspective of Actin Assembly and Its Interactions

Professor Fumio Oosawa, the man who has made the longest continuous contribution to our understanding of the biophysics of actin (Oosawa and Asakura 1975), and who is a living monument in Japan, recalls for us the major publications in the 1940s through the 1960s which led to his consuming interest in actin. In these days of Medline searching for information, it is quite easy to

[1] Institute for Biomedical Research, University of Sydney, Sydney 2006, Australia, and Department of Biochemistry, University of Minnesota Medical School, Minneapolis, MN 55455, USA

Results and Problems in Cell Differentiation, Vol. 32
C. dos Remedios (Ed.): Molecular Interactions of Actin

neglect the huge body of work from these early days. Oosawa's ideas, often considered to be radical, are contained in Chapter 2. He reminds us of how thermal fluctuations can be involved in the biophysical properties of F-actin, of the role played by free energy in determining the flexibility and intrapolymer mobility of actin filaments, and of the potential for F-actin to actively participate in the process of transfer, storage and release of the free energy. Oosawa also raises some difficult questions about the interaction of actin and myosin. For example, how does the effect of one bound myosin head spread over several actin monomers in F-actin? Is energy transferred from the myosin head to the actin filament and, if so, what happens to this stored energy? What kinds of "states" do actin and myosin assume during sliding? Is it possible to follow the states of individual monomers in F-actin during sliding? And finally, is the coupling between F-actin and myosin tight or loose?

Divalent Cations, Nucleotides and Actin Structure

Hanna Strzelecka-Golaszewska provides us with an insightful summary of the binding affinity and kinetic behaviour of actin's ligands (nucleotide and metal cation). These changes are examined in relation to changes in the domain structure of actin. Dr Strzelecka-Golaszewska explains the roles of cations (Ca^{2+} and Mg^{2+}) and nucleotide in stabilizing the structure of actin in terms of binding affinities and exchange kinetics. Polymerization rates, rates of treadmilling of monomers in F-actin, cooperativity of binding of myosin S1 to F-actin are all explained in terms of nucleotide-dependent and bound divalent cation-dependent changes in monomer conformation. This chapter contains a thorough review of the role of conformational changes associated with the binding of these ligands. This chapter should be read by anyone who wants to try their hand at actin-based experiments.

Helical Parameters of F-Actin Precisely Determined from X-Ray Fiber Diffraction of Well-Oriented Sols

Yuichiro Maéda and his colleagues used fiber X-ray diffraction to study highly oriented sols of F-actin to determine the helical parameters of the filaments. The resolution of these diffraction patterns is unprecedented. The angular distribution of filaments is <2°, the pitch of the one-start helix is 59.8 Å. They refined the helical parameters usually applied to F-actin (13 monomers in 6 turns) and show that the filaments actually have 67 monomers per 31 turns of the one-start helix. These data were obtained by aligning the filaments in a strong (13.5 Tesla) magnetic field and are destined to produce an atomic-resolution structure of F-actin. This should forever put an end to the continuing controversy between the inside-out model proposed by the Holmes laboratory and the outside-in model promoted by Schutt et al.

Analysis of Models of F-Actin Using Fluorescence Resonance Energy Transfer Spectroscopy

In this chapter, Pierre Moens and Cris dos Remedios describe a novel and powerful method of analyzing the structure of F-actin using fluorescence resonance energy transfer spectroscopy. The method measures the radial coordinates of known loci on actin monomers in F-actin and, using the known helical geometry of F-actin, it can sense conformational changes by measuring changes in radial coordinates of these loci. Traditional distracting factors such as doubts over the orientation factor in FRET can be disregarded in F-actin because the helical symmetry provides sufficient acceptor orientations relative to the donor emission dipoles to achieve effective randomness. The radial coordinates of actin monomer loci are compared with structures of three models of F-actin.

Microscopic Analysis of Polymerization and Fragmentation of Individual Actin Filaments

Dr. Shin'ichi Ishiwata and his colleagues describe how to directly observe the polymerization-depolymerization dynamics of individual actin filaments. The polymerization and fragmentation processes of individual actin filaments can be visualized by imaging the polymerized filaments through binding of fluorescent rhodamine-phalloidin, using a conventional fluorescence microscope. Thus, the polymerization process at the barbed and pointed ends of actin filaments can be analyzed in the presence of Mg^{2+} or Ca^{2+}. By measuring the rate of polymerization at various G-actin concentrations, these authors confirmed that the depolymerization process was inhibited by the attachment of phallotoxins to the filaments (phallacidin was more effective than phalloidin), but the polymerization rate itself was not significantly affected, such that the critical concentration of polymerization was essentially zero. Also, the fragmentation process of filaments due to the addition of high concentrations of chaotropic salts such as KSCN and KI (except Pi) can be observed, and stabilization of the hydrophobic interactions by Pi was confirmed. Their study demonstrates that the polymerization and fragmentation dynamics of single actin filaments can be examined quantitatively by fluorescence microscopy.

Two Conformations of G-Actin Related to Two Conformations of F-Actin

Ed Egelman and Albina Orlova remind us in Chapter 7 that electron microscopic studies have shown that the actin filament can exist in a number of different conformational states. Since the dynamic properties of actin filaments are likely to be important in a range of processes, from muscle contraction to the control of cell form, these authors have been interested in interpreting the different conformational states of F-actin seen at low resolution in terms of molecular models. They show that two different G-actin structures, an open and a closed state model, demonstrate the opening of the nucleotide-binding cleft seen in yeast actin following ATP hydrolysis and the release of phosphate.

Actin Structure Function Relationships Revealed by Yeast Molecular Genetics

Here, Lisa Belmont and David Drubin summarize the reasoning behind using the yeast to study the structure and function of actin. They review the large range of actin mutants that have been used to: (1) probe the amino acid residues in the nucleotide-binding cleft; (2) test possible movement the hydrophobic putative plug between monomers in opposite strands of the long-pitch helix; (3) compare predictions of the Holmes et al. model of F-actin and Schutt et al. ribbon structure; (4) map the binding sites of actin-binding proteins. This chapter emphasizes the power of mutagenesis studies of yeast actin in studying both the structure and functions of actin in vitro and in vivo.

Actin-Binding Proteins: an Overview

In this overview of actin-binding proteins (Chap. 9), Enrique De La Cruz raises five questions: (1) What are the identities of all actin-binding proteins (Components)? (2) How do they associate with actin and each other (Interactions)? (3) How are their activities modified (Regulation and Dynamics)? (4) What do they look like (Structure)? and (5) What roles do they play in living systems (Cells and Organisms)? Although these questions do not cover all aspects of actin-binding proteins, they provide a rigorous framework for understanding their mechanisms of action at the molecular level and in living systems.

The ADF/Cofilin Family: Accelerators of Actin Reorganization

Amy McGough and her colleagues are amongst the most active groups in investigating the properties of cofilin and its effects on actin. Cofilin belongs to a superfamilty of proteins called the ADF (actin deolymerizing factor) family. The authors discuss the relationships between members of this family and they point out that although cofilin is apparently present in all eukaryotic cells, there are examples of vertebrate cells which contains *both* ADF and cofilin. What is it about vertebrates that apparently requires this protein redundancy? The sequences of members in the ADF family are highly conserved. All contain the sequence: ${}_{30}$KKRKK${}_{34}$, but there are other segments of the sequence which are highly conserved. So far, unlike DNase I, gelsolin and profilin, there are no cocrystals of actin with cofilin. Consequently, mutational studies, chemical cross-linking and peptide mapping have been used to define the sites on cofilin that may interact with actin. The authors discuss the nature of the single actin-binding site on cofilin and how it accomplishes the binding to either G- or F-actin. Cofilin is unique among F-actin-binding proteins in that it changes the angle of rotation between subunits along the genetic helix of F-actin. Finally, the authors discuss how cofilin function is regulated in vivo and how cofilin regulates the turnover of actin in cells. There is an extensive bibliography.

Predicting Interaction Sites Between Glycolytic Enzymes and Cytoskeletal Proteins Employing the Concepts of the Molecular Recognition Theory

Ron Sheedy and Frank Clarke remind us of the molecular recognition theory of Blalock and its potential for predicting, and so assisting in, the identification of interaction sites between glycolytic enzymes and cytoskeletal proteins. This theory represents a radical new approach to determining the binding sites of protein pairs by examining their genetic sequences. This theory has recently been validated for the binding of certain peptides to their receptors. It proposes that the binding sites of proteins have evolved from peptides originally specified by complementary nucleotide sequences. The location of vestigial complementarity between the genes of cognate binding proteins may indicate the location of the interaction sites on the proteins. If the binding sites identified in this investigation are further substantiated by subsequent experimentation, this will provide additional support for the molecular recognition theory. Further interactions between glycolytic enzymes and actin may be delineated, as well as possible interactions between the enzymes themselves. Elucidation of the binding sites may lead to the application of molecular techniques to probe the significance of these interactions both in vitro and in vivo.

Regulation of the Cytoskeleton Assembly: a Role for a Ternary Complex of Actin with Two Actin-Binding Proteins

Cofilin is widely distributed in eukaryotic cells and plays a crucial role in regulating the assembly and disassembly of the actin cytoskeleton. In this chapter, Murat Kekic et al. present evidence that cofilin may not act alone in this regulatory behaviour. Using a model system consisting of actin, DNase I and cofilin, they show that, although these two actin-binding proteins attach to quite separate parts of the actin monomer, never-the-less, DNase I allosterically increases the binding of cofilin to actin and vice versa. The evidence is based on increased stability of the DNase-actin-cofilin complex observed by native polyacrylamide gel electrophoresis. Conformational changes in actin are detected based on changes in the spectral properties of fluorescent probes attached to glutamine-41 and cysteine-374 of actin as well as on fluorescence resonance energy transfer spectroscopy which senses distances between these two loci. We suggest that this in vitro cooperative behaviour may be important in regulating the state of assembly of the actin cytoskeleton.

Structure of Gelsolin

While gelsolin is probably not essential for survival, it does play important roles in regulating the actin cytoskeleton as well as in a number of other functions including apoptosis. The chapter by Les Burtnick et al. contains a description of the complete structure of the six subdomains of gelsolin. They guide the reader through the main structural and function parts of this important actin-binding protein including subdomains 1 and 2 and subdomain 4 which compete for binding to the same site on actin but interact sequentially to sever and then cap the actin filament.

Actin Filament Networks

Tom Stossel and his colleagues describe the biophysics underlying the unusual viscoelastic properties of polymerized actin. These were first recognized more than 50 years ago and have motivated studies of actin filament structure that suggest how the networks formed by these filaments produce the resulting macroscopic rheologic features. From a biological perspective, the relation between single filament structure and network properties is important to predict how activation of specific actin-binding proteins might alter the stiffness of cellular actin networks, and so modulate motility and shape change. From the perspective of material science, F-actin represents a class of semi-flexible polymers that form networks with viscoelastic properties distinct from those of most common synthetic, flexible polymers. The tools developed to visualize actin filaments (e.g., see Chap. by Ishiwata et al.) have enabled studies of single polymer conformation and motion within actin networks that are not yet possible with synthetic polymers.

ARPS: Actin-Related Proteins

Laura Machesky and Robin May review the Arps (*a*ctin-*r*elated *p*roteins) that comprise a recently discovered family of proteins related to actin in sequence and probably in 3D structure. They are multisubunit protein complexes that perform diverse functions. Arp1 is the dynactin complex and is involved in organelle trafficking, in the positioning of the nucleus, and in the assembly and orientation of the mitotic spindle. Arp1 was originally discovered as an activator of dynein-based vesicle motility, hence the name *dyn*ein-*act*ivator, not to be confused with actin which plays no role in this complex. The Arp2/3 complex plays a role in coordinating the positioning and assembly of the actin cytoskeleton. It is also probably capable of nucleating the assembly of cytoplasmic actin as well as cross-linking these filaments. In yeast, the Arp2/3 complex is important for endocytosis, actin patch localization and motility, and membrane growth and polarity. Arp4, Arp7 and Arp9 have been implicated in chromatin structure and function in the nucleus. The functions of the remaining ARPs in this ten-member family are unknown.

Extracellular Signals and the Cytoskeleton

Finally, Michael Hall and his colleagues (Beck et al.) review the mechanisms whereby various signals from outside the cell control the assembly of the internal cytoskeleton. From the above chapters, it is clear that the cytoskeleton is a highly malleable structure that can be rapidly remodeled. This structure can respond to several different types of extracellular signals. This chapter discusses selected examples of such signals and the pathways by which they impinge on the actin cytoskeleton to elicit changes in cell shape or movement. In the yeast *Saccharomyces cerevisiae*, mating pheromones, nutrients, and environmental stress control polarization of the actin cytoskeleton and polarized cell growth. In *Drosophila*, developmental signals control gastrulation and dorsal closure. In mammalian cells, antigens cause T-cell polarization, growth factors control the formation of motile structures as well as cell migration. Also in mammalian cells, extracellularly bound bacterial pathogens can subvert the host actin cytoskeleton to enter cells. In most cases, if not all, the signaling pathways activated by these varied extracellular signals involve Rho-type GTPases.

References

Colombo R (1987) Actin '87 International meeting on the biology of an ubiquitous protein. Monza, 1987, 162 pp

dos Remedios CG, Barden JA (1983) Actin structure and function in muscle and nonmuscle cells. Academic Press, Sydney, 336 pp

Estes JE, Higgins PJ (1994) Actin biophysics, biochemistry, and cell biology. Plenum Press, New York, 236 pp

Oosawa F, Asakura S (1975) Thermodynamics of the polymerization of actin. Academic Press, London, 204 pp

A Historical Perspective of Actin Assembly and Its Interactions

Fumio Oosawa[1]

Introduction

Actin was discovered by F. B. Straub more than 50 years ago as one of the main components of muscle proteins and as a partner of myosin for superprecipitation coupled with the hydrolysis of ATP (Straub 1942). When extracted from dried muscle powder into water, actin was in the state of dispersed monomers, named G-actin, and transformed to fibrous polymers, named F-actin, in a solution of neutral salts. With removal of salts, F-actin returned to G-actin. Later, he noticed that this G-F transformation was coupled with the hydrolysis of ATP bound to G-actin; ADP produced were kept bound to F-actin (Straub and Feuer 1950). The reverse transformation from F to G was not associated with rephosphorylation of bound ADP; G-actin released ADP and bound new ATP in solution.

Albert Szent-Györgyi's superprecipitation, the in vitro contraction, required a combination of two proteins, actin and myosin, both in the state of polymers (Szent-Györgyi 1951). Muscle contraction seemed to be due to a change in the structure formed by two kinds of polymers. In 1954, based on optical microscopic and electronmicroscopic observations of the striated muscle fiber, A. F. Huxley, and H. E. Huxley and J. Hanson proposed the sliding between thick filaments (mainly composed of myosin) and thin filaments (mainly composed of actin) as the mechanism of muscle contraction (Huxley and Niedergerke 1954; Huxley and Hanson 1954).

In the same year, we started to study actin in Department of Physics of Nagoya University. Our questions were: How does the G-F transformation occur? What kind of polymer structure does actin form? What is the function of actin in muscle contraction or sliding? Fortunately, Straub's method of preparation of actin from muscle was not too difficult for us. First, we attempted to make a map on the state of actin in all possible environmental conditions.

Here, I will briefly describe a historical overview of actin research mostly drawn from my own research interests. More details have been published elsewhere (Oosawa and Asakura 1975; Oosawa 1983, 1993, 1995).

[1] Aichi Institute of Technology, Yagusa, Toyota 470-0392, Japan

Results and Problems in Cell Differentiation, Vol. 32
C. dos Remedios (Ed.): Molecular Interactions of Actin

The G-F Transformation

In pure water, all actin molecules are in the state of G-actin, and at physiological salt concentration, all are in the state of F-actin. What happens in the intermediate salt concentration? Is an intermediate structure between G- and F-actins formed? Or is there a mixture of G- and F-actins? Our experiments gave the following results (Oosawa et al. 1959). At low concentrations of actin, no F-actin exists and all actin molecules are in the state of G-actin. At a certain critical concentration of actin, F-actin begins to be formed. With increasing concentration of actin, the amount of F-actin increases, coexisting with G-actin, the concentration of which is kept at the critical concentration. The steady hydrolysis of ATP in this condition demonstrated that actin molecules undergo the cycle between two states, G- and F-actins; that is, G-actin associates to and dissociates from F-actin repeatedly (Asakura and Oosawa 1960). The macroscopic G-F balance is maintained by such microscopic cycling. The critical concentration is high in the absence of salts and very low at physiological salt concentration.

Thus, the G-F transformation has features similar to the gas-liquid condensation or the crystallization. Actually, the transformation was proved to consist of two processes, nucleation and growth (Kasai et al. 1962). Usually, the nucleation by a few actin molecules is rate-limiting. Addition of small fragments of F-actin greatly accelerates the transformation.

We showed that such crystallization-like features of the G-F transformation are theoretically understandable by assuming the helical polymerization of G-actin, where each actin monomer is bound with four neighboring monomers through two kinds of bonds (Oosawa et al. 1961; Oosawa and Kasai 1962). Both thermodynamics and kinetics of the transformation were described well by this theory. In 1963, J. Hanson produced beautiful electron microscopic pictures that showed a two-stranded helical polymer structure of F-actin (Hanson and Lowy 1963) (Fig. 1a).

In 1965, we summarized the role of nucleotides in the G-F transformation as follows (Kasai et al. 1965). In the presence of ATP, G-actin binds ATP and

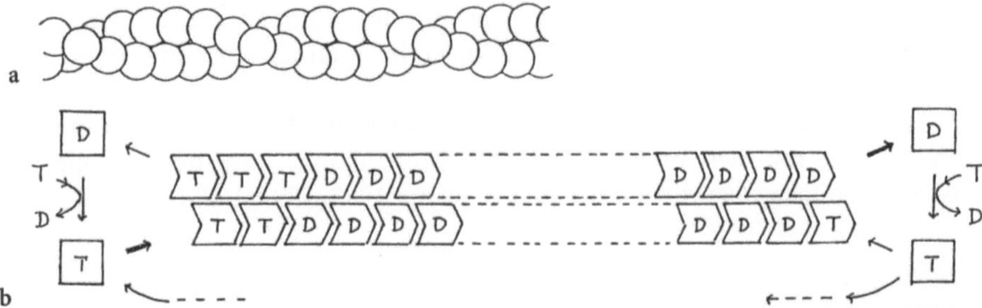

Fig. 1. **a** F-actin as a two stranded helical polymer. **b** Bound nucleotides (T: ATP; D: ADP) near the growing ends of F-actin

when it polymerizes to F-actin, ATP is hydrolyzed. However, G-actin containing ADP instead of ATP also can polymerize, although the rate of its polymerization is much slower than G-actin with ATP. Even G-actin without ATP or ADP can polymerize, if denaturation of this nucleotide-free G-actin is inhibited by a high concentration of sucrose. Bound nucleotides in F-actin are not easily exchangeable with those in the solution. They are a stabilizer of the structure of the actin molecule and an allosteric regulator of the polymerization rate and monomer-monomer bond strength. Divalent cations, calcium ions or magnesium ions, bound to nearby nucleotides are another regulator.

Later, Carlier et al. found a short time lag between the polymerization of G-ATP-actin and the hydrolysis of bound ATP (Carlier et al. 1984; Carlier 1990). In such a case, near the growing ends of F-actin, there are three kinds of actin monomers, those having ATP, those having ADP and inorganic phosphate, and those having ADP alone (Fig. 1b).

F-Actin Structure and Bond Free Energy

F-actin, as a helical polymer, was assumed to be constructed using two kinds of bonds between neighbouring monomers, one along the longitudinal strands and the other between two strands i.e., along the genetic helix. This bonding pattern was finally confirmed by the recent structural analysis at atomic resolution. (Fig. 2) In 1990, the three-dimensional crystal structure of actin complexed with DNase-I was determined by X-ray crystallography in Heidelberg (Kabsch et al. 1990). The actin molecule appeared to be composed of four regions called subdomains. ATP or ADP was found in a deep cleft in the middle of the molecule. Using this molecular structure, K. C. Holmes built up a structure of F-actin that gave the best fit to the X-ray diffraction data from an oriented gel of F-actin (Holmes et al. 1990). Based on this structure, we can now discuss which amino acid residues in the actin molecule are involved in the

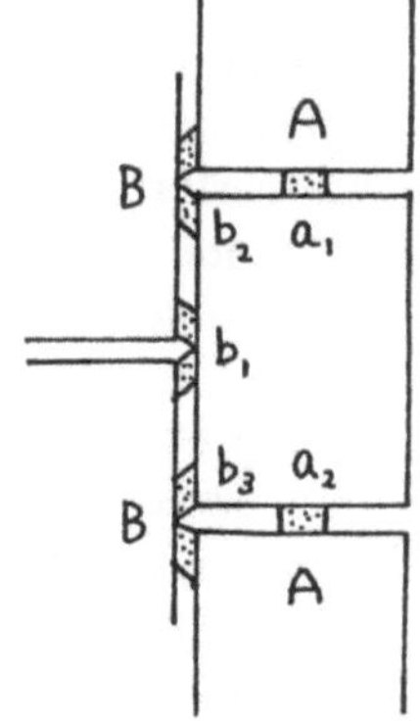

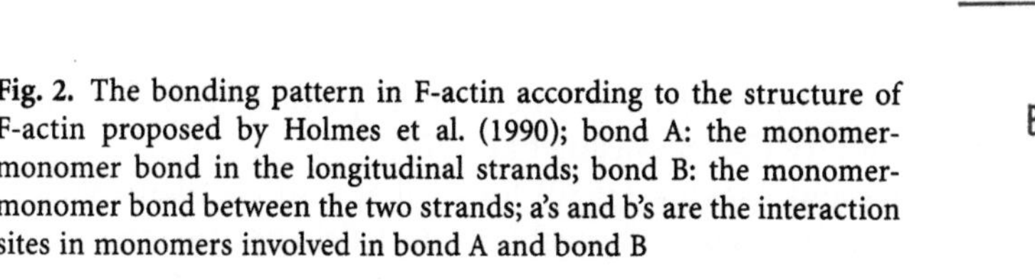

Fig. 2. The bonding pattern in F-actin according to the structure of F-actin proposed by Holmes et al. (1990); bond A: the monomer-monomer bond in the longitudinal strands; bond B: the monomer-monomer bond between the two strands; a's and b's are the interaction sites in monomers involved in bond A and bond B

two kinds of bonds between monomers in F-actin. The role of each amino acid residue can be examined by replacing it with other residues arbitrarily chosen, using gene technology methods.

F-actin has a structural polarity, as was first demonstrated by the way it bound myosin fragments, HMM or S-1 (Huxley 1963). The growth rate of F-actin was found to be much larger at the end named B-end than at the other end named P-end (Kondo and Ishiwata 1976). The depolymerization rate of F-actin was also different at the two ends. However, in the case of polymerization which is not accompanied by the ATP hydrolysis the ratio of the rates of association and dissociation of actin monomer is equal at the two ends and gives the critical concentration of G-actin in equilibrium with F-actin. Then, the critical concentration is related to the free energy of bonds in F-actin, two per monomer. The free energy was estimated to be around −6 to −13 kcal mole^{-1} (Oosawa and Asakura 1975). Its absolute value increases with increasing salt concentration up to the physiological concentration. It depends also on the species of bound nucleotides and divalent cations. There is a strong correlation among bound nucleotides or cations, monomer conformation, and monomer-monomer bonds.

The critical concentration decreases with rising temperature, provided that the temperature is not too high (Asakura et al. 1960). Polymerization is endothermic and the bond free energy comes from a large entropy increase. We do not yet know the structural origin of energy and entropy changes in polymerization, but probably the rearrangement of water molecules around actin monomers is involved in the process.

Dynamics of F-Actin

When we reached the idea of helical polymerization, immediately we imagined that if one of the two kinds of bonds in F-actin was broken or weakened, a large conformational change would occur in F-actin to maintain its structural continuity (Oosawa et al. 1961; Asakaura et al. 1963); (Fig. 3a,b). The monomer-monomer bond and also the monomer structure in F-actin must have thermal fluctuations.

The first successful experiment to detect thermal movements in F-actin was performed in 1970 by S. Fujime using the quasielastic light-scattering technique (Fujime 1970). The thermal bending movement of F-actin made remarkable contributions to the frequency broadening of the light scattered from an F-actin solution. F-actin was a flexible filament and the average amplitude of bending was estimated to be about 40 nm for F-actin of length 1 μm. The curved shape of F-actin in electron microscopic pictures was also taken as evidence of its flexibility (Takebayashi et al. 1977).

In 1980, Nagashima and Asakura (1980) found that a single filament of F-actin, if it was long enough, was directly visible by dark field optical microscopy

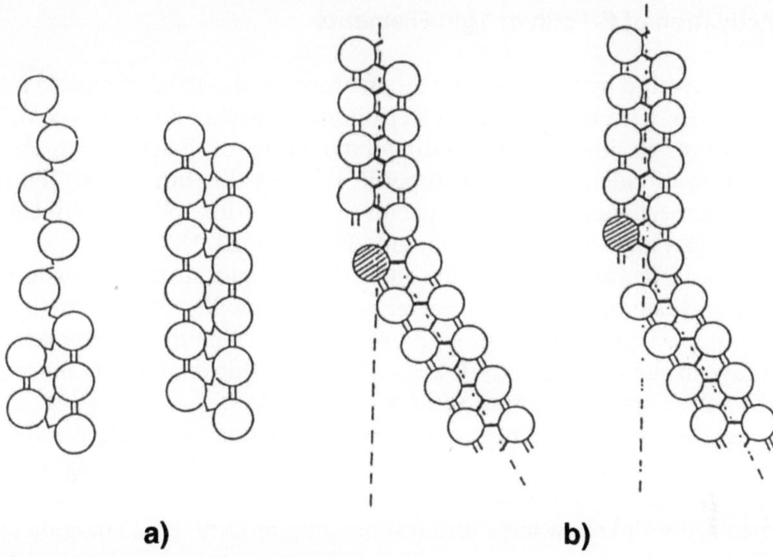

Fig. 3a,b. Hypothetical conformational changes in F-actin produced by breaking or weakening of one of the two kinds of bonds. (Oosawa et al. 1961; Asakura et al. 1963)

after decoration with myosin fragments. As expected, F-actin showed thermal bending movements. Later, by labeling with fluorescent molecules, particularly with fluorescent phalloidin, Yanagida et al. (1984) made F-actin easily observable by fluorescence microscopy. All previous results on flexibility were confirmed by direct observations.

The rigidity decreased or flexibility increased by raising the temperature in contrast to endothermic polymerization (Takebayashi et al. 1977). The recovery force from bending was due to an energy drop. The short-range elasticity and the long-range interaction between actin monomers in F-actin have different characters. The energy and entropy are complex functions of the monomer-monomer distance and orientation.

In 1988, Kishino and Yanagida (1988) captured the ends of an F-actin filament on thin glass needles and stretched the filament to determine the force necessary to break the monomer-monomer bonds in F-actin. This was the first experiment to use micromanipulation of single F-actin filaments in solution. F-actin was extensible as expected from its flexibility (Kojima et al. 1994). X-ray diffraction of the thin filaments in a muscle fiber showed that F-actin or the thin filaments were appreciably stretched when the fiber generated a tension isometrically (Huxley et al. 1994; Wakabayashi et al. 1994).

It should be noted that, because of thermal fluctuations, actin monomers in F-actin are not always in the same physical state.

Activation of F-Actin or Thin Filaments

As discovered by S. Ebashi, the thin filament in the striated muscle is a complex of F-actin with tropomyosin and troponin (Ebashi and Endo 1968). Troponin-tropomyosin inhibits the sliding interaction of F-actin with myosin, and binding of calcium ions to troponin releases this inhibition. The steric hindrance mechanism was proposed for this inhibition, although conclusive structural data have not yet been obtained (Huxley 1973).

It was demonstated by the quasielastic light scattering experiment that tropomyosin and troponin decreased the flexibility of F-actin (Ishiwata and Fujime 1972; Oosawa et al. 1973). Addition of calcium ions recovered the flexibility to the level of pure F-actin. Thus, the thin filament is more flexible in the active than in the resting state.

As soon as it was made possible to directly observe single F-actin filaments, we attempted to see what happens when soluble myosin fragments, S-l, and ATP are added to a solution of F-actin. F-actin did not exhibit sliding movements, but did show large and fast bending movements (Yanagida et al. 1984). Their amplitude was increased and the apparent period was shortened by adding S-l and ATP. This means that the bending movements of F-actin were activated using the free energy liberated by the ATP hydrolysis. In the case of the F-actin-tropomyosin-troponin complex, such activation occurred only in the presence of calcium ions.

In some versions of the sliding filament mechanism, F-actin was treated as if it were a rigid rod. For instance, if myosin molecules walk on F-actin during sliding, F-actin is assumed to be rigid. For walking, solid ground would be better. However, the above results indicated that the flexibility or intrapolymer mobility of F-actin increased upon activation. Why?

In a molecular machine of sliding, the input free energy of ATP hydrolysis is not very much larger than the energy of thermal fluctuation. Under such conditions, it is likely that a thermal fluctuation in the machine is positively incorporated into the mechanism of free energy conversion.

Sliding of F-Actin on Myosin

In 1985, 30 years after the proposal of the sliding mechanism, Yanagida et al. (1985) made direct investigations of the input-output relation in the molecular machine of sliding. They used glycerinated crab leg muscle fibers which had long thick and thin filaments aligned in parallel. After addition of ATP, the sliding velocity of thin filaments into the thick filaments was measured under a fluorescence microscope and ATP hydrolysis was followed chemically. From these results, we estimated the sliding distance of a thin filament produced by a myosin head during hydrolysis of one ATP molecule. It was about 60 nm, which was, surprisingly, much longer than either the size of a myosin head or an actin monomer. The free energy of ATP hydrolysis is about 20 kT under the ordinary conditions. The long sliding distance per ATP molecule means that the free energy used for sliding 5 nm, the distance between actin monomers, is of the same order as the average energy of the thermal fluctuation.

The sliding distance of F-actin produced by the hydrolysis of one ATP molecule on a myosin head seems to be widely variable, from 0 nm under a heavy load to 60 nm under no load. Thus, the input-output coupling in sliding is not tight but loose.

This work raised a previously unrecognized and serious question concerning input-output coupling during sliding. To solve the coupling mechanism, it was necessary to observe directly the sliding behavior of single F-actin filaments and myosin molecules.

In an internodal cell of *Nitella*, we can see fast circulation of cytoplasm caused by interaction between F-actin fixed on the gel layer and myosin molecules in the cytoplasm (Kamiya and Kuroda 1956). Later, Higashi-Fujime (1980) observed under an optical microscope that a bundle of F-actin removed from this cell onto a glass slide undergoes fast translational and rotational movements in an artificial solution containing ATP. In this experiment, however, localization of myosin was not confirmed.

Then, in 1986, Spudich and Kron (1986) invented a method convenient for constructing the unitary sliding machine. They fixed myosin heads on a glass plate and applied a solution of F-actin. Upon addition of ATP, fluorescently-labelled single F-actin filaments showed unidirectional sliding movements. If myosin heads were suitably treated fixed to a surface, the sliding velocity of F-actin on them was comparable to that in a muscle fiber. (Harada et al. 1990) Yanagida et al. attached one end of F-actin to a very thin glass needle and placed it on the fixed myosin heads. (Ishijima et al. 1994). After addition of ATP, the sliding of F-actin of nanometers and the force for sliding of piconewtons were simultaneously measured. Single events of sliding of F-actin on one myosin head were observed.

Then, an optical microscopic technique was developed to detect single, fluorescently labeled myosin molecules and single fluorescent ATP-analogue molecules bound to myosin heads (Funatsu et al. 1995). The binding of one ATP-analogue to a myosin head and dissociation of the product after hydrolysis from that head were made observable. Combining these techniques, recent experiments have provided direct evidence for a loose coupling between the ATP hydrolysis and the sliding (Kitamura et al. 1999). During or after the hydrolysis of one ATP molecule, a myosin head slides over several actin monomers along F-actin. (For more detailed discussion of this approach see Ishii et al. 2nd Vol. this Series).

Pathway of Free Energy Conversion

Loose coupling means that the molecular machine that produces sliding is not a rigid mechanical device but a flexible, statistical mechanical machine. The free energy of ATP hydrolysis depends on the concentrations of ATP and ADP. What happens if this free energy is decreased by decreasing the ATP concentration and increasing the ADP concentration? Large unidirectional sliding movements of F-actin are expected to approach to small bi-directional thermal movements. However, this kind of experiment is not easy in practice. If this scenario is correct, it is unlikely that the sliding movement of F-actin is directly caused by a conformational change of a myosin head coupled to a definite step in the ATP hydrolysis reaction (Oosawa 1995).

The long-distance sliding between F-actin and a myosin head produced by hydrolysis of one ATP molecule suggests that the free energy of ATP hydrolysis is transferred and stored somewhere in the myosin head (and/or F-actin), then divided into small pieces of arbitrary size and released either gradually or stepwise depending on the mechanical work needed. Actually, recent experiments showed that there is often a time lag between the start of sliding after the product release of ATP hydrolysis on myosin head (Ishijima et al. 1998). Does the myosin head store the free energy of ATP hydrolysis (Huxley 1998)? Or does F-actin also actively participate in the process of transfer, storage and release of the free energy?

Several years ago, we (Vale and Oosawa 1990) proposed a thermal ratchet mechanism for sliding. A localized high temperature state and an asymmetric flexible structure were assumed in the machine to transform thermal fluctuation to unidirectional sliding. This mechanism can explain the observed relations among the sliding velocity, the load and the ATP hydrolysis. If this kind of mechanism does work, the question is how and where the free energy of ATP hydrolysis is transferred and stored to realize an effective high temperature state in a part of the machine. Up to now, various kinds of thermal ratchets have been discussed in relation to the sliding (Magnesco 1993; Faucheux et al. 1995).

The State of F-Actin During Sliding

Conformational differences in F-actin have been detected, depending on the species of bound nucleotides and divalent cations (Orlova and Egelman 1995; Orlova et al. 1995; also see Egelman and Orlova, this Vol.). Binding of myosin heads to F-actin changes the conformation of F-actin (Oosawa 1983). The effect of one bound head spreads over several actin monomers in F-actin. As noticed before, it is also probable that each actin monomer in F-actin is not in a fixed conformation but thermally fluctuates among different conformations.

What happens when F-actin interacts with free myosin heads during hydrolysis of ATP? Bending movements of F-actin are activated. A hypothetical temperature of the bending freedom of F-actin was estimated to be three to four times higher than room temperature. The free energy of ATP hydrolysis was partially transferred to the bending movements and it was not instantaneously consumed as heat. The twisting movements of F-actin were also

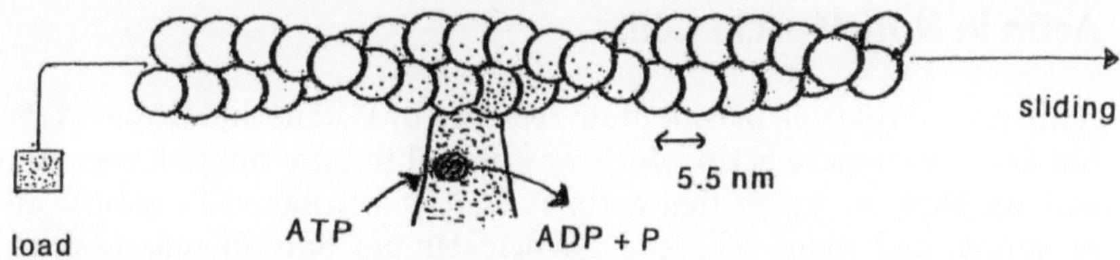

Fig. 4. Sliding of an F-actin filament on a myosin head coupled with the ATP hydrolysis; the state of each actin monomer may change during sliding

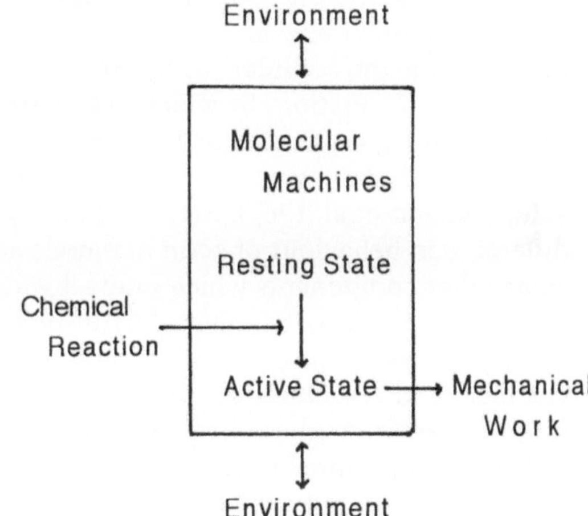

Fig. 5. The input-output (the chemical reaction to the mechanical event) relation in the sliding machine, where the output is controlled by the internal state of the machine

activated by interaction with free myosin heads hydrolyzing ATP. Therefore, the transfer of free energy may occur to various degrees of freedom in F-actin. In the case of sliding on myosin heads fixed on a plate, the free energy must be transferred and stored in a form convenient to be utilized for translational movements.

In a molecular sliding machine, the relation between the chemical reaction and a structural change and force generation is not straightforward (Oosawa 1995). To understand the mechanism of chemomechanical free energy conversion, it seems useful to introduce the concept of "state". Let us consider, for example, the case of superconductivity. Even if the structure of a high temperature superconductor was determined at atomic resolution, the mechanism of superconductivity is not readily understandable. We have to know what kind of state is established in the superconductor for superconductivity.

Similarly, our question is: what kind of states do myosin heads and actin molecules in F-actin assume during sliding? New active states are expected to appear in both. Is it possible to follow the changes of the states of individual actin molecules in F-actin successively sliding on myosin heads?

Actin in Non-Muscle Cells

Actin from *Physarum* plasmodium reported by Hatano and Oosawa (1966) was the first non-muscle actin which was proved to have full G-F transformation activity. Now we know that actin is widely distributed in almost all kinds of animal and plant cells and participates not only in muscle contraction and other forms of cell motility but also in various cellular activities and their regulation.

In general, actin molecules are in a dynamic balance between G- and F-actins. In a muscle fiber, the critical concentration of actin for the G-F transformation is very low and F-actin is stable. In other nonmuscle cells like *Amoeba*, the intracellular structure is reversibly changed by the so-called sol-gel transformation, in which a G-F transformation probably takes part. The critical concentration of *Physarum* actin for the G-F transformation was approximately the same under the same environmental condition as muscle actin (Hatano et al. 1967). Actin is a highly conserved protein. Therefore, the difference in behaviour of actin in muscle and nonmuscle cells must be due to some other components which interact with actin.

Since then, a large number of actin-binding proteins have been discovered (Hatano 1994; see also the Actin-Binding Protein section, this Vol.) and biochemical and structural studies have defined their functions. Some of them bind to G-actin to inhibit its polymerization. Some bind to one of the two ends of F-actin to promote nucleation or inhibit its growth. Some break or weaken the monomer-monomer bonds in F-actin. Others bind to the side of F-actin, causing cross-linking or bundling of F-actin. Using these proteins, cells control the G-F transformation, the amount, length, localization, flexibility and conformation of F-actin, and the assembly of F-actin.

In nonmuscle cells also, intracellular movements are often due to sliding between F-actin and myosin molecules or thick filaments of myosin. However, in those cells we can detect a motility of actin other than sliding. The polymerization and depolymerization at the ends of F-actin provides a force to move intracellular structures or membranes (Tilney et al. 1992). F-actin itself can translate by treadmilling. If the strength of the two kinds of bonds in F-actin is selectively controlled, a large conformational change of F-actin may take place. Formation and destruction of crosslinks of F-actin results in movements of the network structure (see also Janmey et al. this Vol.). Actin is involved in various kinds of motility and morphological changes in living cells.

Future of Actin Research

Since the three-dimensional structure of the actin molecule was determined, many experiments have attempted to find the structural basis of various properties of actin. The structure of the actin-myosin complex has not yet been solved at atomic resolution. However, the analysis of binding of a myosin head to F-actin seen by electron microscopy has been based on the separately determined structures of the myosin head and a model of F-actin. One of the principal lines of actin research is the structural study on the interaction of actin with myosin, tropomyosin, and many other actin-binding proteins.

Concerning the motility of actin, we need to know not only the structure, but also its dynamics and energetics. Protein molecules in a molecular machine are not like parts of a macroscopic machine. Energy and entropy in parts of the machine may be transiently incorporated into the process of free energy conversion. In this sense, thermodynamic and statistical mechanical studies on the state of actin in the machine must proceed in parallel to structural studies.

New experimental techniques have been developed to directly observe the behavior of single protein molecules in solution. As described above, these techniques have made it possible to make simultaneous measurement of mechanical and chemical events at the level of single molecules. Instead of a huge number of molecules in solution, we can now treat single molecules and follow their dynamics. This opens a new, very powerful line of actin research to understand the mechanism of actin-based motility.

To end this chapter, the great advances in our knowledge under the influence of actin research must be mentioned. (1) Thermodynamic and kinetic studies of the G-F transformation of actin have provided a standard procedure to treat the formation of fibrous polymers of various proteins. (2) Studies of the dynamic properties of F-actin have given fundamental data to an understanding of the rigidity of globular protein molecules and the bonds between them. (3) Actin is also a nice example of how the three-dimensional structure of a protein molecule can be used for interactions with many different kinds of proteins. (4) The input-output coupling in the sliding mechanism of myosin and F-actin is loose. Living cells have many kinds of sliding or rotary machines composed of different pairs of proteins. Therefore, the same question should be asked (Oosawa and Hayashi 1986): is the input-output coupling tight or loose?

References

Asakura S, Oosawa F (1960) Dephosphorylation of ATP in actin solutions at low concentration of Mg ions. Arch Biochem Biophys 87:273–285

Asakura S, Kasai M, Oosawa F (1960) The effect of temperature on the equilibrium state of actin solutions. J Polym Sci 44:35–49

Asakura S, Taniguchi M, Oosawa F (1963) Mechanochemical behavior of F-actin. J Mol Biol 7:55–69

Carlier M-F (1990) Actin polymerization and ATP hydrolysis. Adv Biophys 26:51–73

Carlier M-F, Pantaloni D, Korn E (1984) Evidence for an ATP cap at the ends of actin filaments and its regulation of the F-actin steady state. J Biol Chem 259:9983–9986

Ebashi S, Endo M (1968) Calcium ions and muscle contraction. Prog Biophys Mol Biol 18:123–183

Faucheux L, Bourdieu L, Kaplan P, Libchaber A (1995) Optical thermal ratchet. Phys Rev Lett 74:1504–1507

Finer J, Simmons R, Spudich JA (1994) Single myosin molecule mechanics; picoNewton forces and nanometer steps. Nature 369:113–119

Fujime S (1970) Quasielastic light scattering from solutions of macromolecules, II Doppler broadening of light scattered from solutions of semi-flexible polymers, F-actin. J Phys Soc Jpn 29:751–759

Funatsu T, Harada Y, Tokunaga M, Saito K, Yanagida T (1995) Imaging of single fluorescent molecules and individual ATP turnovers by single myosin molecules in aqueous solution. Nature 374:555–559

Hanson J, Lowy J (1963) The structure of F-actin and actin filaments isolated from muscle. J Mol Biol 6:46–60

Harada Y, Sakurada K, Aoki T, Thomas DD, Yanagida T (1990) Mechanochemical coupling in actomyosin energy transduction studied by in vitro motility assay. J Mol Biol 216:49–68

Hatano S (1994) Actin-binding proteins in cell motility. Int Rev Cytol 156:199–273

Hatano S, Oosawa F (1966) Isolation and characterization of plasmodium actin. Biochim Biophys Acta 127:488–498

Hatano S, Totsuka T, Oosawa F (1967) Polymerization of plasmodium actin. Biochim Biophys Acta 140:109–122

Higashi-Fujime S (1980) Active movement in vitro of bundles of microfilaments isolated from *Nitella* cell. J Cell Biol 87:569–578

Holmes K, Popp D, Gebhard W, Kabsch W (1990) Atomic model of the actin filament. Nature 347:44–49

Huxley AF (1998) Biological motors: Energy storage in myosin molecules. Curr Biol 8:485–488

Huxley AF, Niedergerke R (1954) Structural changes in muscle during contraction. Nature 173:971–973

Huxley H, Stewart A, Sosa H, Irving T (1994) X-ray diffraction measurements of the extensibility of actin and myosin filaments in contracting muscle. Biophys J 67:2411–2421

Huxley HE (1963) Electronmicroscopic studies on the structure of natural and synthetic protein filaments from striated muscle. J Mol Biol 7:281–308

Huxley HE (1973) Cold Spring Harbor Symp. Quant Biol 37:361–376

Huxley HE, Hanson J (1954) Changes in the cross-striation of muscle during contraction and stretch and their structural interpretation. Nature 173:973–975

Ishijima A, Harada Y, Kojima H, Funatsu T, Higuchi H, Yanagida T (1994) Single molecule analysis of the actomyosin motor using nano-manipulation. Biochem Biophys Res Commun 199:1057–1063

Ishijima A, Kojima H, Funatsu T, Tokunaga M, Higuchi H, Tanaka H, Yanagida T (1998) Simultaneous observation of individual ATPase and mechanical events by a single myosin molecule during interaction with actin. Cell 92:161–171

Ishiwata S, Fujime S (1972) Effect of calcium ions on the flexibility of reconstituted thin filament of muscle studied by quasielastic light scattering of laser light. J Mol Biol 68:511–522

Kabsch W, Mannherz H, Suck D, Pai E, Holmes K (1990) Atomic structure of the actin-DNase I complex. Nature 347:37–43

Kamiya N, Kuroda K (1956) Velocity distribution of protoplasmic streaming in *Nitella* cells. Bot Mag 69:544–554

Kasai M, Asakura S, Oosawa F (1962) Cooperative nature of G-F transformation of actin. Biochim Biophys Acta 57:22–31

Kasai M, Nakano E, Oosawa F (1965) Polymerization of actin free from nucleotides and divalent cations. Biochim Biophys Acta 94:494–503

Kishino A, Yanagida T (1988) Force measurements by micromanipulation of a single actin filament by glass needle. Nature 334:74–76

Kitamura K, Tokunaga M, Iwane A, Yanagida T (1999) A single myosin head moves along an actin filament with regular steps of 5.3 nm. Nature 397:129–134

Kojima H, Ishijima A, Yanagida T (1994) Direct measurement of stiffness of single actin filaments with and without tropomyosin by in vitro nanomanipulation. Proc Natl Acad Sci USA 91:12962–12966

Kondo H, Ishiwata S (1976) Unidirectional growth of F-actin. J Biochem 79:159–171
Kron S, Spudich JA (1986) Fluorescent actin filaments move on myosin fixed on a glass surface. Proc Natl Acad Sci USA 83:6272–6276
Magnesco M (1993) Forced thermal ratchet. Phys Rev Lett 71:1477–1481
Nagashima H, Asakura S (1980) Dark field light microscopic study of the flexibility of F-actin complexes. J Mol Biol 136:169–182
Oosawa F, Asakura S (1975) Thermodynamics of the polymerization of protein. Academic Press New York
Oosawa F, Hayashi S (1986) The loose coupling mechanism in molecular machines of living cells. Adv Biophys 22:151–183
Oosawa F, Kasai M (1962) Theory of linear and helical polymerization of macromolecules. J Mol Biol 4:10–21
Oosawa F, Asakura S, Ooi T (1961) Physical chemistry of muscle protein, actin. Prog Theor Phys suppl 17:14–34
Oosawa F, Asakura S, Hotta K, Imai N, Ooi T (1959) G-F transformation of actin as a fibrous condensation. J Polym Sci 37:323–336
Oosawa F, Fujime S, Ishiwata S, Mihashi K (1973) Dynamic property of F-actin and thin filament. Cold Spring Harbor Symp Quant Biol 37:277–286
Oosawa F (1983) Macromolecular assembly of actin In: Stracher A (ed)., Muscle nonmuscle motility, Academic Press New York, 152–216
Oosawa F (1993) Physical chemistry of actin: past, present and future. Biophys Chem 47:101–111
Oosawa F (1995) Sliding and ATPase. J Biochem 118:863–870
Orlova A, Egelman E (1995) Structural dynamics of F-actin I. J Mol Biol 245:582–597
Orlova A, Prochniewicz E, Egelman EH (1995) Structural dynamics of F-actin II. J Mol Biol 245:598–607
Straub FB, Feuer G (1950) Adenosinetriphosphate, the functional group of actin. Biochim Biophys Acta 4:455–470
Straub FB (1942) Actin. Studies Med Inst Szeged 2:3–15
Szent-Györgyi A (1951) Chemistry of muscular contraction Academic Press, New York
Takebayashi T, Morita Y, Oosawa F (1977) Electronmicroscopic investigation of the flexibility of F-actin. Biochim Biophys Acta 492:357–363
Tilney L, DeRosier D, Tilney M (1992) How *Listeria* exploits host cell actin to form its own cytoskeleton. J Cell Biol 118:71–81
Vale R, Oosawa F (1990) Protein motors and Maxwell's demons: Does mechanochemical transduction involve a thermal ratchet? Adv Biophys 26:97–131
Wakabayashi K, Sugimoto Y, Tanaka H, Ueno Y, Takezawa Y, Amemiya Y (1994) X-ray diffraction evidence for the extensibility of actin and myosin filaments during muscle contraction. Biophys J 67:2422–2435
Yanagida T, Nakase M, Nishiyama N, Oosawa F (1984) Direct observation of motion of single F-actin filaments in the presence of myosin. Nature 307:58–60
Yanagida T, Arata T, Oosawa F (1985) Sliding distance of actin filament induced by a myosin cross-bridge during one ATP hydrolysis cycle. Nature 316:366–369

Divalent Cations, Nucleotides, and Actin Structure

Hanna Strzelecka-Gołaszewska[1]

Introduction

Actin has one high-affinity site for a divalent cation, with a K_d for Ca^{2+} and Mg^{2+} in the nanomolar range. This binding site is located at the bottom of the cleft between the two domains of the molecule (Fig. 1). The cation is coordinated not only by amino acid residues but also by the oxygens of the γ- and/or β-phosphate groups of the nucleotide, ADP or ATP respectively, which is bound further up in the cleft (Valentin-Ranc and Carlier 1989; Kabsch et al. 1990). The tightly bound cation of G-actin exchanges with other divalent cations by a simple competitive mechanism.

Consistent with the high ratio of free Mg^{2+} to Ca^{2+} concentration under physiological conditions, Mg^{2+} is most likely the tightly-bound cation in vivo. In actin purified in vitro by commonly-used procedures, this ion is replaced by Ca^{2+} present in buffer solutions. The replacement is facilitated by the affinity of ATP-G-actin about five fold higher for Ca^{2+} than for Mg^{2+}. For the same reason, a complete back exchange of G-actin-bound Ca^{2+} for Mg^{2+} requires the presence of a selective Ca^{2+} chelator such as EGTA, because the high free Mg^{2+} concentrations required to compensate for the lower affinity of Mg^{2+} would induce polymerization of actin, which, in turn, dramatically reduces the exchangeability of the tightly bound cation (reviewed by Estes et al. 1992). On the contrary, with ADP as the bound nucleotide, it is difficult to convert Mg-G-actin into Ca-G-actin because ADP-G-actin binds Mg^{2+} more strongly than Ca^{2+} (Kinosian et al. 1993 and references therein).

Beside the single high-affinity site for a divalent cation, actin contains multiple sites of lower affinity that can bind both divalent and monovalent cations. Sites of moderate or intermediate affinity (initially termed low-affinity sites) bind Ca^{2+} and Mg^{2+} with K_d in the range of 10^{-4} M and K^+ with affinity 2 orders of magnitude lower. Their saturation initiates polymerization of actin. Sites presently termed low-affinity sites bind Ca^{2+} and Mg^{2+} with a K_d in the range of 10^{-2} M, and their saturation correlates with an assembly of actin filaments into paracrystals (Strzelecka-Gołaszewska et al. 1978; Carlier et al. 1986a; Estes et al. 1992 and references therein).

[1] Department of Muscle Biochemistry, Nencki Institute of Experimental Biology, 3 Pasteur Street, 02-093 Warsaw, Poland

Results and Problems in Cell Differentiation, Vol. 32
C. dos Remedios (Ed.): Molecular Interactions of Actin

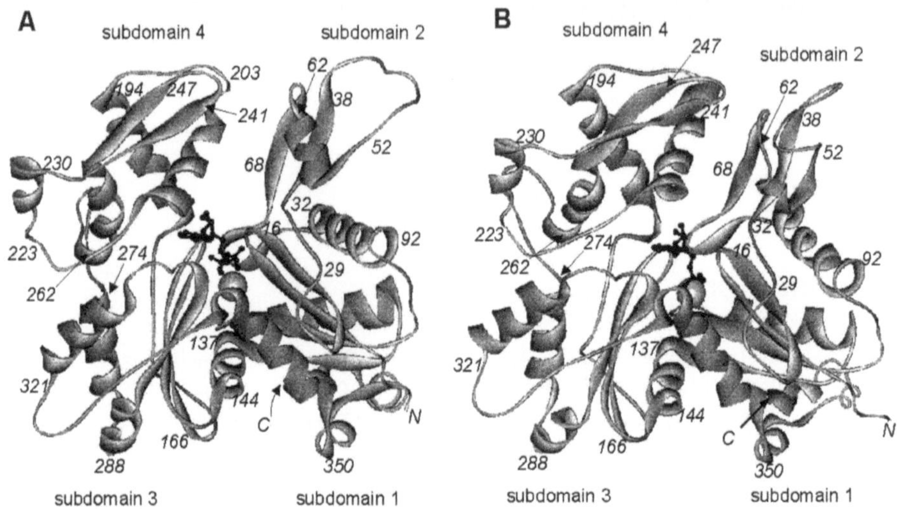

Fig. 1A,B. Diagrams of the actin monomer structure as determined by Kabsch et al. (1990) (**A**) and of the F-actin protomer according to Lorenz et al. (1993) (**B**). The coordinates for **A** were taken from the Brookhaven Protein Data Bank (file name 1ATN), and those for **B** were obtained from <http://indy.mpimf-heidelberg.mpg.de/~michael/ftp.html>. The diagrams were prepared using WebLab ViewerLite. The adenine nucleotide (ATP) is shown as a ball-and-stick model, and the divalent cation (Ca^{2+}) as a van der Waal's sphere

Effects of Cations on the State and Properties of Actin

The primary role of the tightly bound divalent cation is to stabilize, in its complex with the bound nucleotide, the native conformation of the actin monomer. The nucleotide bridges the two domains of actin (Fig. 1) and probably restricts their slow collective motions relative to one another (Tirion and ben-Avraham 1993). The importance of the tightly bound cation for this function is clearly seen from its increasing the binding constant for nucleotide by 3–4 orders of magnitude. The dissociation of cation (slow) is rate-limiting for nucleotide dissociation from the cation-nucleotide-actin complex.

The affinity of cation-free G-actin for ATP is still high (K_d of about 10^{-6} M), but the rate constant for nucleotide release from this form of actin is also high, and nucleotide-free actin undergoes a fast irreversible denaturation. The kinetic relationships between nucleotide binding (ATP or ADP) and divalent cation (Ca^{2+} or Mg^{2+}) binding in the cleft have been best described by Estes and colleagues (Estes et al. 1992; Kinosian et al. 1993). These authors show that nucleotide binding to actin is less affected by Mg^{2+} than by Ca^{2+} and is regulated in different concentration ranges of these cations (higher for Mg^{2+}) because the rate constant for cation association with ATP-G-actin is two orders of magnitude lower for Mg^{2+} than for Ca^{2+}. The dissociation rate constant of Mg^{2+} (from ATP-G-actin) is also lower than that of Ca^{2+}, but only about ten fold. This means that at the same free cation concentration, actin equilibrated with

Mg^{2+} spends a longer time in the cation-free state than actin equilibrated with Ca^{2+}. As a result, there is an increased probability of nucleotide dissociation and denaturation of the nucleotide-free protein for Mg-G-actin.

From these kinetic considerations it follows that at free divalent cation concentrations of 50–100 μM and 200 μM ATP usually present in buffer solutions used to store G-actin, Ca^{2+} and Mg^{2+} should equally well protect actin from denaturation (Kinosian et al. 1993). However, this is not so because of the ability of MgATP-G-actin to slowly hydrolyze the bound ATP. This Mg^{2+}-dependent G-actin ATPase activity, estimated at about 0.2 h^{-1} at pH 7.5 and 20–27°C (Brenner and Korn 1981; Mossakowska et al. 1993; Kasprzak 1994), leads to a conversion of MgATP-G-actin into MgADP-G-actin as free ATP is consumed and eventually to actin denaturation since ADP-G-actin is unstable even at 0°C (Gershman et al. 1989; Pollard et al. 1992).

These effects should be kept in mind while interpreting data obtained with actin that had been stored in the Mg^{2+}-bound form before measurements. Prolonged storage of Mg-G-actin is inadvisable also because this actin form is known to slowly associate into dimers and larger aggregates (Mozo-Villarias and Ware 1985; Newman et al. 1985; Goddette et al. 1986) and, at high protein concentrations, even to polymerize in low ionic strength solutions (Borejdo et al. 1981; Attri et al. 1991).

According to Estes and colleagues, the different effects of Ca^{2+} and Mg^{2+} on nucleotide binding to actin are entirely accounted for by the differences in the association and dissociation rate constants for Ca^{2+} and Mg^{2+} respectively, and these are due to different hydration properties of Ca^{2+} and Mg^{2+} ions in aqueous solutions (Estes et al. 1992; Kinosian et al. 1993). The stimulation of ATP hydrolysis on G-actin by the presence of Mg^{2+} at the high-affinity site, however, can only be explained in terms of a difference in the conformation of the nucleotide site between CaATP-G-actin and MgATP-G-actin. This effect is no surprise in view of the binding of the cation and nucleotide as a complex. More extensive conformational differences between G-actin in Ca^{2+}- and Mg^{2+}-bound form are apparent from studies on actin polymerization.

Salt-induced polymerization of actin, initiated by cation binding to sites of intermediate affinity (see Introduction), has been described by Oosawa and his colleagues as a cooperative two-step process involving a slow, rate-limiting nucleation followed by a fast polymer elongation (see this Vol. Oosawa). Polymerization ensues when actin concentration exceeds a critical concentration (Cc) that depends on solvent conditions (type and concentrations of salts, nucleotides, pH, temperature) and corresponds to the concentration of monomers coexisting with the polymer at the steady state of polymerization (Oosawa et al. 1959; Kasai et al. 1962a, 1962b; Oosawa 1983). Further studies confirmed that the nucleus is a trimer (the smallest oligomer with helical structure) and provided various lines of evidence for there being a step preceding nucleation, the so-called monomer activation for polymerization by a change in the protein conformation (see dos Remedios and Barden 1983; Carlier 1991). A relation between the activation of monomers and conformational changes induced by the binding of cations to various classes of sites in G-actin will be discussed below.

There is an agreement that the kinetics of polymerization is largely influenced by the type of tightly bound divalent cation and is little dependent on the type of cation saturating the sites of intermediate affinity. MgATP-G-actin polymerizes faster than CaATP-G-actin, mostly because of a higher rate of its nucleation (Maruyama 1981; Tobacman and Korn 1983; Gershman et al. 1984; Carlier et al. 1986b; Attri et al. 1991). The Cc of MgATP-G-actin is five-to-ten fold lower than the Cc of CaATP-G-actin, which reflects a lower relative rate constant of depolymerization (*k*–) and a higher rate constant of filament elongation (k+) of the MgATP-form (Cc = k–/k+) (Selden et al. 1986). In contrast, MgADP-G-actin and CaADP-G-actin are remarkably similar to each other in their polymerization properties. They largely differ from MgATP-G-actin (Cc is 20–30-fold higher for ADP-actins) and less from CaATP-G-actin, which means that a nucleotide-dependent change in the monomer conformation underlying the difference between polymerization properties of ADP-G-actin and ATP-G-actin requires the presence of Mg^{2+} at the high-affinity site (Gershman et al. 1989 and references therein).

Not only the polymerization kinetics but also the rate of ATP hydrolysis accompanying polymerization of ATP-actins is influenced by the type of cation bound at the high-affinity site. With CaATP-actin, ATP hydrolysis lags behind the polymer growth, whereas with MgATP-actin it more closely follows polymerization (reviewed in Carlier 1991). This effect of Mg^{2+}, together with the large difference between Cc values for MgATP-actin and MgADP-actin, have important consequences for the dynamics of actin filaments. ATP hydrolysis and the subsequent release of Pi generate a difference between critical concentrations for the fast growing barbed end of the actin filament, capped by newly incorporated ATP-monomers (low Cc), and for slowly growing pointed end carrying ADP-subunits (high Cc). This difference supports a continuous dissociation of the ADP-subunits from the pointed ends (followed by exchange of G-actin-bound ADP for free ATP present in solution) at steady state, compensated for by addition of ATP-monomers to the barbed ends, which results in a monomer flux through the filaments known as treadmilling (Wegner 1976).

How this principle of actin polymerization-depolymerization at steady state may be used in regulation of turnover and redistribution of actin filaments in the cell by capping proteins, actin depolymerizing factor (cofilin), profilin and perhaps other actin-binding proteins, is discussed in subsequent chapters in this Volume (for example see McGough et al., this Volume). With tightly bound Ca^{2+}, treadmilling does not occur because the hydrolysis of ATP of the newly incorporated monomers is slow and Cc values for the two filament ends are similar (Pollard and Mooseker 1981).

It is common knowledge that the activation of the myosin ATPase by actin, which is a key event in the mechanism of force generation by actin-myosin systems, is not influenced by the type of cation bound at the high-affinity site in actin. However, with the introduction of the in vitro motility assay, it was observed that certain modifications of actin may impair the sliding movement of the actin filaments without having a comparable effect on the ATPase rate or the affinity of actin for myosin in the presence of ATP (Prochniewicz and Yanagida 1990; Schwyter et al. 1990; Kwon et al. 1994). These and other studies emphasized the role of F-actin structure and, in particular, its flexibility in the mechanism of motility (reviewed in Egelman and Orlova 1995; see also Egelman and Orlova, this Volume). Therefore, the large difference in flexural rigidity between the filaments assembled from CaATP-G-actin and from MgATP-G-actin revealed in an electron microscopic study by Orlova and Egelman (1993) has been met with a considerable interest. However, this difference has not been confirmed by a similar study by Steinmetz et al. (1997). Measurements of flexural rigidity of actin filaments in solution failed to detect any significant difference between Ca-F-actin and Mg-F-actin (Isambert et al. 1995; Scharf and Newman 1995; Yasuda et al. 1996), whereas torsional rigidity of Ca-F-actin was found to be about three fold larger than that of Mg-F-actin (Yasuda et al. 1996).

An interesting property of Ca-F-actin, not shared by Mg-F-actin, is a cooperativity of rigor binding of heavy meromyosin (HMM) at low HMM:actin ratios. Although this effect of the tightly bound cation of actin does not seem to be of physiological importance, it suggests that one component of the widely studied cooperativity in the interaction with myosin imposed on F-actin by the regulatory proteins is within actin itself (Orlova and Egelman 1997).

Tightly Bound Cation-Dependent Conformational Changes in G-Actin

Although the high-affinity cation-binding site in actin under in vivo conditions is probably always occupied by Mg^{2+}, exploration of conformational differences between Mg- and Ca-actin is of considerable interest for at least two reasons: (1) the atomic structure has been solved for G-actin with either Ca^{2+} (Kabsch et al. 1990; McLaughlin et al. 1993; Chik et al. 1996) or Sr^{2+} at the high-affinity site (Schutt et al. 1993), and the atomic models of F-actin are based on the Ca-monomer structure and X-ray fiber diffraction pattern of Ca-F-actin (Holmes et al. 1990; Lorenz et al. 1993; Tirion et al. 1995); and (2) in view of the large differences in the kinetics of polymerization of CaATP-G-actin and MgATP-G-actin, identification of conformational differences between these actins seems to be a good approach to better understand structure-function relationships in actin.

The tightly bound cation-dependent effects on polymerization might be exerted either by a propagation of conformational changes from the cation-binding site in the interdomain cleft to the surface loops involved in the intermonomer interactions, or through changes in the orientation of the two domains of actin and/or their subdomains relative to each other. The latter possibility is consistent with a prediction from a normal mode analysis of G-actin that Gln-137, which is involved in coordination of the tightly bound Ca^{2+} (Kabsch et al. 1990), is a hinge point for a twisting motion of the two domains of actin. It was also predicted that a scissor-type movement of the domains, which opens and closes the cleft, is mediated by the α-helix 137–144 acting as a spring (Tirion and ben-Avraham 1993). Conformational changes in this segment of the polypeptide chain accompanying rotation of the two domains relative to one another have been confirmed by comparison of two different crystal structures of β-actin complexed to profilin, "tight" and "open", the latter having the interdomain cleft more widely open (Page et al. 1998). Refinements of the original atomic model of F-actin to optimize its fit to the fiber diffraction data suggest that a change in a relative orientation of either the two domains (Tirion et al. 1995) or subdomains 2 and 4 (Lorenz et al. 1993) resulting in a narrowing of the interdomain cleft (see Fig. 1) accompanies polymerization of actin. Narrowing of the cleft in F-actin has been confirmed by fluorescence resonance energy transfer (FRET) measurements (Miki and Kouyama 1994).

Other changes in the monomer structure predicted by the refinements of the F-actin model include a rearrangement of the DNase-I-binding loop (residues 38–52) on the top of subdomain 2 and a shift of the C-terminus (Lorenz et al. 1993; Tirion et al. 1995; also see Egelman and Orlova, this Vol.). Proposed alterations in these areas of the monomer during polymerization are based on comparison of three-dimensional reconstructions of the filament from electron micrographs with the Holmes et al. atomic model of F-actin constructed on the assumption that the structure of free G-actin in solution is the same as in actin:DNase-I crystals and does not substantially change upon incorporation into the filament (Bremer and Aebi 1992; Orlova and Egelman 1992). Yet another deformation, a reorientation of the hydrophobic loop 262–274 has been postulated by Holmes et al. (1990) and Lorenz et al. (1993) to allow insertion of this loop into a hydrophobic pocket formed by two adjacent actin subunits on the other strand of the two-start F-actin helix and thus fill a gap between the strands that is not seen in the electron microscopic reconstructions (see Belmont and Drubin this Vol.). This "hydrophobic plug" hypothesis has been questioned by Schutt et al. (1993), but received considerable support from biochemical studies on yeast actin mutants (Chen et al. 1993; Feng et al. 1997).

The first experimental evidence for long-range conformational effects of Ca^{2+}/Mg^{2+} replacement in G-actin was an enhancement by Mg^{2+} of the fluorescence of acetyl-N′-(5-sulfo-1-naphthyl)-ethylenediamine (AEDANS) covalently attached to the penultimate Cys-374 residue (Frieden et al. 1980; Frieden 1982). This change does not occur when bound ATP is replaced with ADP (Frieden and Patane 1985). The same pattern of changes near the C-terminus of actin has been revealed by limited proteolytic digestion. Tryptic cleavage at Lys-373 and Arg-272 in MgATP-G-actin proceeds four-to-six times more slowly than in CaATP-G-actin, and conversion of MgATP-G-actin into MgADP-G-actin largely diminishes the protective effect of Mg^{2+} (Strzelecka-Gołaszewska et al. 1993). Cation-dependent changes are also sensed by the fluorescent label 7-chloro-nitrobenzene-2-oxa-1,3-diazole (NBD) attached to Lys-373 (Carlier et al. 1986b). The fluorescence lifetime measurements on AEDANS-actin and EPR measurements on maleimide spin-labeled actin suggest an increased rigidity of the environment of Cys-374 in MgATP-G-actin compared to CaATP-G-actin (Nyitrai et al. 1997).

Another effect of conversion of CaATP-G-actin into MgATP-G-actin is a strong protection of trypsin cleavage sites at Arg-62 and Lys-68 in subdomain 2. Also this effect is eliminated by replacement of the bound ATP with ADP in Mg-G-actin, confirming that MgADP-G-actin is conformationally similar to CaATP-G-actin (Strzelecka-Gołaszewska et al. 1993).

The effect of Mg^{2+} on ATP-G-actin closely resembles the effect of polymerization. It is well known that F-actin is highly resistant to proteolysis. Direct comparison of tryptic fragmentation patterns of CaATP- and MgATP-G-actins before and after polymerization revealed that the inhibitory effect of polymerization on cleavages within segment 61–69 can only be observed with Ca^{2+} as the tightly bound cation. With Mg^{2+}, these sites are nearly equally well protected in the monomer and polymer state (Strzelecka-Gołaszewska et al. 1996).

As shown in Fig. 2, segment 61–69 is on a side of the interdomain cleft, with the side chain of Arg-62 pointing into the cleft. Lys-68, located at the base of subdomain 2, was shown to act as a hinge for a rotation of this subdomain (Page et al. 1998). Such a location of the residues recognized by trypsin

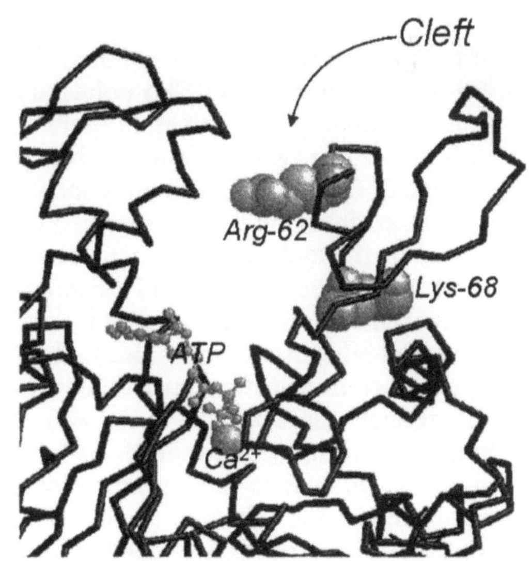

Fig. 2. Diagram depicting the orientation of the side chains of Arg-62 and of Lys-68 with respect to the interdomain cleft in G-actin. The diagram is based on the G-actin model of Kabsch et al. (1990) and was constructed using Rasmol 2.5 with the coordinates as in Fig. 1A

suggests that stabilization of the monomer conformation with the cleft narrowed is a common structural basis for the diminished accessibility of these residues to proteolysis in F-actins and in MgATP-G-actin. A Mg^{2+}-induced rotation of subdomain 2, presumably contributing to the cleft closure, might be mediated by a conformational change around Lys-68 (Strzelecka-Gołaszewska et al. 1993, 1996).

Using antibody binding as a probe it was demonstrated that segment 18–29 in subdomain 1 of actin is sensitive to Ca^{2+}/Mg^{2+} replacement in G-actin. It was suggested that this segment may participate in propagating structural perturbations from the cation-binding site to subdomain 2 (Adams and Reisler 1994).

The hypothesis of a Mg^{2+}-induced cleft closure in G-actin by a hinged movement or rotation of subdomains 2 and 4 received a support from molecular dynamics simulations of Ca- and Mg-G-actin structures (Wriggers and Schulten 1997), although not all results of this study are consistent with this view. Cleft closure relative to the Kabsch et al. crystal structure, albeit less pronounced than in Mg-G-actin, was also observed with bound Ca^{2+}, and substitution of Mg^{2+} for Ca^{2+} in the simulated Ca-G-actin structure (i.e., after relaxation from the crystal structure) did not produce any further change. In contrast to the results of proteolytic digestion experiments, the molecular dynamics simulations did not reveal any substantial effect of ATP/ADP replacement on the simulated structures of Ca- and Mg-G-actin.

As it was mentioned above, experimental proof that G-actin can adopt either an open or tight conformation by domain and subdomain movements was obtained by comparing the structures of β-actin:profilin crystals produced under different solvent conditions (Chik et al. 1996). Interestingly, the open state of β-actin (with Ca^{2+} bound) shows the interdomain cleft more widely open than it is in the crystal structure of α-actin-DNase-I which was used to model the F-actin structure. This points to the possibility that free Ca-G-actin in solution adopts a conformation more widely open than in actin:DNase-I crystals where the two domains are bridged by DNase. Consequently, polymerization of Ca-G-actin might be associated with an even larger change than was predicted from the F-actin models. However, this is contrary to what was suggested by the molecular dynamics simulations.

Chick et al. (1996) also showed a significant effect of transition between the open and tight states on the environment of Cys-374, supporting the suggestion that domain rotation may also underlie transmission of conformational changes between the cation/nucleotide site and the C-terminus (Crosbie et al. 1994; Strzelecka-Gołaszewska et al. 1996).

Both the C-terminus and the 38–52 loop at the "top" of subdomain 2 have been implicated in actin-actin contacts by the molecular models of the F-actin structure and by the effects of chemical or proteolytic modifications of these regions on polymerization and on the filament stability (reviewed in dos Remedios and Moens 1995). These areas are not uniquely oriented in the refined F-actin models (Lorenz et al. 1993; Tirion et al. 1995). The arrangement of the 38–52 loop is critical for the number of atomic interactions along the filament, which, in turn, influences the dynamic and elastic properties of the filament (ben-Avraham and Tirion 1995). Therefore, it was of interest to examine whether the conformational effects of substituting Mg^{2+} for the nonphysiological Ca^{2+} extend to this loop. The literature data on this issue are largely confusing.

A profound change in the environment of Gln-41 upon conversion of CaATP-G-actin into MgATP-G-actin was observed as a change in the fluorescence of dansyl ethylenediamine (DED) covalently attached to this residue (Kim et al. 1995). This change was not seen with another dansyl probe, dansyl cadaverine (DC) (Moraczewska et al. 1996). Similarly, an inhibition of subtilisin cleavage at Met-47 in MgATP-G-actin was observed by Kim et al. which contrasted with an earlier reported insensitivity of proteolytic cleavages in this area to Ca^{2+}/Mg^{2+} replacement (Strzelecka-Gołaszewska et al. 1993).

These controversies seem to be resolved by the recent finding that the changes in the fluorescence of both DED- and DC-labeled actin which had been attributed to an effect of the cation replacement represent slow spontaneous changes following substitution of Mg^{2+} for Ca^{2+} in G-actin which result from the denaturation-driven dissociation of the bound nucleotide. Either DED- or DC-labeling of Gln-41 accelerates the Mg^{2+}-dependent hydrolysis of ATP on G-actin and diminishes (by 1 order of magnitude) the affinity of actin for nucleotide. The resulting accelerated denaturation of the protein is accompanied by a dramatic change in the fluorescence of the label (Moraczewska et al. 1999). Partial conversion of MgATP-G-actin into MgADP-G-actin during prolonged storage of actin in the Mg-bound form might also underlie the reported inhibition of subtilisin cleavage within loop 38–52 that has not been observed with preparations examined shortly after replacement of Ca^{2+} with Mg^{2+} (Strzelecka-Gołaszewska et al. 1993).

The absence of substantial cation-dependent conformational transitions within subdomain 2 is consistent with FRET measurements, showing that Ca^{2+}/Mg^{2+} replacement is not accompanied by any significant change in the distances between probes on Cys-374 in subdomain 1 and on Tyr-69 (Miki 1991), Gln-41 (Moraczewska et al. 1996; although see Kekic et al. this Vol.), or Lys-61 (Nyitrai et al. 1998) in subdomain 2. Do these data preclude the postulated domain or subdomain rotation?

Neither a hinged rotation of subdomain 4 (Wriggers and Schulten 1997), nor a shift of the two domains without alteration of the intradomain structure as in the Tirion et al. (1995) model of F-actin (but without the rearrangement of the 38–52 loop) would have any influence on the interprobe distances under consideration. Likewise, a rotation of subdomain 2 resulting in a shift of the probes located in this subdomain in a plane nearly perpendicular to the line joining the donor-acceptor pair would have a negligible effect on the interprobe distance.

The insensitivity to Ca^{2+}/Mg^{2+} exchange contrasts with a profound change in the conformation of the 38–52 loop upon replacement of ATP with ADP in Mg-G-actin repeatedly observed as an increased resistance of this loop to proteolytic digestion (Strzelecka-Gołaszewska et al. 1993; Muhlrad et al. 1994) and as a large change in the fluorescence of DED- or DC-labeled actin (Kim et al. 1995; Moraczewska et al. 1996, 1999). Important functional implications result from a rearrangement of loop 38–52 upon release of Pi from ADP-Pi intermediate of the ATP hydrolysis accompanying polymerization. This rearrangement was revealed by comparison of three-dimensional reconstructions of actin filaments in an F-ADP-BeFx state – an analogue of either the F-ADP-Pi or the transient F-ATP state – with those in the final F-ADP state (Orlova and Egelman 1992) and confirmed by limited proteolysis experiments (Muhlrad et al. 1994) and fluorescence measurements on DC-labeled actin (Moraczewska et al. 1999). The observed changes suggest weakening of the longitudinal intermonomer contacts involving the 38–52 loop upon transformation of ADP-Pi-F-actin into ADP-F-actin. As originally proposed by Orlova and Egelman (1992), this effect may underlie the earlier demonstrated destabilization of F-actin upon the Pi release (reviewed in Carlier 1991). The corresponding change observed in G-actin on replacement of ATP with ADP may contribute to the diminished polymerizability of ADP-G-actin.

So far, little is known of possible cation-dependent changes in the large domain of actin. A change in the antigenic reactivity of the region around Val-201 on the top of subdomain 4 in response to Ca^{2+}/Mg^{2+} exchange in G-actin (Méjean et al. 1988) reports a conformational transition that may be relevant to polymerization. According to the atomic models, this region participates in the longitudinal contacts in the filament (Holmes et al. 1990; Lorenz et al. 1993; Tirion et al. 1995).

Molecular dynamics simulations (Wriggers and Schulten 1997) revealed that the rotation of subdomain 4 associated with the closure of the interdomain cleft in MgATP-G-actin is accompanied by a detachment of the 262–274 loop from the surface of subdomain 4 and its shift toward the position it has in the Lorenz et al. F-actin model (Fig. 1B). In terms of the hydrophobic plug hypothesis, this change primes the monomer for interactions along the genetic helix of F-actin. It might be crucial for the Mg^{2+} acceleration of the nucleation of actin because this step of polymerization critically depends on stabilization of the lateral interactions. The conformational coupling between the 262–274 loop and the interdomain cleft in G-actin has been confirmed by studies on a mutant yeast actin with replacements expected to promote the detachment of this loop from the protein surface (Kuang and Rubenstein 1997).

Effects of Polymerizing Salts on G-Actin Conformation: Monomer Activation

The idea of monomer activation as a preliminary step in actin polymerization was born from findings which indicated that addition of a polymerizing salt induces a rapid change in G-actin transforming the monomer to a state that is conformationally similar to the F-actin subunit. Such postulated changes were based on a diminished rate of proteolytic digestion by subtilisin or pronase (Rich and Estes 1976), and a change in the UV absorption spectrum of actin at subcritical concentrations (Rouayrenc and Travers 1981; Pardee and Spudich 1982). This experimental evidence for a novel monomer state was questioned by Fisher et al. (1983) who were unable to reproduce the KCl-dependent change in UV absorbance or to find any effect of salt on the rate of digestion with chymotrypsin when the temperature was lowered to 0°C to slow down the nucleation. These authors concluded that the inhibition of proteolysis of G-actin by KCl is due to a charge-screening effect of ionic strength on either actin or the protease. Thus, the influence on the protease-actin interaction could occur without any change in the actin conformation. In support of this conclusion, KCl had little or no influence on the ^{1}H-NMR and near-UV circular dichroism spectra of G-actin at subcritical concentrations (Barden et al. 1983).

Meanwhile, the first reports of conformational changes in G-actin upon Ca^{2+}/Mg^{2+} exchange at the high-affinity site appeared (Frieden et al. 1980; Frieden 1982; Barden and dos Remedios 1985), and monomer activation by the tightly-bound Mg^{2+} was postulated to explain the acceleration by Mg^{2+} of the nucleation step of polymerization (Cooper et al. 1983; Frieden 1983; Gershman et al. 1984; Barden and dos Remedios 1985). This view has been disputed and the original idea of monomer activation by polymerizing salt was revived by Carlier et al. (1986a). Their evidence was based on a salt-induced rapid increase in the fluorescence of AEDANS-labeled G-actin in both Ca- and Mg-bound form, contrasting with a slow fluorescence enhancement following substitution of Mg^{2+} for tightly bound Ca^{2+}.

This kinetic argument against a contribution of the Mg^{2+}-induced change to the monomer activation lost its value once it was documented that, contrary to an earlier view, the exchange of tightly-bound cation is slow and rate -limiting for the conformational change sensed by the AEDANS label (reviewed in Estes et al. 1992). Moreover, MgATP-G-actin was shown to associate (albeit slowly) into oligomers and, at higher protein concentrations, even to polymerize without addition of salt (Attri et al. 1991). These data clearly show that Mg^{2+} binding at the high-affinity site transforms G-actin into a species capable of polymerizing when the repulsive electrostatic forces between monomers are overcome by high protein concentrations or neutralized by addition of salt.

From the currently available information it appears that at least two regions of the actin molecule, the C-terminal segment and the 61–69 segment or their environments, undergo a conformational change preceding nucleation. These changes are induced by substitution of Mg^{2+} for Ca^{2+} at the high-affinity site as well as by binding of either divalent of monovalent cations to the moderate-affinity sites. As discussed above, the Mg^{2+}-dependent change in the proteolytic susceptibility of the 61–69 segment resembles the effect of polymerization.

Actually, most of the change observed on polymerization of CaATP-G-actin can be accounted for by a salt-induced change preceding polymer formation.

Addition of 0.1 M KCl under nonpolymerizing conditions (low protein concentration, 0°C, time limited to 90 s) diminishes the accessibility of tryptic cleavage sites in segment 61–69 even more strongly than does replacement of the bound Ca^{2+} with Mg^{2+} at low temperature used in these experiments (Khaitlina et al. 1996). The absence of any significant oligomer formation within the time of digestion was confirmed by the lack of any inhibition of subtilisin cleavage within the DNase-I-binding loop, the region strongly protected in F-actin.

The recent observation that the 61–69 segment is more strongly protected from proteolysis in F-actin polymerized with 0.1 M KCl than when polymerized with 2 mM $MgCl_2$ or $CaCl_2$ (Strzelecka-Gołaszewska et al. 1996) might argue for the interpretation of the KCl-dependent change as a charge-screening effect on the enzyme-protein interaction. However, the cleavage of $MgCl_2$-polymerized actin is completely inhibited (as in KCl-polymerized actin) by beryllium fluoride at concentrations which are equimolar to actin (Muhlrad et al. 1994). This is difficult to explain in terms of a nonspecific charge-screening effect. Similarly, the insensitivity of chymotryptic digestion of G-actin to the presence of salt, apparently arguing against the salt-induced conformational change in actin (Fisher et al. 1983), in the light of more recent data can be explained in line with the conformational effect of salt.

Chymotrypsin sequentially splits two peptide bonds in subdomain 2 of actin: between residues 44 and 45 in the loop 38–52 and, subsequently, between residues 67 and 68 in 61–69 segment (Konno 1987). Using other enzymes, it was shown that the conformational flexibility of segment 61–69 is lost or largely diminished when the continuity of the loop 38–52 is broken (Strzelecka-Gołaszewska et al. 1993). It is quite possible that a disruption of local intramolecular charge interactions or, alternatively, neutralization of electrostatic repulsive forces by cation binding to the charged residues (or by screening effect of elevated ionic strength) produces a change in the local conformation of the protein. Thus, the interdomain cleft closure is a plausible explanation for the protection of the 61–69 segment from proteolysis in KCl-treated G-actin.

Cation-Dependent Differences in F-Actin Structure

In general, the F-actin structure is less sensitive to the type of tightly bound cation than the conformation of G-actin. This is connected with the fact that, as discussed in the preceding sections, certain changes induced in G-actin by replacement of Ca^{2+} with Mg^{2+} (and preserved in Mg-F-actin) are mimicked by salt-dependent changes accompanying the polymerization of Ca-G-actin. One can also expect attenuation of the cation-dependent differences by ATP hydrolysis and Pi release from the polymer, by analogy with the effects of ATP for ADP replacement in G-actin. Thus, the tightly bound cation-dependent differences in proteolytic susceptibility of the C-terminal segment and, in particular of the interdomain cleft area, largely diminish after polymerization (Strzelecka-Gołaszewska et al. 1996).

The cation-dependent difference in antibody binding to the segment 18–29 in G-actin (Adams and Reisler 1994) is preserved in F-actins (Méjean et al. 1988; Adams and Reisler 1994), but no quantitative comparison between the G- and F-forms has been made. In line with these data, ^{1}H NMR spectra show a larger mobility of the 1–22 segment in Mg-F-actin compared with Ca-F-actin (Slósarek et al. 1994; Heintz et al. 1996). These observations are of interest in view of the involvement of negatively charged residues in sequences 1–4 and 18–29 of actin in the myosin binding.

Differences between Ca-F-actin and Mg-F-actin suggesting a change in the orientation of subdomain 2, and the presence of a bridge of density connecting the two long-pitch strands in Ca-F- but not in Mg-F-actin, were observed by Orlova and Egelman (1993, 1995) in electron microscopic reconstructions of negatively stained filaments. The authors documented that the shift of mass resulting in development of the interstrand connectivity involves the C-terminal segment. However, in an apparently similar study, Steinmetz et al. (1997) have not found any significant structural difference between Ca-F- and Mg-F-actin. The discrepancy might at least in part be due to different polymerizing salt concentrations used by these two groups. Limited proteolysis showed that the accessibility of the interdomain cleft area (segment 61–69) and of the C-terminal segment in F-actin is more influenced by changes in the ionic strength of the solution than by Ca^{2+}/Mg^{2+} replacement at the high-affinity site (Strzelecka-Gołaszewska et al. 1996). The collection of three-dimensional reconstructions of F-actins polymerized with different combinations of salts, reported by Orlova and Egelman (1995), also suggests ionic strength-dependent structural changes, confirming that the changes observed with limited proteolysis reflect true conformational differences. Interesting, although difficult to explain, is the electron microscopic observation that supplementing of Mg-F-actin in 50 mM KCl with 0.1 mM $CaCl_2$ transforms apparently flexible filaments into rigid ones, with a corresponding structural change that has been interpreted as a bending of subdomain 2 toward the cleft (Orlova and Egelman 1993). No rapid replacement of the tightly bound cation under these conditions can be expected.

A conformational effect of either Ca^{2+} or Mg^{2+} added to KCl-polymerized Mg-actin, but at concentrations sufficiently high (2 mM) to compete with K^{+} for the moderate-affinity sites in actin, has been observed by measuring linear dichroism spectra of fluorescein mercuric acetate (FMA)-labeled actin filaments oriented by flow (Mihashi et al. 1979). This effect, reporting a change at or near Cys-374, could not be observed with Ca-F-actin, in line with the view that the C-terminal region is in a different, more flexible conformation in Mg-F-actin than in Ca-F-actin. Contrary to these data, but in agreement with limited proteolysis experiments (Strzelecka-Gołaszewska et al. 1996), recent measurements of steady-state fluorescence anizotropy of AEDANS attached to Cys-374 suggest that the microenvironment of this residue is more flexible in Ca-F-actin than in Mg-F-actin (Nyitrai et al. 1999). Clearly, more work is needed to reconcile the disparate findings and elucidate how cations bound at various classes of sites in actin modulate the actin filament structure.

Conclusions

Spectroscopic and biochemical studies consistently show that the conformation of G-actin in solution is modulated at several sites by the type of tightly bound divalent cation, as well as by cation binding at the multiple moderate-affinity sites. Although the exact nature of these changes is not yet known, current data point to the possibility that a common effect of exchange of tightly bound Ca^{2+} for Mg^{2+} in ATP-G-actin, and of neutralization (or screening) of certain charged residues by an added salt, may represent a transition from an open conformation characteristic of CaATP-G-actin to a closed (or tight) conformation resembling that of the F-actin subunit. This transition is best explained as a ligand-dependent stabilization of one of different conformations that the monomer can adopt upon thermal motions of its domains and/or subdomains (see Oosawa this Vol.). Stabilization of the closed conformation, although still hypothetical, is likely to be the main event in the initial monomer activation step of the G-F transformation because, as shown by the refinements of the F-actin model, the domain or subdomain rotation whereby the interdomain cleft narrows is necessary to optimize the intermonomer contacts in the polymer (see Egelman and Orlova, this Vol.).

Another common effect of replacement of tightly bound Ca^{2+} with Mg^{2+} and of addition of a polymerizing salt to CaATP-G-actin is a local change(s) in the environment of the C-terminus. Although the nature of the changes generated by these different treatments may not be exactly the same, as a salt-dependent change is also observed in MgATP-G-actin, both of them correlate with a transformation of G-actin into a polymerizable species (monomer activation). On the other hand, there is no compelling evidence against possible contribution of the Mg^{2+}-dependent change around the C-terminus to the acceleration of actin nucleation in the presence of salt (Carlier et al. 1986b). The recently reported reorientation of loop 262–274 in subdomain 4 in Mg-G-actin (Wriggers and Schulten 1997), presumably facilitating the lateral interactions between the monomers, is most likely to play a role in the Mg^{2+}-dependent acceleration of the nucleation step in salt-induced polymerization. This latter conformational effect of Mg^{2+} has been suggested to also explain a faster condensation of G-actin : myosin subfragment 1 complexes into oligomers when Mg^{2+} rather than Ca^{2+} is the bound cation (Fievez et al. 1997).

In view of the role of the DNase-binding loop on the top of subdomain 2 in establishing both longitudinal and lateral contacts in F-actin, it comes as a surprise that there is no indication of a change in the conformation of this loop prior to polymerization (however, see Kekic et al. this Vol.). Quite contrary, the available evidence suggests that the rearrangement of this loop, predicted by the refinements of the F-actin model, occurs subsequently to polymerization and results in destabilization of the intermonomer contacts. One should, however, keep in mind that the reference point for the F-actin models was the monomer conformation in actin:DNase-I crystals. The structure of actin in complex with gelsolin segment 1 (McLaughlin et al. 1993) suggests that in free G-actin in solution the DNase-I-binding loop is highly mobile. This structural flexibility appears to be more favorable for establishing the inter-monomer contacts during polymerization than is stabilization of this loop, e.g., folded back onto subdomain 2, which is likely to be a common feature of ADP-G-actin (Strzelecka-Gołaszewska et al. 1993; Moraczewska et al. 1996, 1999) and of the final ADP-F-actin state (Lorenz et al. 1993).

Knowledge of the tertiary structures of different conformational states of G- and F-actin is necessary to fully understand the mechanism of actin polymerization and its interaction with myosin and other proteins. The general conclusion from current data is that in vivo (in nonmuscle cells) the tightly bound Mg^{2+} and ATP, and physiological salt concentrations keep actin in the conformation most favorable for polymerization, enabling rapid changes in the monomer/polymer equilibrium in response to intracellular and extracellular signals transmitted through numerous actin-binding proteins.

Acknowledgment. The author is supported by a grant to the Nencki Institute from the State Committee for Scientific Research.

References

Adams SB, Reisler E (1994) Sequence 18–29 on actin: antibody and spectroscopic probing of conformational changes. Biochemistry 33:14426–14433

Attri AK, Lewis MS, Korn ED (1991) The formation of actin oligomers studied by analytical ultracentrifugation. J Biol Chem 266:6815–6824

Barden JA, dos Remedios CG (1985) Conformational changes in actin resulting from Ca^{2+}/Mg^{2+} exchange as detected by proton NMR spectroscopy. Eur J Biochem 146:5–8

Barden JA, Wu C-SC, dos Remedios CG (1983) Actin monomer conformation under polymerizing conditions studied by proton nuclear magnetic resonance and circular dichroism spectroscopy. Biochim Biophys Acta 748:230–235

ben-Avraham D, Tirion MM (1995) Dynamic and elastic properties of F-actin: a normal-modes analysis. Biophys J 68:1231–1245

Borejdo J, Muhlrad A, Leibovich SJ, Oplatka A (1981) Polymerization of G-actin by hydrodynamic shear stresses. Biochim Biophys Acta 667:118–131

Bremer A, Aebi U (1992) The structure of the F-actin filament and the actin molecule. Curr Opin Cell Biol 4:20–26

Brenner SL, Korn ED (1981) Stimulation of actin ATPase activity by cytochalasins provides evidence for a new species of monomeric actin. J Biol Chem 256:8663–8670

Carlier M-F (1991) Actin: protein structure and filament dynamics. J Biol Chem 266:1–4

Carlier M-F, Pantaloni D, Korn ED (1986a) Fluorescence measurements of the binding of cations to high-affinity and low-affinity sites on ATP-G-actin. J Biol Chem 261:10778–10784

Carlier M-F, Pantaloni D, Korn ED (1986b) The effects of Mg^{2+} at the high-affinity and low-affinity sites on the polymerization of actin and associated ATP hydrolysis. J Biol Chem 261:10785–10792

Chen X, Cook RK, Rubenstein PA (1993) Yeast actin with a mutation in the "hydrophobic plug" between subdomains 3 and 4 (L266D) displays a cold-sensitive polymerization defect. J Cell Biol 123:1185–1195

Chik JK, Lindberg U, Schutt CE (1996) The structure of an open state of β-actin at 2.65 Å resolution. J Mol Biol 263:607–623

Cooper JA, Buhle EL, Jr, Walker SB, Tsong TY, Pollard TD (1983) Kinetic evidence for a monomer activation step in actin polymerization. Biochemistry 22:2193–2202

Crosbie RH, Miller C, Cheung P, Goodnight T, Muhlrad A, Reisler E (1994) Structural connectivity in actin: effect of C-terminal modifications on the properties of actin. Biophys J 67:1957–1964

dos Remedios CG, Barden JA (1983) in Actin Structure and Function in Muscle and Nonmuscle Cells. Academic Press, Sydney pp. 1–336

dos Remedios CG, Moens PDJ (1995) Actin and the actomyosin interface: a review. Biochim Biophys Acta 1228:99–124

Egelman ED, Orlova A (1995) New insights into actin filament dynamics. Curr Opin Struct Biol 5:172–180

Estes JE, Selden LA, Kinosian HJ, Gershman LC (1992) Tightly-bound divalent cation of actin. J Muscle Res Cell Motil 13:272–284

Feng L, Kim E, Lee W-L, Miller CJ, Kuang B, Reisler E, Rubenstein PA (1997) Fluorescence probing of yeast actin subdomain 3/4 hydrophobic loop 262–274. J Biol Chem 272:16829–16837

Fievez S, Carlier M-F, Pantaloni D (1997) Kinetics of myosin subfragment-1-induced condensation of G-actin into oligomers, precursors in the assembly of F-actin-S1. Role of the tightly bound metal ion and ATP hydrolysis. Biochemistry 36:11843–11850

Fisher AJ, Curmi PMG, Barden JA, dos Remedios CG (1983) A reinvestigation of actin monomer conformation under polymerizing conditions based on rates of enzymatic digestion and ultraviolet difference spectroscopy. Biochim Biophys Acta 748:220–229

Frieden C (1982) The Mg^{2+}-induced conformational change in rabbit skeletal muscle G-actin. J Biol Chem 257:2882–2886

Frieden C (1983) Polymerization of actin: mechanism of the Mg^{2+}-induced process at pH 8 and 20°C. Proc Natl Acad Sci USA 80:6513–6517

Frieden C, Patane K (1985) Differences in G-actin containing bound ATP or ADP: the Mg^{2+}-induced conformational change requires ATP. Biochemistry 24:4192–4196

Frieden C, Lieberman D, Gilbert HR (1980) A fluorescent probe for conformational changes in skeletal muscle G-actin. J Biol Chem 255:8991–8993

Gershman LC, Newman J, Selden LA, Estes JE (1984) Bound-cation exchange affects the lag phase in actin polymerization. Biochemistry 23:2199–2203

Gershman LC, Selden LA, Kinosian HJ, Estes JE (1989) Preparation and polymerization properties of monomeric ADP-actin. Biochim Biophys Acta 995:109–115

Goddette DW, Uberbacher EC, Bunick GJ, Frieden C (1986) Formation of actin dimers as studied by small angle neutron scattering. J Biol Chem 261:2605–2609

Heintz D, Kany H, Kalbitzer HR (1996) Mobility of the N-terminal segment of rabbit skeletal muscle F-actin detected by ^{1}H and ^{19}F nuclear magnetic resonance spectroscopy. Biochemistry 35:12686–12693

Holmes KC, Popp D, Gebhard D, Kabsch W (1990) Atomic model of the actin filament. Nature 347:44–49

Isambert H, Venier P, Maggs AC, Fattoum A, Kassab R, Pantaloni D, Carlier M-F (1995) Flexibility of actin filaments derived from thermal fluctuations. J Biol Chem 270:11437–11444

Kabsch W, Mannherz HG, Suck D, Pai EF, Holmes KC (1990) Atomic structure of the actin: DNase I complex. Nature 347:37–44

Kasai M, Asakura S, Oosawa F (1962a) The G-F-equilibrium in actin solutions under various conditions. Biochim Biophys Acta 57:13–21

Kasai M, Asakura S, Oosawa F (1962b) The cooperative nature of G-F transformation of actin. Biochim Biophys Acta 57:22–31

Kasprzak AA (1994) Myosin subfragment 1 activates ATP hydrolysis on Mg^{2+}-G-actin. Biochemistry 33:12456–12462

Khaitlina S, Wawro B, Pliszka B, Strzelecka-Gołaszewska H (1996) Conformational changes associated with the monomer activation step of actin polymerization. J Muscle Res Cell Motil 17:122–123

Kim E, Motoki M, Seguro K, Muhlrad A, Reisler E (1995) Conformational changes in subdomain 2 of G-actin: fluorescence probing by dansyl ethylenediamine attached to Gln-41. Biophys J 69:2024–2032

Kinosian HJ, Selden LA, Estes JE, Gershman LC (1993) Nucleotide binding to actin. Cation dependence of nucleotide dissociation and exchange rates. J Biol Chem 268:8683–8691

Konno K (1987) Functional, chymotryptically split actin and its interaction with myosin subfragment 1. Biochemistry 26:3582–3589

Kuang B, Rubenstein PA (1997) The effects of severely decreased hydrophobicity in a subdomain 3/4 loop on the dynamics and stability of yeast G-actin. J Biol Chem 272:4412–4418

Kwon H, Hardwicke PMD, Collins JH, Zhao X, Szent-Györgyi AG (1994) Myosin filament ATPase is enhanced by intramolecularly cross-linked actin. J Muscle Res Cell Motil 15:555–562

Lorenz M, Popp D, Holmes KC (1993) Refinement of the F-actin model against X-ray fiber diffraction data by the use of a direct mutation algorithm. J Mol Biol 234:826–836

Maruyama K (1981) Effects of trace amounts of Ca^{2+} and Mg^{2+} on the polymerization of actin. Biochim Biophys Acta 667:139–142

McLaughlin PJ, Gooch JT, Mannherz H-G, Weeds AG (1993) Structure of gelsolin segment 1-actin complex and the mechanism of filament severing. Nature 364:685–692

Méjean C, Hué HK, Pons F, Roustan C, Benyamin Y (1988) Cation binding sites on actin: a structural relationship between antigenic epitopes and cation exchange. Biochem Biophys Res Commun 152:368–375

Mihashi K, Nakabayashi M, Yoshimura H, Ohnuma H (1979) Absorption, fluorescence, and linear dichroism spectra of fluorescein mercuric acetate (FMA) bound to F-actin. J Biochem 85:359–366

Miki M (1991) Detection of conformational changes in actin by fluorescence resonance energy transfer between tyrosine-69 and cysteine-374. Biochemistry 30:10878–10884

Miki M, Kouyama T (1994) Domain motion in actin observed by fluorescence resonance energy transfer. Biochemistry 33:10171–10177

Moraczewska J, Strzelecka-Gołaszewska H, Moens PDJ, dos Remedios CG (1996) Structural changes in subdomain 2 of G-actin observed by fluorescence spectroscopy. Biochem J 317:605–611

Moraczewska J, Wawro B, Seguro K, Strzelecka-Gołaszewska H (1999) Divalent cation-, nucleotide-, and polymerization-dependent changes in the conformation of subdomain 2 of actin. Biophys J 77:373–385

Mossakowska M, Moraczewska J, Khaitlina S, Strzelecka-Gołaszewska H (1993) Proteolytic removal of three C-terminal residues of actin alters the monomer-monomer interactions. Biochem J 289:897–902

Mozo-Villarias A, Ware BR (1985) Actin oligomers below the critical concentration detected by fluorescence photobleaching recovery. Biochemistry 24:1544–1548

Muhlrad A, Cheung P, Phan BC, Miller C, Reisler E (1994) Dynamic properties of actin: structural changes induced by beryllium fluoride. J Biol Chem 269:11852–11858

Newman J, Estes JE, Selden LA, Gershman LC (1985) Presence of oligomers at subcritical actin concentrations. Biochemistry 24:1538–1544

Nyitrai M, Hild G, Belágyi J, Somogyi B (1997) Spectroscopic study of conformational changes in subdomain 1 of G-actin: influence of divalent cations. Biophys J 73:2023–2032

Nyitrai M, Hild G, Belágyi J, Somogyi B (1999) The flexibility of actin filaments as revealed by fluorescence resonance energy transfer. The influence of divalent cations. J Biol Chem 274:12996–13001

Nyitrai M, Hild G, Lakos ZS, Somogyi B (1998) Effect of Ca^{2+}-Mg^{2+} exchange on the flexibility and/or conformation of the small domain in monomeric actin. Biophys J 74:2474–2481

Oosawa F (1983) Macromolecular assembly of actin. In: Stracher A (ed) Muscle and nonmuscle motility, vol 1 Academic Press, New York, pp. 151–216

Oosawa F, Asakura S, Hotta K, Imai N, Ooi T (1959) G-F transformation of actin as a fibrous condensation. J Polymer Sci 37:323–336

Orlova A, Egelman EH (1992) Structural basis for the destabilization of F-actin by phosphate release following ATP hydrolysis. J Mol Biol 227:1043–1053

Orlova A, Egelman EH (1993) A conformational change in the actin subunit can change the flexibility of the actin filament. J Mol Biol 232:334–341

Orlova A, Egelman EH (1995) Structural dynamics of F-actin. Changes in the C terminus. J Mol Biol 245:582–597

Orlova A, Egelman EH (1997) Cooperative rigor binding of myosin to actin is a function of F-actin structure. J Mol Biol 265:469–474

Page R, Lindberg U, Schutt CE (1988) Domain motions in actin. J Mol Biol 28:463–473

Pardee LD, Spudich JA (1982) Mechanism of K^+-induced actin assembly. J Cell Biol 93:648–654

Pollard TD, Mooseker MS (1981) Direct measurement of actin polymerization rate constants by electron microscopy of actin filaments nucleated by isolated microvillus cores. J Cell Biol 88:654–659

Pollard TD, Goldberg I, Schwarz WH (1992) Nucleotide exchange, structure, and mechanical properties of filaments assembled from ATP-actin and ADP-actin. J Biol Chem 267:20339–20345

Prochniewicz E, Yanagida T (1990) Inhibition of sliding movement of F-actin by crosslinking emphasizes the role of actin structure in the mechanism of motility. J Mol Biol 216:761–772

Rich SA, Estes JE (1976) Detection of conformational changes in actin by proteolytic digestion: evidence for a new monomeric species. J Mol Biol 104:777–792

Rouayrenc J-F, Travers F (1981) The first step in polymerisation of actin. Eur J Biochem 116:73–77

Scharf RE, Newman J (1995) Mg- and Ca-actin filaments appear virtually identical in steady-state as determined by dynamic light scattering. Biochim Biophys Acta 1253:129–132

Schutt CE, Myslik JC, Rozycki MD, Goonesekere NCW, Lindberg U (1993) The structure of crystalline profilin-actin. Nature 365:810–816

Schwyter DH, Kron SJ, Toyoshima YY, Spudich JA, Reisler E (1990) Subtilisin cleavage of actin inhibits in vitro sliding movement of actin filaments over myosin. J Cell Biol 111:465–470

Selden LA, Gershman LC, Estes JE (1986) A kinetic comparison between Mg-actin and Ca-actin. J Muscle Res Cell Motil 7:215–224

Slósarek G, Heinz D, Kalbitzer HR (1994) Mobile segments in rabbit skeletal muscle F-actin detected by ^1H nuclear magnetic resonance spectroscopy. FEBS Lett 351:405–410

Steinmetz MO, Goldie KN, Aebi U (1997) A correlative analysis of actin filament assembly, structure, and dynamics. J Cell Biol 138:559–574

Strzelecka-Gołaszewska H, Moraczewska J, Khaitlina SY, Mossakowska M (1993) Localization of the tightly bound divalent-cation-dependent and nucleotide-dependent conformation changes in G-actin using limited proteolytic digestion. Eur J Biochem 211:731–742

Strzelecka-Gołaszewska H, Prochniewicz E, Drabikowski W (1978) Interaction of actin with divalent cations. I. The effect of various cations on the physical state of actin. Eur J Biochem 88:219–227

Strzelecka-Gołaszewska H, Wozniak A, Hult T, Lindberg U (1996) Effects of the type of divalent cation, Ca^{2+} or Mg^{2+}, bound at the high-affinity site and of the ionic composition of the solution on the structure of F-actin. Biochem J 316:713–721

Tirion MM, ben-Avraham D (1993) Normal mode analysis of G-actin. J Mol Biol 230:186–195

Tirion MM, ben-Avraham D, Lorenz M, Holmes KC (1995) Normal modes as refinement parameters for the F-actin model. Biophys J 68:5–12

Tobacman LS, Korn ED (1983) The kinetics of actin nucleation and polymerization. J Biol Chem 258:3207–3214

Valentin-Ranc C, Carlier M-F (1989) Evidence for the direct interaction between tightly bound metal ion and ATP on actin. J Biol Chem 264:20871–20880
Wegner A (1976) Head to tail polymerization of actin. J Mol Biol 108:139–150
Wriggers W, Schulten K (1997) Stability and dynamics of G-actin: back door water diffusion and behavior of subdomain 3/4 loop. Biophys J 73:624–639
Yasuda R, Miyata H, Kinosita K, Jr (1996) Direct measurement of the torsional rigidity of single actin filaments. J Mol Biol 263:227–236

Valentin-Ranc C, Carlier M-F (1989) Evidence for the direct interaction between tightly bound divalent cation and ATP on actin. J Biol Chem 264: 20871–20880

[illegible]

[illegible] of actin [illegible]

Woody R [illegible] Tinoco I Jr (1967) [illegible] J Mol Biol 29: [illegible]

The Helical Parameters of F-Actin Precisely Determined from X-Ray Fiber Diffraction of Well-Oriented Sols

Toshiro Oda[1,3], Kouji Makino[1,2,3], Ichiro Yamashita[1], Keiichi Namba[1,4], and Yuichiro Maéda[1,3]

Structural Analysis of F-Actin

Actin plays important roles in many biological events such as cell division, muscle contraction, signal transduction and cell motility. To understand the molecular mechanism in these events, it is crucial to know the atomic structure of not only G-actin – for which multiple numbers of crystal structures have been solved (Kabsch et al. 1990; McLaughlan et al. 1993; Shutt et al. 1993; Chik et al. 1996) – but also of F-actin, the functional cellular filamentous form of actin. However, in spite of much effort, F-actin structure has not been solved at high resolution.

The F-actin structure has been studied by electron microscopy and X-ray fiber diffraction. In the early days of electron microscopic studies, negatively stained F-actin filaments showed that there are 13 monomers in 6 turns of the one-start helix with a pitch of 59 Å and a diameter of 80 Å (Hanson and Lowy 1963). Using these helical parameters, a low-resolution model of F-actin was built by the three-dimensional image reconstruction technique of DeRosier and Klug (1968; Moore et al. 1970).

In the 1980s, nonstained F-actin filaments in a frozen-hydrated state were examined by electron cryomicroscopy and a three-dimensional image was reconstructed (Trinick et al. 1986). Even today, the resolution limit of EM image analysis techniques is still only 20–30 Å (e.g. Milligan et al. 1990). However, recent advances in the EM technologies, such as field emission guns, energy filters and back projection analysis, may further improve this resolution limit (e.g. Stowell et al. 1998).

On the other hand, X-ray fiber diffraction from well-oriented filament sols is a potentially powerful tool for atomic resolution structural analysis of fibrous assemblies of macromolecules with helical symmetries. In the case of tobacco mosaic virus (TMV), the atomic structure was determined at 2.9 Å

[1] International Institute for Advanced Research, Matsushita Electric Industrial Ltd. 3-4, Hikaridai, Seika, Kyoto, 619-0237, Japan

[2] Faculty of Science, Department of Molecular Biology, Nagoya University, Furou, Chikusa, Nagoya, 464-8602, Japan

Present address:

[3] Laboratory for Structural Biochemistry, RIKEN Harima Institute at Spring-8, 1-1-1 Kouto, Mikazuki, Sayo, Hyogo, 679-5198, Japan

[4] Protonic Nano-machine project, ERATO, JST, 3-4, Hikaridai, Seika, Kyoto, 619-0237, Japan

Results and Problems in Cell Differentiation, Vol. 32
C. dos Remedios (Ed.): Molecular Interactions of Actin

resolution using isomorphous replacement and stereochemically restrained least-squares refinement techniques in a way similar to X-ray crystallography (Namba et al. 1989a). However, the method has not been applicable to the analysis of F-actin structure because of difficulties in precisely extracting the layer-line intensities from X-ray diffraction patterns. The intervals between layer-lines are much narrower due to the helical symmetry, and layer-line reflections are more diffuse due to larger filament disorientation, as compared with those in TMV sols.

As an alternative approach, F-actin structure was modeled using the atomic structure of the monomeric actin and the model was evaluated or further refined based on X-ray fiber diffraction patterns from well-oriented F-actin sols (Holmes et al. 1990; Lorenz et al. 1993; Tirion et al. 1995). Even for this purpose, X-ray fiber diffraction patterns need to be of high quality. Well-resolved layer-lines are crucially dependent on the orientation of the filaments in F-actin sols (Makowski 1991; Holmes 1995). Various procedures to make well-oriented F-actin sols have been developed to date (Cohen and Lowy 1953; Gillis and O'Brien 1975; Popp et al. 1987). Recently, we have proposed an improved procedure (Oda et al. 1998), based on the method developed for orienting the bacterial flagellar filaments (Yamashita et al. 1998a).

In the present chapter, we summarize the strategy to prepare well-oriented F-actin sols. From the diffraction patterns of the oriented F-actin sols, we have precisely determined the helical symmetry of the monomer arrangement within F-actin. We also describe the effect of phalloidin on the structure of F-actin. Finally, we discuss the conditions under which we can draw a quantitative comparison between two diffraction patterns obtained from F-actin sols with different degrees of orientation.

Procedures for Recording and Analyzing X-Ray Fiber Diffraction Patterns from Well-Oriented F-Actin Sols

In the following sections, the procedures are briefly described. For further details, we refer the reader to our recent publication (Oda et al. 1998).

Sample Preparation

Actin was extracted from acetone powder and further purified by G-150 gel filtration (Spudich and Watt 1971; MacLean-Fletcher and Pollard 1980). Gelsolin was purified from bovine serum (Kurokawa et al. 1990). Gelsolin was added to G-actin solution at various molar ratios of actin to gelsolin, typically 100 : 1, to control the filament length of F-actin (Janmey et al. 1986). After 1 h incubation on ice, KCl was added to a final concentration of 60 mM to polymerize the actin. The final actin concentration was 2 mg ml^{-1}. Solvent conditions were adjusted for various measurements by overnight dialysis. The standard solvent condition was: 30 mM NaCl, 10 mM Tris-acetate (pH = 8), 1 mM $CaCl_2$, 0.5 mM ATP, 1 mM NaN_3, and 1 mM 2-mercaptoethanol. The F-actin filaments were spun down at ~11 000 g for ~3 days. The resulting soft pellet was drawn into a quartz capillary with a diameter of 0.7 mm. The sols in the capillary were centrifuged at 100–2600 g to further concentrate the filaments. Finally, a strong magnetic field of 13.5 Tesla was applied to the sols in the capillary.

X-Ray Diffraction

X-ray fiber diffraction patterns were recorded on 20 × 25 cm Fuji imaging plates (IP) using a Rigaku rotating anode X-ray RU-200 generator operated at 40 kV and 30 mA. Ni-coated double mirror optics were used to focus the beam which had a dimension of 100 × 200 μm on IP. An X-ray wavelength of 1.5418 Å was used. The specimen-to-film distance (~166 mm) was determined from powder patterns of $CaSO_4 \cdot 2H_2O$ crystals (a_0 = 6.2846 Å, b_0 = 15.2011 Å, c_0 = 5.6737 Å, β = 114.10°) placed at the same position as F-actin sols. Reflections at (020), (021) and (040) were used (Yamashita et al. 1998b). Diffraction patterns from the F-actin sols were recorded out to a resolution of 3.4 Å. The exposure time was about 10 h. The exposed IPs were processed with a Fuji BA100 system using 0.1 mm rasters. The image data were transferred to a VAX 4500 computer and displayed on a D-SCAN GR4416 graphic terminal. The diffraction intensity distribution in the Cartesian coordinate of the flat IP was converted into the polar coordinate of the reciprocal space, and a circular symmetrical background was subtracted.

Analysis of X-Ray Fiber Diffraction Patterns

The angular distributions of the filament orientation in F-actin sols were determined by the two-dimensional profile fitting procedures partially described previously (Oda et al. 1998; Yamashita et al. 1998b). Detailed explanations will be given elsewhere (K. Hasegawa, I. Yamashita and K. Namba in prep.).

An ideal fiber diffraction pattern would be obtained from sols containing filaments with a perfect orientation, which would show layer-lines with their intensities confined on straight lines perpendicular to the filament axis. In reality, because of finite disorientation of the filaments, each point of the layer-lines is smeared out into an arc. Angular spread functions of the intensity can be calculated using the equations deduced by Holmes and Barrington-Leigh (1974), who assumed that the angular distribution of the filaments can be approximated by a Gaussian: $f(\theta) \propto \exp(-\theta^2/2\sigma^2)$ where σ is the standard deviation of the distribution.

Given the standard deviation and layer-line distribution, the intensity profile of diffraction patterns can be reconstructed by summing up the point spread functions weighted by the intensities. The overall residuals between the observed and the reconstructed profiles can thus be calculated and minimized by adjusting the intensities in an iterative manner to achieve the fitting of the two profiles. Local residuals on some layer-line peaks are also monitored to find the best-fitting values of the variable parameters, such as the standard deviation of the filament orientation, the pitch of the one-start helix and the helical symmetry.

The resolution range used for the profile fitting was 0.012–0.05 Å^{-1}. Local residuals were monitored in two areas, one on the main peak of the 59 Å layer-line (curved box area A in Fig. 1) and the other on the main peak of the 51 Å layer-line (curved box area B in Fig. 1). A set of best-fitting parameters was determined on the basis of the local residuals by varying the parameters one by one in the following sequences (Fig. 2): (1) the pitch of one-start helix (P) was first determined from the axial position of the 59 Å layer-line by minimizing the residual in the area A; (2) the standard deviation (σ) of the angular distribution of the filaments was deduced from the intensity spread of 59 Å layer-line; and finally (3) the helical symmetry was obtained from the axial position of the 51 Å layer-line as determined in the area B and of the 59 Å layer-line. Based on these parameters, the diffraction intensities were extracted. In other words, the reconstructed profiles were obtained which bring about an overall fitting to the observed pattern.

Spencer (1969) showed that the Bragg spacing of interference peak on the equator is dependent on the actin concentration and that the concentration can be calculated from the spacing. Hence, the interfilament distance was calculated from the spacing (d) of the first interference peak on the equator as $1.114 \times d$, which was deduced from the two-dimensional gas model.

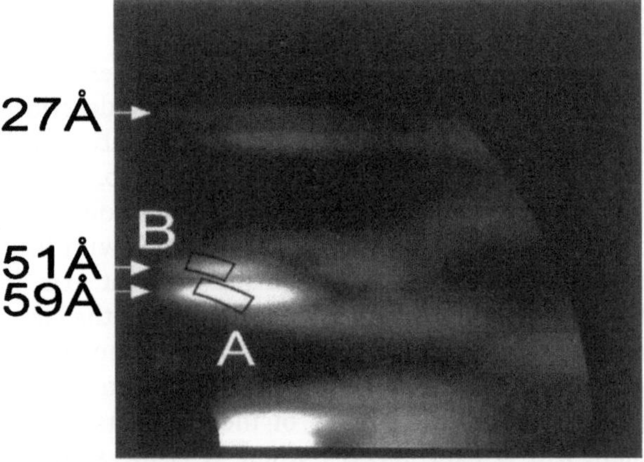

Fig. 1. Areas in the diffraction pattern where local residuals were calculated. Curved box area *A* lies across the *59 Å* layer-line. Curved box area *B* covers a section of *51 Å* layer-line. *White arrows* indicate the axial positions of the layer lines of *27*, *51* and *59 Å*

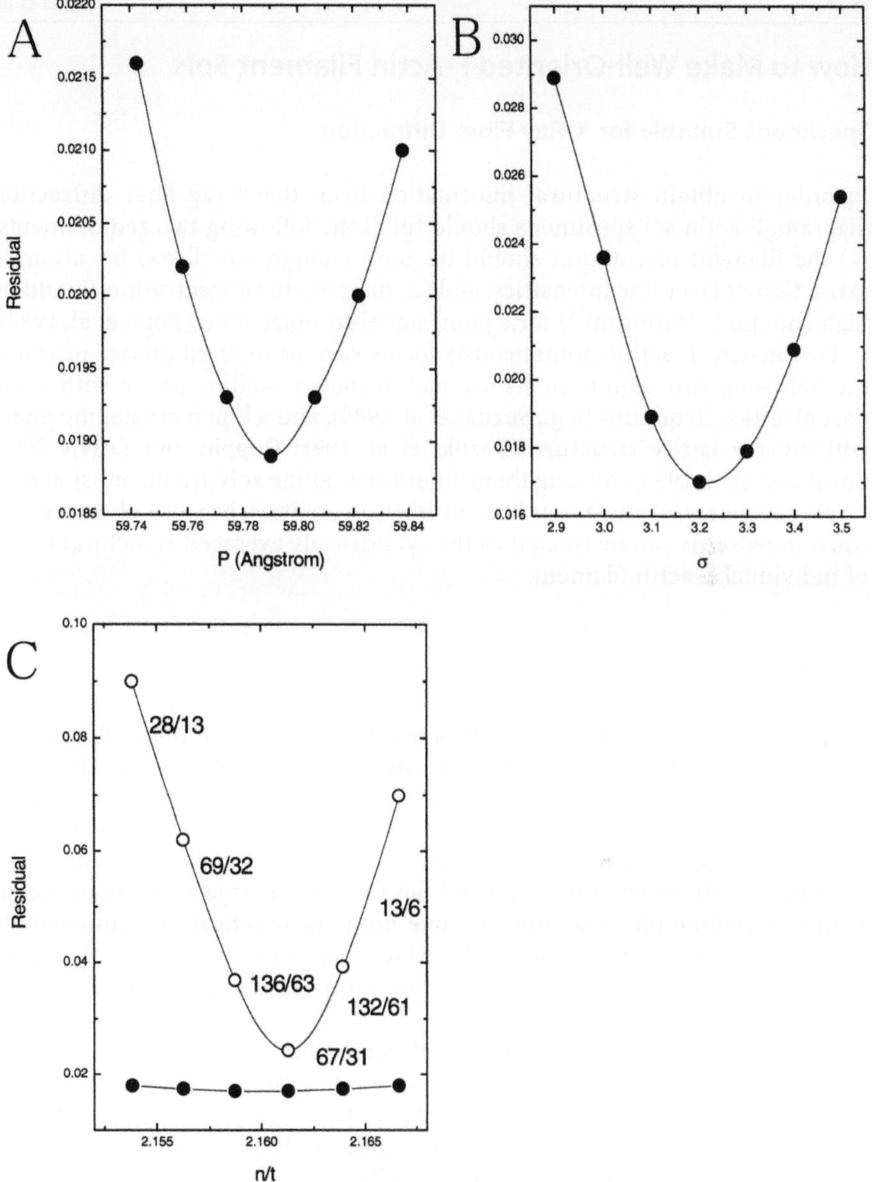

Fig. 2A–C. Residual plots for: **A** the axial position of the 59 Å layer-line which corresponds to the pitch of the one-start helix of the F-actin, P; **B** the standard deviation of filament orientation, σ; and **C** the helical symmetry n/t, the number of subunits per turn. Final values were determined from these plots as those which bring about a minimal value of the local residuals. The local residuals were obtained either in the curved box A (*filled circles*) or in the curved box B (*open circles*) of the diffraction pattern in Fig. 1. In this particular series, all the parameters were obtained from a single diffraction pattern, and the parameters were determined in the following sequence. Firstly, in A, residuals were calculated for various values of P, the axial position of the 59 Å layer-line, as indicated on the abscissa. The minimal residual was obtained with P = 59.79 Å. Then, in B, the best fit σ was estimated to be 3.22°. Finally, in C, the local residuals were calculated in the curved box B for various values of n/t (by *open circles*), clearly indicating that the helical symmetry is close to 67/31 with high accuracy

How to Make Well-Oriented F-Actin Filament Sols

Specimens Suitable for X-Ray Fiber Diffraction

In order to obtain structural information from the X-ray fiber diffraction diagram, F-actin sol specimens should fulfill the following two requirements: (1) the filament orientation should be good enough (small σs) for accurate extraction of layer line intensities; and (2) the protein concentration should be high enough (>100 mg ml^{-1}) for a good signal-to-noise ratio (Popp et al. 1987).

Fortunately, F-actin spontaneously forms various ordered phases in which the following two requirements are met. A paracrystalline phase with some partial lattice structures (e.g., Suzuki et al. 1989), and a liquid crystalline phase without any lattice structure (Suzuki et al. 1991; Coppin and Leavis 1992; Furukawa et al. 1993). Among them, liquid crystalline sols are the most appropriate specimen for the X-ray fiber diffraction analysis, because observed diffraction patterns can be treated as the cylindrically averaged structural factor of individual F-actin filament.

Effect of the Filament Length

To prepare F-actin sols at protein concentrations higher than 100 mg ml^{-1}, filaments in solutions were spun down by slow centrifugation at 11 000 g for approximately 60 h. The resulting soft F-actin sols were drawn into quartz capillaries with diameters of about 0.7 mm. Sols of F-actin filaments polymerized without gelsolin showed mosaic color patterns under a polarization microscope. These small domains did not fuse into a large and homogeneous liquid crystalline phase within the time limits of practical experimentation, namely 2–3 days. This may be attributed to a high viscosity of the sols caused by close packing of the long filaments to a protein concentration as high as 100 mg ml^{-1}.

The viscosity was reduced by controlling the filament length. By polymerizing G-actin in the presence of gelsolin at an actin/gelsolin molar ratio of 100 : 1, the average length of F-actin was ~350 nm. The F-actin sols with this short average length formed a large homogeneous domain spontaneously only several hours after being drawn into a capillary.

This suggests that F-actin sols with shorter average length would form liquid crystals spontaneously in a fairly short time even at a very high filament concentration. On the other hand, Onsager's formulation on predicted liquid crystals is that sols with longer filaments should give rise to better orientation in the liquid crystalline phase (Onsager 1949). We therefore searched for the optimal filament length that compromises the two requirements – that filaments should be short enough to form liquid crystalline phase within a realistic period of time, and the filaments should be long enough to result in better orientation.

The interfilament distance and the standard deviation of angular distribution of the filament orientation were measured from X-ray diffraction patterns from a series of F-actin sols, in which the average lengths of the filament were controlled by the amount of gelsolin added. The results showed that the filament orientation was improved in proportion to the filament length, confirming the Onsager's prediction. The orientation reached the highest limit at the average length of ~0.6 μm, which may be accounted for in terms of the flexibility of F-actin filaments. It was concluded that the optimal filament length for making well-oriented sols is around 0.6 μm.

Effect of Salt Concentration and pH

Electrostatic interactions between charged rod-like particles, being controlled by the salt concentration and the pH of solvent, must strongly influence the formation of the liquid crystalline phase. Therefore, we studied the effect of varying the pH (6–8.5) and ionic strength (30, 60 and 90 mM NaCl) of the solvent on the filament orientation in the sols. Relatively short F-actin filaments (polymerized in the presence of gelsolin at a molar ratio of 100:1) were used for these experiments. The interfilament distances and the standard deviations of the filament orientation were measured from diffraction patterns.

The result indicated that, at any interfilament distance, the smallest standard deviation was obtained at 30 mM NaCl, and the largest at 90 mM NaCl. On the other hand, filament orientation did not depend in any systematic way on the solvent pH between 6 and 8.5.

As described above, lower ionic strength results in better orientation. On the other hand, the complete removal of salt caused depolymerization of F-actin filaments, which clearly increased small-angle scattering from actin monomers. Therefore we needed to find the lowest possible salt concentrations under which the depolymerization of F-actin can be suppressed. One option for further decreasing the salt concentration was to use phalloidin because it inhibits the depolymerization of F-actin filaments.

Effect of Various Kinds of Anions and Cations

Generally, lowering of protein solubility and the resulting precipitation (protein coalescence) is controlled by preferential hydration of the protein. This change is controlled by a balance between cosolvent binding and exclusion from the domain of the protein (Timasheff 1993).

The relative effectiveness in inducing the preferential hydration of proteins is identical to the order of the Hofmeister series, namely the anion series: SO_4^{2-} > CH_3COO^- > Cl^- > NO_3^- > SCN^- (Cacace et al. 1977). Thus, ionic specification depend on the charge density of ion, i.e., the size and charge (Collins 1997). The ions at the left in the series have high charge densities and thus bind water molecules strongly (strongly hydrated ions). The ones to the right of the series have low charge densities and consequently bind water molecule weakly (weakly hydrated ions) relative to the strength of water-water interaction in bulk solution.

We then studied the effect of SO_4^{2-}, CH_3COO^-, Cl^-, NO_3^-, SCN^- as well as $C_6H_5COO^-$ and glutamic acid on the filament orientation in liquid crystalline sols. We used sodium salts of these anions as a replacement of NaCl in the standard solvent conditions. The filament orientations attained in the F-actin sols following concentration using the low speed centrifugation and application of the high magnetic field of 13.5 Tesla, were independent of the anion species. Standard deviations were 3–3.5°.

We also examined the effect of various kinds of cations on the filament orientation in the liquid crystalline sols. For cations, the Hofmeister series are $K^+ > Na^+ > Li^+ > Mg^{2+} > Ca^{2+}$ (Cacace et al. 1977). The results indicated that the attained filament orientation was independent of these monovalent anions; Cs^+, K^+, Na^+, Li^+. However, by adding a divalent cation, for instance, 5 mM $CaCl_2$, the fluidity of the F-actin sols decreased and thus homogeneous domains were not formed spontaneously. It is therefore concluded that filament orientation in the F-actin sols is independent of species of anions and monovalent cations, with exceptions for divalent cations, and is strongly dependent only on the ionic strength.

Practical Procedures – Centrifugation and Magnetic Field

Our measurements indicated that the filaments are better oriented when they are packed more closely. Therefore, we would expect that the higher the filament concentration, the better the filament orientation. Actually, reduction of the interfilament distance by a slow, long centrifugation of sols in capillaries (such as 1400 g for 3 days and 2500 g for 3 days) improved the filament orientation. However, diffraction patterns from some closely packed sols showed lattice sampling at the 51, 59 and 360 Å layer-lines as well as the equator, indicating that these sols were not suitable for the structural analysis. Therefore, there is apparently a practical limit to the improvement of filament orientation by concentrating F-actin filaments.

Strong magnetic fields have been used to obtain highly oriented specimens of biological assemblies including F-actin and its paracrystals (Torbet and Dickens 1984), microtubules (Bras et al. 1998), and flagellar filaments (Yamashita et al. 1998a). We therefore applied a magnetic field of 13.5 Tesla to the F-actin sols and found that the filaments were oriented parallel to the magnetic field. This procedure is a practical and effective way of improving the orientation of F-actin filaments.

Strategy

In conclusion, we have established a procedure that reproducibly achieves highly oriented F-actin sols. The average length of the filaments was controlled at ~0.6 μm by adding gelsolin during the polymerization of the filament. The ionic strength of the solvent should be as low as possible, typically less than 30 mM NaCl, the filaments should be concentrated by very slow and long centrifugation, and the sols should be exposed to a strong (13.5 Tesla) magnetic field.

X-Ray Fiber Diffraction Pattern

Figure 3 illustrates a typical diffraction pattern from a well-oriented F-actin sol prepared by the above procedures. The pattern has a limiting resolution of

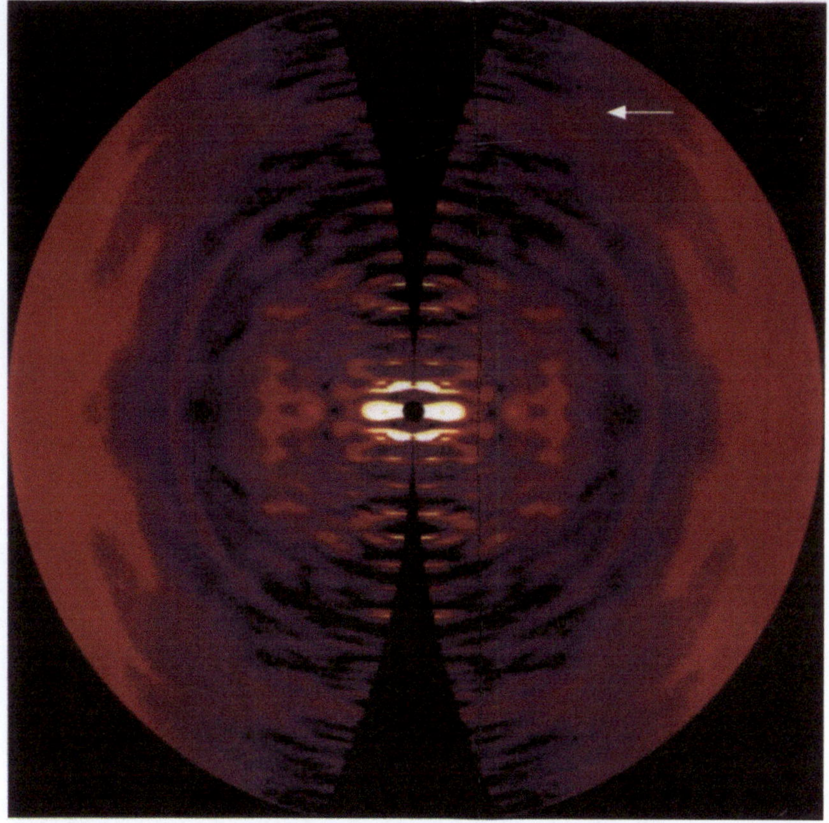

Fig. 3. An X-ray fiber diffraction pattern from well-oriented F-actin sols up to the resolution limit of 3.4 Å. *White arrow* indicates reflections in the region of 5 Å spacing

3.4 Å with a standard deviation of filament orientation of <2°. The diffraction pattern may be divided into two regions and a short explanation is given for each of them.

Small-Angle Region (up to the Resolution of 8 Å) of the X-Ray Diagram

From a small to medium angle region below 0.125 Å^{-1} (1/8 Å), the layer lines are well separated, along which the intensity modulation is clearly observed (Fig. 4). Holmes and co-workers used X-ray fiber diffraction patterns from F-actin sols with a resolution limit of 8 Å (Holmes et al. 1990; Lorenz et al. 1993; Tirion et al. 1995). Unlike ours, they added phalloidin to their F-actin. The stoichiometric binding of phalloidin stabilized the F-actin structure which improved the tolerance of F-actin to damage by the X-ray beam. In the present

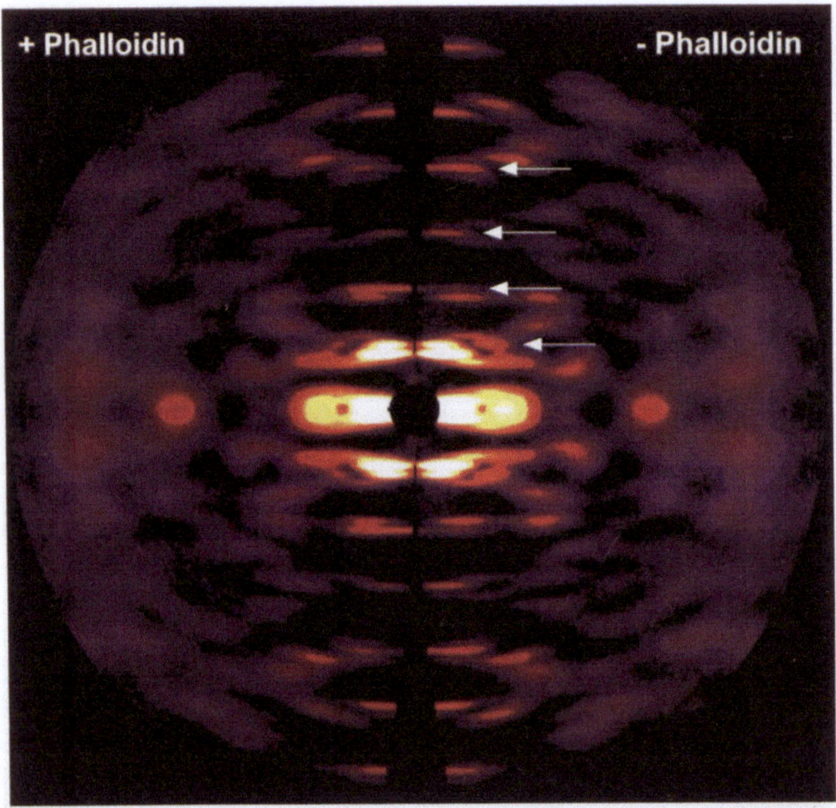

Fig. 4. The small to medium angle region (up to the resolution limit of 8 Å) of X-ray fiber diffraction patterns of F-actin sols with (*left-hand half*) and without (*right-hand half*) phalloidin. The color gradients of the two halves were adjusted by eye for easy comparison. *White arrows* indicate the layer-lines at 51, 27, 18 and 14 Å, *from bottom to top*

work, we compared the diffraction patterns from F-actin sols in the presence (Fig. 4, left half) and absence (Fig. 4 right half) of phalloidin.

Substantial differences were observed between the two patterns, particularly in the intensities of the meridional/near-meridional reflections at the 27, 18 and 14 Å layer-lines. In the presence of phalloidin, the 27 Å reflection was much stronger than those at 18 and 14 Å, whereas in the absence of phalloidin, the 27 Å reflection was much weaker than the others (cf. Zheleznaya et al. 1980).

Since the intensity distribution along the meridian is the Fourier transform of the mass distribution projected onto the filament axis, these differences may be caused by either the additional mass of phalloidin itself or a conformational change of actin subunits. It is also noted that the second, subsidiary peak along the 51 Å layer-line is weakened upon addition of phalloidin.

High-Angle Region (from 8 to 3.4 Å) of the Diffraction Pattern

In the outer part of the pattern (the high–angle region beyond 0.125 $Å^{-1}$) layer-lines were separated only near the meridian, as seen in Fig. 3. However, even if the layer-lines cannot be deconvoluted, some structural information can also be extracted from this region. For instance, the strong layer-line intensities observed at a spacing of ~5 Å as in the case of flagellar filaments (Yamashita et al. 1998a) indicate that α-helices are more or less aligned along the filament axis of F-actin.

Structural Information Obtained from X-Ray Fiber Diffraction Patterns

Helical Structure of F-Actin Filament

To characterize the subunit arrangement in the F-actin filament, we refined two helical parameters: the helical symmetry and the axial pitch (P) of the genetic helix. The helical symmetry is expressed as n/t where n subunits are included in t turns of the one-start helix. As is well known, the helical symmetry of F-actin approximates to 13/6 and the pitch of the one-start, left-handed (genetic) helix is about 59 Å (Hanson and Lowy 1963). Starting with these values, we refined these parameters of F-actin based on the diffraction data.

Pitch of the One-Start Helix

Firstly, we precisely measured the axial position of the 59 Å layer-line which directly gives rise to the pitch of the one-start helix (P), by applying a profile fitting technique to the observed diffraction patterns. A series of intensity profiles were reconstructed using layer-line patterns with various values of P but with a fixed (assumed) helical symmetry of 136/63, and a local residual between the reconstructed and the observed intensities within the curved box area A in Fig. 1 was plotted against P (Fig. 2A). The best–fitting P that produces a minimal residual was obtained by interpolation of the plot. The result is independent of the assumed helical symmetry.

Under the standard solvent condition used in the present work, the pitch, P, was 59.8(2) ± 0.06 Å (average ± standard deviation where n = 44). The pitch of the thin filament in live resting frog sartorius muscle, obtained by Huxley and Brown (1967), corresponds to 59.3 ± 0.03 (or 59.6 ± 0.05) Å, after correcting for the X-ray wavelength of Cu Kα, of 1.537 Å, which was previously used instead of the corrected one, 1.5418 Å. From anterior byssus retractor muscle of *Mytilus edulis*, a molluskan smooth muscle in the live resting state, the pitch of the thin filament was measured to be 59.8 Å (Tajima et al. 1983). The pitch obtained from F-actin in the present study is closer to that of the molluskan smooth muscle thin filament than to that of frog skeletal muscle thin filaments. The pitch of the one-start helix may depend on the isoform of the actin and/or on the binding status of actin-associated proteins.

Helical Symmetry

We then determined the helical symmetry of F-actin as follows. A series of layer line patterns was generated with the position of 59 Å layer-line fixed as determined above and with one of the following helical symmetries: 13/6 (= 2.167), 132/61 (= 2.164), 67/31 (= 2.161), 136/63 (= 2.159), 69/32 (= 2.156) and 28/13 (= 2.154). The values in brackets represent the number of units per turn, n/t. A local residual between the reconstructed and the observed profiles within the curved box area B in Fig. 1 was plotted as a function of n/t. This resulted in the best fit for the axial position of the 51 Å layer-line. The results indicated that a symmetry of 67/31 (= 2.161) was the best match to recorded diffraction patterns (Fig. 2C).

F-actin filaments in the liquid crystal sols do not make strong contacts with each other because the typical interfilament distances of 150 Å in the sols are much larger than the maximum diameter of filament ~90 Å (Lorenz et al. 1993). Where there are strong filament contacts, such as in hydrated Mg^{2+}-paracrystals, the interfilament distance was observed to be ~70 Å (Matsudaira et al. 1983). Therefore the 67/31 parameter is probably from stress-free F-actin filaments.

Extraction of Layer-Line Intensities

Finally, using the determined helical parameters, the intensity profiles of the layer-lines were extracted from the diffraction patterns in the region between

0.012 and 0.05 Å^{-1}, i.e., basically the Bessel terms below the sixth order. In the region near the origin, the intensities along 360 Å and 180 Å layer-lines, as well as on the equator, were shadowed by the beam stop, therefore these intensities could not be determined. Also the meridional intensities were not measured because we did not tilt the specimen.

The overall intensity profile as well as first peak positions of the layer-line intensities are similar to those (in the region below 0.04 Å^{-1}) obtained by a computer Fourier transform of cryo-EM pictures of isolated F-actin filaments (Trinick et al. 1986; Milligan et al. 1990; Lepault et al. 1994). This suggests that at low resolution, X-ray diffraction and electron microscopy may "see" F-actin in the same way, despite the fact that the physics of the interaction of materials with X-rays and electrons is different, and that a 3D model of F-actin filament can be constructed by combination of X-ray diffraction intensities (amplitudes) and phase information by cryo-EM (Namba et al. 1989b). A study of flagellar filaments from *Salmonella* (Mimori et al. 1995) confirmed the similar intensity profiles by X-ray and cryo-EM up to at least 20 Å resolution.

Relation Between F-Actin Filament Disorienation and the Quality of Diffraction Patterns

Even using our F-actin preparative procedures, filament orientation was variable from capillary to capillary and for different spots irradiated by the X-ray beam. In order to quantitatively compare two patterns of different degrees of orientation, it was desirable to estimate the reliability of the comparison. For this purpose, we calculated the R-factors (the consistency between the intensity profiles) in the region between 0.012 and 0.042 Å^{-1} and compared them with a standard diffraction pattern with a filament disorientation of 3.22°. The R-factor is defined as:

$$R = \sum \left| I_b^{1/2} - k \cdot I_a^{1/2} \right| / \sum I_b^{1/2} \times 100\,(\%)$$

where I_a is the layer-line intensities of data set a, I_b is the layer-line intensities of the standard pattern, and k is the scaling factor.

The R-factor is plotted against the filament orientation, as shown in Fig. 5. It is highly dependent on the disorientation of filament. This indicates that, as described in the previous sections, the quality of diffraction pattern (therefore the amount of information extracted from the diffraction pattern) is strongly dependent on filament orientation. Moreover, in this particular case, the R-factor reaches the minimal value 4.8 ± 0.7% at a disorientation of 3.5°. This is close to the disorientation (3.22°) of the standard pattern. It is therefore concluded that the two diffraction patterns could be compared quantitatively only if the difference in the filament disorientation is within ±0.3°.

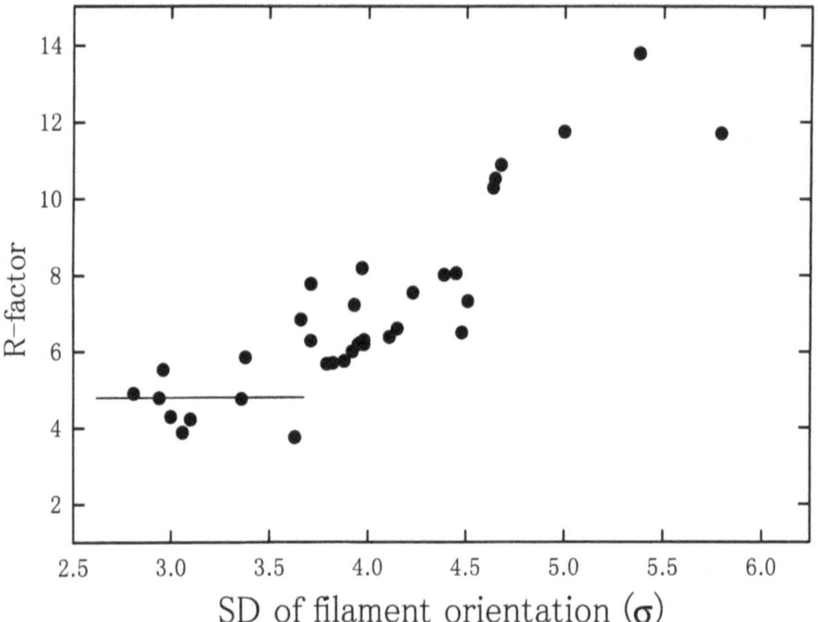

Fig. 5. Relationship between the R-factor and the standard deviation (σ) of the angular distribution of filament orientation. The R-factors were calculated for the layer-line intensities below the resolution limits of 20 Å, against a standard intensity profile. The standard intensity profile used in this particular case was extracted from the diffraction pattern having a standard deviation of 3.22°

Conclusions

The procedures for preparing well-oriented F-actin sols suitable for X-ray fiber diffraction have been established. Using such F-actin sols, the parameters have been precisely determined which specify the helical arrangement of actin subunits within F-actin. The influence of solvent conditions, such as pH and salts, on the structure of F-actin is now under way with respect to helical symmetry and subunit conformation (T. Oda et al. in prep.). The X-ray fiber diffraction diagrams from the F-actin sols should also be useful in obtaining the 3D structure of F-actin filament at a high resolution.

Acknowledgments. We thank Dr. K. Hasegawa for much technical help in recording and processing the diffraction patterns.

References

Bras W, Diakun GP, Diaz JF, Maret G, Kramer H, Bordas J, Medrano FL (1998) The susceptibility of pure tubulin to high magnetic fields: a magnetic birefringence and X-ray fiber diffraction study. Biophys J 74:1509–1521

Cacace M, Landau EM, Ramsden JJ (1977) The Hofmeister series: salt and solvent effects on interfacial phenomena. Q Rev Biophys 30:241–277

Chik JK, Lindberg U, Schutt C (1996) The structure of an open state of β-actin at 2.65 Å resolution. J Mol Biol 263:607–623

Cohen C, Lowy J (1953) An X-ray diffraction study of F-actin. Biochim Biophys Acta 21:177–178

Collins KD (1997) Charge density-dependent strength of hydration and biological structure. Biophys J 72:65–76

Coppin CM, Leavis PC (1992) Quantitation of liquid-crystalline ordering in F-actin solutions. Biophys J 63:794–807

DeRosier DJ, Klug A (1968) Reconstruction of three dimensional structure from electron micrographs. Nature 217:130–134

Furukawa R, Kundra R, Fechheimer M (1993) Formation of liquid crystals from actin filaments. Biochemistry 32:12346–12352

Gillis JM, O'Brien EJ (1975) The effect of calcium ions on the structure of reconstituted muscle thin filament. J Mol Biol 99:445–459

Hanson J, Lowy S (1963) The structure of F-actin and of actin filaments isolated from muscle. J Mol Biol 6:46–60

Holmes KC (1995) Solving the structure of macromolecular complexes with the help of X-ray fiber diffraction diagrams. J Struct Biol 115:151–158

Holmes KC, Barrington Leigh J (1974) The effect of disorientation on the intensity distribution of non-crystalline fibers. I. Theory. Acta Crystallogr A30:635–638

Holmes KC, Popp D, Gebhard W, Kabsch W (1990) Atomic model of the actin filament. Nature 347:44–49

Huxley HE, Brown W (1967) The low-angle X-ray diagram of vertebrate striated muscle and its behavior during contraction and rigor. J Mol Biol 30:383–434

Janmey PA, Peetermans J, Zaner KS, Stossel TP, Tanaka T (1986) Structure and mobility of actin filaments as measured by quasielastic light scattering, viscometry, and electron microscopy. J Biol Chem 261:8357–8362

Kabsch W, Mannherz HG, Suck D, Pai EF, Holmes KC (1990) Atomic structure of the actin: DNase I complex. Nature 347:37–44

Kurokawa H, Fujii W, Ohmi K, Sakurai T, Nonomura Y (1990) Simple and rapid purification of brevin. Biochem Biophys Res Commun 168:451–457

Lepault J, Ranck J-L, Erk I, Carlier M-F (1994) Small angle X-ray scattering and electron cryomicroscopy study of actin filaments: role of the bound nucleotide in the structure of F-actin. J Struct Biol 112:79–91

Lorenz M, Popp D, Holmes KC (1993) Refinement of the F-actin model against X-ray fiber diffraction data by the use of a directed mutation algorithm. J Mol Biol 234: 826–836

MacLean-Fletcher SD, Pollard TD (1980) Identification of a factor in conventional muscle actin preparations which inhibits actin filament self-association. Biochem Biophys Res Commun 96:18–27

Makowski L (1991) An estimate of the number of structural parameters measurable from a fiber diffraction pattern. Acta Crystallogr A47:562–567

Matsudaira P, Mandelkow E, Renner W, Hesterberg LH, Weber K (1983) Role of fimbrin and villin in determining the interfilament distances of actin bundles. Nature 301:209–214

McLaughlin PJ, Gooch JT, Mannherz HG, Weeds AG (1993) Structure of gelsolin segment 1-actin complex and the mechanism of filament severing. Nature 364:685–692

Milligan RA, Whittaker M, Safer D (1990) Molecular structure of F-actin and location of surface binding sites. Nature 348:217–221

Mimori Y, Yamashita I, Murata K, Fujiyoshi Y, Yonekura K, Toyoshima C, Namba K (1995) The structure of the R-type straight flagellar filament of *Salmonella* at 9 Å resolution by electron cryomicroscopy. J Mol Biol 29:69–87

Moore PB, Huxley HE, DeRosier DJ (1970) Three-dimensional reconstruction of F-actin, thin filaments and decorated thin filaments. J Mol Biol 50:279–295

Namba K, Pattanayek R, Stubbs G (1989a) Visualization of protein-nucleic acid interactions in a virus. Refined structure of intact tobacco mosaic virus at 2.9 Å resolution by X-ray fiber diffraction. J Mol Biol 208:307–325

Namba K, Yamashita I, Vondervist F (1989b) Structure of core and central channel bacterial flagella. Nature 342:648–654

Oda T, Makino K, Yamashita I, Namba K, Maéda Y (1998) Effect of the length and effective diameter of F-actin on the filament orientation in liquid crystalline sols measured by X-ray fiber diffraction. Biophys J 75:2672–2681

Onsager L (1949) The effects of shape on the interaction of colloidal particles. Ann N Y Acad Sci 51:627–659

Popp D, Lednev VV, Jahn W (1987) Methods of preparing well-orientated sols of F-actin containing filaments suitable for X-ray diffraction. J Mol Biol 197:679–684

Schutt CE, Myslik JC, Rozycki MD, Goonesekere NCW, Lindberg U (1993) The structure of crystalline profilin-β-actin. Nature 365:810–816

Spencer M (1969) Low-angle X-ray diffraction from concentrated sols of F-actin. Nature 223:1361–1362

Spudich JA, Watt S (1971) The regulation of rabbit skeletal muscle contraction. I. Biochemical studies of the interaction of the tropomyosin-troponin complex with actin and the proteolytic fragments of myosin. J Biol Chem 246:4866–4871

Stowell MHB, Miyazawa A, Unwin N (1998) Macromolecular structure determination by electron microscopy: new advances and results. Curr Opin Struct Biol 8:595–600

Suzuki A, Yamazaki M, Ito T (1989) Osmoelastic coupling in biological structures: Formation of parallel bundles of actin filaments in a crystalline-like structure caused by osmotic stress. Biochemistry 25:6513–6518

Suzuki A, Maéda T, Ito T (1991) Formation of liquid crystalline phase of actin filament solutions and its dependence on filament length as studied by optical birefringence. Biophys J 59:25–30

Tajima Y, Kamiya K, Seto T (1983) X-ray structure analysis of thin filaments of a molluscan smooth muscle in the living relaxed state. Biophys J 43:335–343

Timasheff SN (1993) The control of protein stability and association by weak interactions with water: how do solvents affect these processes? Ann Rev Biophys Biomol Struct 22:67–97

Tirion M, ben-Avraham D, Lorenz M, Holmes KC (1995) Normal modes as refinement parameters for the F-actin model. Biophys J 68:5–12

Torbet J, Dickens MJ (1984) Orientation of skeletal muscle actin in strong magnetic fields. FEBS Lett 173:403–406

Trinick J, Cooper J, Seymour J, Egelman EH (1986) Cryo-microscopy and three-dimensional reconstruction of actin filaments. J Microsc 141:349–360

Yamashita I, Suzuki H, Namba K (1998a) Multiple-step method for making exceptionally well oriented liquid-crystalline sols of maromolecular assemblies. J Mol Biol 278:609–615

Yamashita I, Hasegawa K, Suzuki H, Vondervist F, Mimori-Kiyosue Y, Namba K (1998b) Structure and switching of bacterial flagellar filament studied by X-ray fiber diffraction. Nat Struct Biol 5:125–132

Zheleznaya LA, Mevkh NG, Gherasimov VS (1980) X-ray diffraction study of F-actin induced by phalloidin. Naturwissenschaften 67:363–364

Analysis of Models of F-Actin Using Fluorescence Resonance Energy Transfer Spectroscopy

Pierre D. J. Moens[1] and Cristobal G. dos Remedios[2]

Introduction

Fluorescence resonance energy transfer spectroscopy (FRET) has been widely used to determine distances ranging from 10 to 100 Å between amino acids within proteins. However, FRET is not very good at measuring atomic distances because it measures the distance between two probes attached to an amino acid but not the position of the amino acid itself. Therefore, one has to take into consideration the size of the probes used to interpret the FRET data. FRET compensates for that deficit by being particularly sensitive to changes in distance and therefore is suitable to study conformational change in proteins (dos Remedios and Moens 1995a, 1995b).

In 1981, Taylor et al. (1981) used FRET to determine the radial coordinate of Cys-374 of actin and study the assembly of actin filament. Subsequently, several authors have used this method to determine the radii of Gln-41 (Kasprzak 1988) the nucleotide binding site (Miki et al. 1986b; Kasprzak 1988) Cys-10 (Miki et al. 1986) and Cys-374 (Kasprzak 1988; Moens et al. 1994). When myosin S1 binds to actin filaments, the radial coordinate of Gln-41 increases by 3 Å (Kasprzak 1988) and we showed that the radius of Cys-374 increases by approximately 4.5 Å (Moens et al. 1997).

Radial coordinate determinations when used to study conformational changes in a filament are limited in the sense that a large change in the position of a probe and hence of the labeled residue is not necessarily detected. Indeed, measurements of radial coordinates do not detect rotations around the filament axis or translations along that axis. Therefore, to study conformational changes in detail, we need to localize the probe in the three dimensions.

Miki et al. (1986b) devised a novel method to determine the radial coordinate of a residue in F-actin, knowing the radial coordinate of a second site and the intramonomer spatial relationship between the two residues. Therefore, if we inverse the problem, i.e., if we know the radii for the two residues, we should be able to get information about their spatial relationship. Three parameters

[1] Department of Genetics & Molecular Biology, 1960 East-West Road, A209 Honolulu, Hawaii 96822

[2] Muscle Research Unit, Department of Anatomy and Histology, Institute for Biomedical Research, University of Sydney, Sydney 2006, Australia

Results and Problems in Cell Differentiation, Vol. 32
C. dos Remedios (Ed.): Molecular Interactions of Actin

define this relationship: the intramonomer distance (R_{intra}); the height (h); and the angle (ϕ) between the two residue (Fig. 1a). If two of these parameters are known, we can deduce the third.

Here, we use the nucleotide binding site as the reference point in the monomer with a radial coordinate for the fluorescent ADP analogue of 25 Å (Miki et al. 1986a,b) and a radius of 13.5 Å for the probe fixed to Cys-374 (Moens and dos Remedios 1997). We designed two types of experiment where the efficiency of transfer is measured between the nucleotide and the probe bound to Cys 374 in F-actin, but one is particularly sensitive to the R_{intra} distance. By combining the results obtained and the radial coordinate distances, we can position the probe in 3D. These coordinates are then compared with the refined F-actin models of Lorenz et al. (1993), Tirion and Benavraham (1993), and the original model of F-actin published by Holmes et al. (1990).

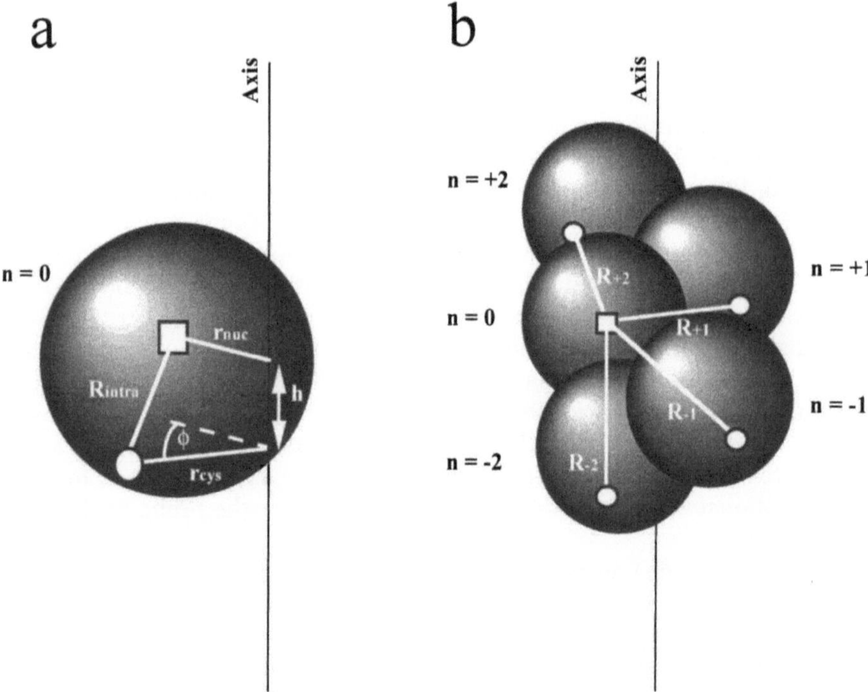

Fig. 1a,b. Spatial relationship and distances between two nonhomologous sites in actin filaments. The monomers are represented as *shaded spheres*. **a** The spatial relationship between the nucleotide-binding site (□) and Cys-374 (○) in one monomer is defined by the radial coordinates of the two sites (r_{nuc} and r_{cys}), the intramonomer distance (R_{intra}), the angle (ϕ), and the height (h) between the two sites. **b** The FRET efficiency in a filament will be a function of the distances (R_{-2}, R_{-1}, R_{+1}, R_{+2}) separating a donor labeled monomer (n = 0) to acceptors bound to up to four adjacent monomers ($n = -2, -1, +1, +2$)

Protein Preparation

Actin was prepared from an acetone powder of rabbit skeletal muscle by the method of Spudich and Watt (1971) with minor modifications (Barden and dos Remedios 1984). The concentrations of unlabeled G-actin was determined spectrophotometrically using an extinction coefficients of $\varepsilon = 6.3\,cm^{-1}$ (0.1%, 290 nm).

Labeling of Cys-374

The labeling of Cys-374 with an acceptor probe, N-(4-dimethyl-amino-3,5-dinitrophenyl)-maleimide (DDPM), from Aldrich (Milwaukee, Wisconsin) was performed in accordance to the methods of Miki and Mihashi (1978). Unbound labels were removed from actin by passing it through a polymerization-depolymerization cycle followed by dialysis in the presence of 1% dimethylformamide, clarified at 130000 g for 60 min at 4°C, and then passed through a G-25 Sephadex column (2 × 10 cm).

Labeling of the Nucleotide-Binding Site of Unlabeled and DDPM-Labeled Actin

In order to replace the bound ATP with ε-ATP (Molecular Probes Inc., Eugene, Oregon) unlabeled and DDPM-actin ($\simeq 2\,mg\,ml^{-1}$) were passed through a Dowex-1 column (1/2 vol Dowex over actin vol). A ten fold molar excess of ε-ATP pH 8 was immediately added and the actin was incubated overnight at 4°C.

Experimental Design

We designed two types of experiments that are both dependent on the angle (ϕ) and the intramonomer distance (R_{intra}). However, one is very sensitive to R_{intra}. We further refer to them as angle-R_{intra} determination and R_{intra}-angle determination (for the one sensitive to that distance).

Actin Polymerization and Determination of the Angle-R_{intra}

Just prior to polymerization, the DDPM-labeled actin and ε-ATP-actin were passed through, Dowex-1 columns twice to remove free ATP in the DDPM-actin and the excess ε-ATP from the ε-ATP-actin samples. Following the Dowex treatment, the concentration of DDPM-labeled actin was determined spectrophotometrically using an exctinction coefficient $\varepsilon_{440} = 3.0 \times 10^3$ (Miki and

Table 1. Determination of the acceptor molar fraction for various actin samples

DDPM-actin vol (ml)	ε-ATP-actin vol (ml)	DDPM-actin conc. (μM)	DDPM conc. (μM)	ε-ATP-actin conc. (μM)	Final DDPM conc. (μM)	Total actin conc. (μM)	Acceptor molar fraction
0	1	–	–	55.3	0	55.3	0
0.3	0.7	36.2	23.1	56.3	6.9	50.3	0.14
0.4	0.6	38	28	56.4	11.2	49	0.23
0.5	0.5	34.4	31.7	54.9	15.8	44.6	0.35
0.6	0.4	38.2	26.6	44.7	15.9	40.8	0.39
0.7	0.3	44.3	31.5	45.3	22	44.6	0.49
0.9	0.1	47	31.7	34.2	28.5	45.7	0.62

The final volume for each sample was 1 ml. The concentration (conc.) of proteins and labels were determined for each sample after the Dowex-1 treatment.

Mihashi 1978). A small sample of each solution was kept for a later determination of protein concentration. The samples were mixed to obtained different acceptor molar fractions as described in a previous paper (Moens and dos Remedios 1997) and shown in Table 1. The samples were then added to the polymerization buffer containing 20 mM phosphate buffer pH 7, 50 mM KCl, and 2 mM $MgCl_2$ and a twofold molar excess of phalloidin (Molecular Probes Inc., Eugene, Oregon). All the steps from the Dowex treatment to the polymerization are executed in less than 3 min to avoid denaturation of actin monomers.

Actin Polymerization for R_{intra}-Angle Determination

The same procedure was applied for the R_{intra}-angle determination but different acceptor molar fractions were obtained by mixing the double-labeled DDPM-ε-ATP-actin with unlabeled actin.

Actin Concentration After Dowex-1

The protein concentration was measured for each sample after the Dowex treatment and determined from the absorbance at 595 nm using the Coomassie protein assay reagent (Bradford 1976). A range of unlabeled actin concentration was used to construct a standard curve. Actin concentrations ranged from 20.4 to 55.3 μM. Following the fluorescence measurements, we checked for free labels and the presence of denatured actin by sedimentation experiments. In all samples, less than 1 μM of probes was detected in the supernatant and the actin concentrations were beyond the detection level of the Bradford assay (data not shown).

Fluorescence Measurements

The fluorescence lifetimes of the donor probe were recorded using a photon counting EEY nanosecond fluorimeter. The excitation wavelength was selected using a filter centered at 350 nm with bandpass of 70 nm and the emission wavelength was selected using a bandpass filter centered at 415 nm having a width of 40 nm. Samples were temperature controlled at 16°C. The lifetime data were analyzed using a sum of two exponentials (Moens and dos Remedios 1997) and the efficiency of transfer E_{obs} is given by:

$$E_{obs} = 1 - (\tau_{da}/\tau_d), \tag{1}$$

with τ_{da} being the lifetime of the donor probe in an actin filament in the presence of acceptors and τ_d the lifetime of the donor probe in an actin filament without acceptors.

Theoretical Calculations

FRET in actin filaments can occur between a donor labeled monomer (n = 0) and the four adjacent monomers when they are labeled with acceptor probes (n = −2, −1, +1, +2). Monomers above or below these four adjacent monomers do not significantly participate in the FRET efficiency because the distance is larger than 100 Å (i.e. more than twice the R_0 value). The FRET efficiency is determined by the distance separating the donor and acceptor probe and the Förster distance (R_0):

$$E = R_0^6/(R_0^6 + R^6). \tag{2}$$

R_0 is the distance at which the efficiency of energy transfer is 50%. This distance is a function of the probe pair and is 31 Å for ε-ATP:DDPM probe pair (Miki & Mihashi, 1978). However, Eq. (2) is only valid when the distance is measured between a donor and a single acceptor probe. In a filament, there can be up to four acceptors for one donor probe and the FRET efficiency for each arrangement of acceptors around a donor is given by:

$$E_k = R_0^6 \left\{ R_0^6 + \left(1 \middle/ \left[\sum_N R_n^{-6} \right] \right) \right\}^{-1} \tag{3}$$

where k is one of the 16 possible arrangements of acceptors around a donor, N is the number of acceptors for the kth arrangement and n is the position of the acceptor labeled monomers (n = −2, −1, +1, +2) in relation with the donor-labeled monomer (n = 0). Because of the helical symmetry of the actin filament (a monomer is translated by 27.5 Å and rotated by −166.15° relative to the previous one in the genetic helix), the four distances R_{-2}, R_{-1}, R_1 and R_2 (Fig. 1b) separating the donor probe (□: ε-ATP at the nucleotide binding site) and the acceptor probe (O:DDPM bound to Cys-374) are given by:

$$R_n^2 = (27.5n + h)^2 + r_{nuc}^2 + r_{cys}^2 - 2r_{nuc}r_{cys}\cos(-166n + \phi) \quad (4)$$

and

$$h^2 = R_{intra}^2 - r_{nuc}^2 - r_{cys}^2 + 2r_{nuc}r_{cys}\cos\phi \quad (5)$$

where r_{nuc}, r_{cys} are the radial coordinates of ε-ATP and DDPM, respectively, and h, ϕ and R_{intra} are the height, angle and intramonomer distance between ε-ATP and DDPM bound to the protein, respectively.

For a given R_{intra} and angle ϕ, we can calculate using Eq. (3) the efficiency of transfer for each arrangement of acceptors around a donor probe. Then, the FRET efficiency in a filament (E_t) is the sum of the efficiency of transfer for each kth arrangement E_k multiplied by its probability δ_k. The sum is taken over all possible arrangements (K) and is given by:

$$E_t = \sum_{k=1}^{K} \delta_k E_k \quad (6)$$

When this equation is computed for each acceptor molar ratio (0, 0.01, . . .1), one can build a series of theoretical curves of transfer efficiency versus acceptor molar ratio for different R_{intra} and angles. The experimental FRET data are then plotted together with the theoretical curve and aχ^2 value is calculated for each R_{intra} and angle ϕ using:

$$\chi^2 = \sum_i \left[(E_{obs} - E_t)^2 / E_t \right] \quad (7)$$

where i is the number of FRET data, E_{obs} is the FRET data at a given acceptor molar fraction and E_t is the theoretical efficiency of transfer at that acceptor molar fraction. The best fit corresponds to the smallest χ^2 value.

Determination of Cys-374 Radial Coordinate

Figure 2 shows the variation of χ^2 as a function of the radii of the probes bound to Cys-374 using the FRET data published by Moens and dos Remedios (1997). Low χ^2 value were obtained for radial coordinates ranging from 12.5 to 15.5 Å. The radial coordinates of the probes were re-calculated using the FretLab software developed in our laboratory and available at: http://perso.libertysurf.fr/PierreMoens/index.htm The mean radius of the probe bound to Cys-374 is 13.7 ± 0.2 Å (mean ± sem). Figure 3 is a plot of the FRET data (•) with the mean radius (solid line).

Determination of the Angle-R_{intra}

In these experiments, FRET occurs from ε-ATP-labeled monomers (donor) to different monomers labeled with DDPM (acceptors). A lifetime of 32.2 ns

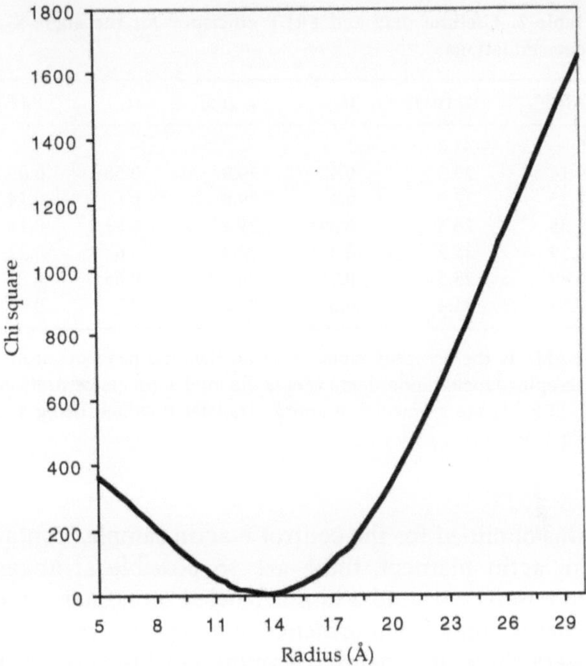

Fig. 2. Variations of χ^2 for Cys-374 radial coordinate as a function of the radius. Low χ^2 values are obtained for radii from 12.5 to 15.5 Å

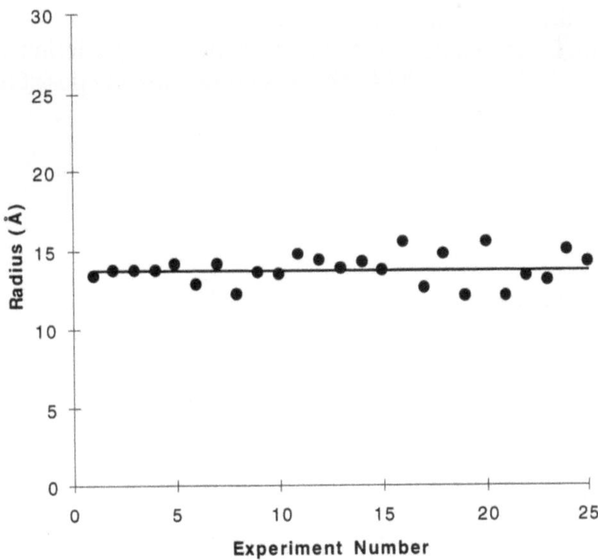

Fig. 3. Cys-374 radii calculated using the FretLab software. Each *filled circle* represents a separate experimental point. *Solid line* is the mean radius of 13.5 Å

Table 2. Lifetime data and FRET efficiency for the angle-R_{intra} determinations

AMF[a]	τ_1 (ns)	α_1	τ_2 (ns)	α_2	FRET[b]
0	32.2	1	–	–	–
0.14	29.3	0.42	30.9	0.58	0.09
0.23	27.8	0.4	29.8	0.6	0.14
0.35	26.5	0.39	29.4	0.61	0.18
0.39	25.2	0.4	27.4	0.6	0.22
0.49	23.5	0.35	26.8	0.65	0.27
0.62	21.4	0.5	25.8	0.5	0.33

[a] AMF is the acceptor molar fraction (i.e., the concentration of acceptor-labeled monomers versus the total actin concentration).
[b] FRET is the efficiency of energy transfer obtained using τ_1 in Eq. (1).

was obtained for the control F-actin samples containing only donor probe. In an actin filament, there are 16 possible arrangements of acceptor-labeled monomers around a donor-labeled monomer. As a consequence, the lifetime curves obtained in presence of acceptors are a sum of 16 exponentials. These were therefore analyzed assuming a bi-exponential curve using the smaller lifetime to calculate the FRET efficiency (Moens & dos Remedios, 1997). Table 2 shows an example of lifetimes for different acceptor molar fractions. As expected from Eq. (6), the efficiency of transfer increases as a function of the acceptor molar fraction. Using a radial coordinate of 25 Å for the ε-ATP probe at the nucleotide binding site (Miki et al. 1986), we calculated χ^2 (Fig. 4a) for angles (ϕ) from −90° to 90° and allowed a variation in the intramonomer distance from 25 to 36 Å. There is more than one possible fit of the FRET data and this is illustrated in Fig. 4b. The best fits for a particular angle ϕ and R_{intra} distance are cross-hatched. It is clear from this figure that if the R_{intra} distance is restricted from 26 to 34 Å, each angle ϕ has an R_{intra} value which fits the FRET data (white areas).

Determination of the R_{intra}-Angle

In order to restrict the possible number of fits of the FRET data, we double-labeled actin monomers and mixed the ε-ATP-DDPM actin with unlabeled actin. Under these conditions, energy transfer occurs between the ε-ATP and the DDPM probes within the monomer but also with up to four DDPM probes bound to adjacent monomers. Therefore Eq. (3) has to be rewritten as:

$$E_k = R_0^6 \left\{ R_0^6 + 1 \Big/ \left[R_{intra}^{-6} + \sum_N R_n^{-6} \right] \right\}^{-1} \qquad (8)$$

a

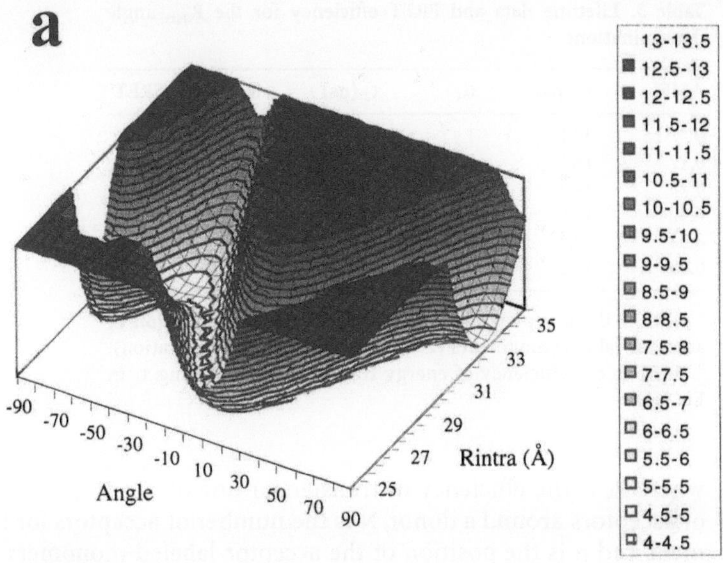

b

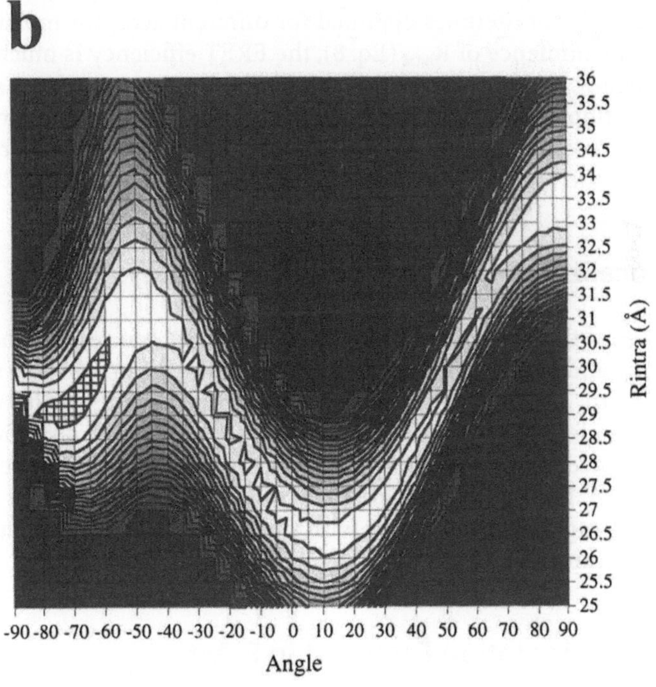

Fig. 4a,b. Surface map of the χ^2 variations for the angle-R_{intra} determination as a function of the angle and intramonomer distances. **a** Values of χ^2 greater than 13.5 have not been plotted. **b** Top view of **a**, the lowest χ^2 values are highlighted by *cross-hatching*

Table 3. Lifetime data and FRET efficiency for the R_{intra}-angle determinations

AMF[a]	τ_1 (ns)	α_1	τ_2 (ns)	α_2	FRET[b]
0	32.8	1	–	–	–
0.17	14.7	0.35	21.8	0.65	0.55
0.28	14.1	0.3	19.7	0.7	0.57
0.40	12.9	0.3	17.9	0.7	0.60
0.54	12.6	0.3	17.5	0.7	0.61
0.73	12.2	0.3	16.6	0.7	0.63

[a] AMF is the acceptor molar fraction (i.e., the concentration of acceptor-labeled monomers versus the total actin concentration).
[b] FRET is the efficiency of energy transfer obtained using τ_1 in Eq. (1).

where E_k is the efficiency of transfer for one of the 16 possible arrangements of acceptors around a donor, N is the number of acceptors for that k^{th} arrangement and n is the position of the acceptor-labeled monomers (n = −2, −1, +1, +2) in relation with the donor-labeled monomer (n = 0). Table 3 shows some examples of lifetimes obtained for different acceptor molar fractions. Because of the influence of R_{intra} (Eq. 8), the FRET efficiency is much higher than in the angle-R_{intra} determination (Table 2). Figure 5a represents a surface plot of χ^2 in function of R_{intra} and the angle ϕ The best fits obtained for these experiments are cross-hatched in Fig. 5b. Because these experiments are more sensitive to R_{intra}, the range of possible R_{intra} values is limited to 27 to 29 Å.

Localization of the Probe Bound to Cys-374

Even if there is a good fit for every angle, we can restrict the possible values of R_{intra} and ϕ by combining the results obtained from the two sets of experiments (Fig. 6). A low χ^2 value is obtained in both experiments for an R_{intra} of 27 Å and an angle ϕ of −70° to −60°, −20° to −10° and 35° to 40° (Fig. 6). The best fit is obtained for an angle of −15°. Using these values in Eq. 5, we calculated a height of 24 Å. Figure 7a and b are plots of the FRET data (□) for the angle-R_{intra} and R_{intra}-angle determinations with the theoretical curves calculated with the best fit parameters, respectively.

The Orientation Factor κ^2 in F-Actin

The distances measured by FRET and therefore the localization of the probes in F-actin rely on the assumption that $\kappa^2 = 2/3$. Censullo et al. (1992) demonstrate that for the determination of radial coordinates in actin filaments, the assumption that $\kappa^2 = 2/3$ can be justified. This is because of the averaging of

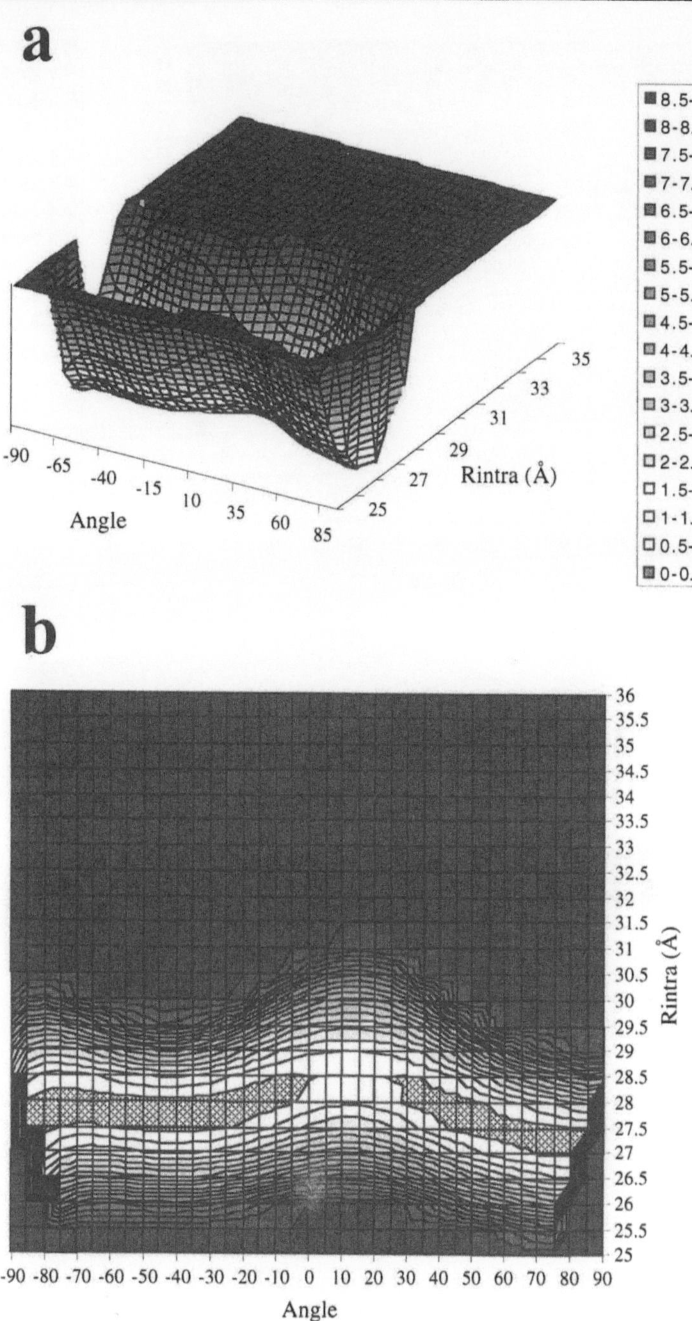

Fig. 5a,b. Surface map of the χ^2 variations for the R_{intra}-angle determination as a function of the angle and intramonomer distances. **a** Values of χ^2 greater than 9 have not been plotted. **b** Top view of **a**, the lowest χ^2 values are highlighted by *cross-hatching*

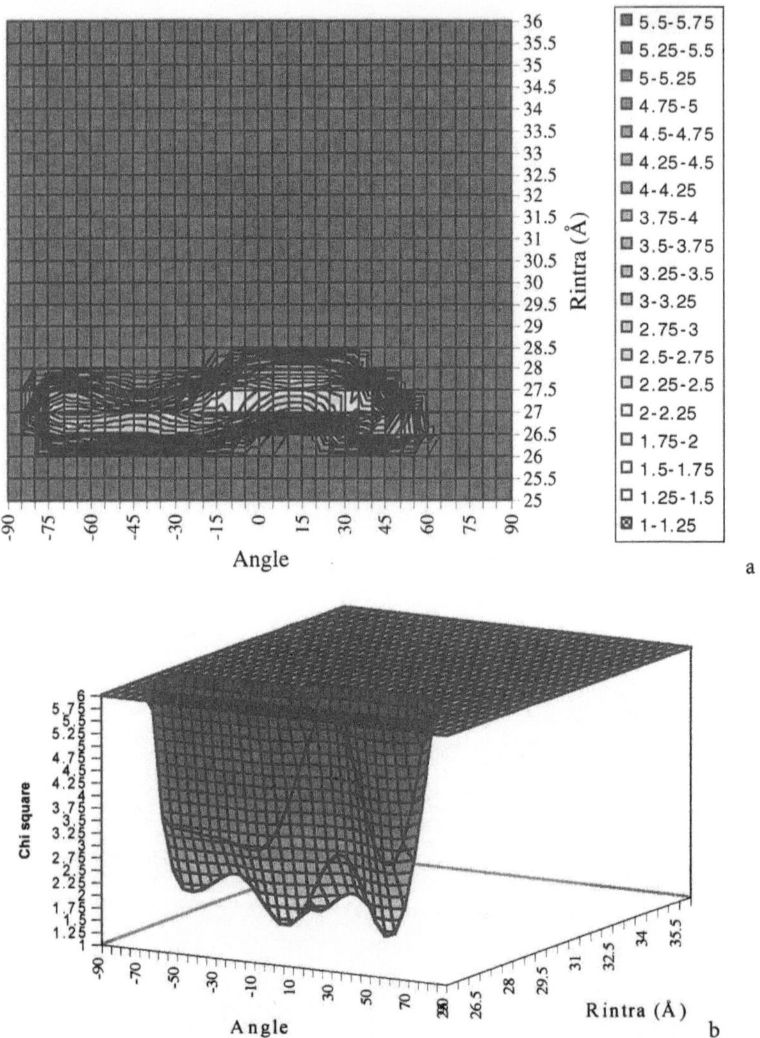

Fig. 6a,b. Surface map of the normalized χ^2 variations for the combined angle-R_{intra} and R_{intra}-angle determinations as a function of the angle and intramonomer distances. Values of χ^2 greater than 5.75 have not been plotted. The lowest χ^2 values are for an R_{intra} of 27 Å and an angle ϕ of –70° to –60°, –20° to –10° and 35° to 40°

κ^2 due to the different orientations of the acceptor dipoles in the four adjacent donor-labeled monomers. This is probably also true in our experiments. However, in radial coordinate measurements, the distances separating the donor and the acceptors are equal two by two. Using a radial coordinate of the probe of 13.5 Å, R_{-2} and R_{+2} = 55.4 Å and R_{-1}, R_{+1} = 38.4 Å. In the experiments described here, the energy transfer occurs between probes bound to

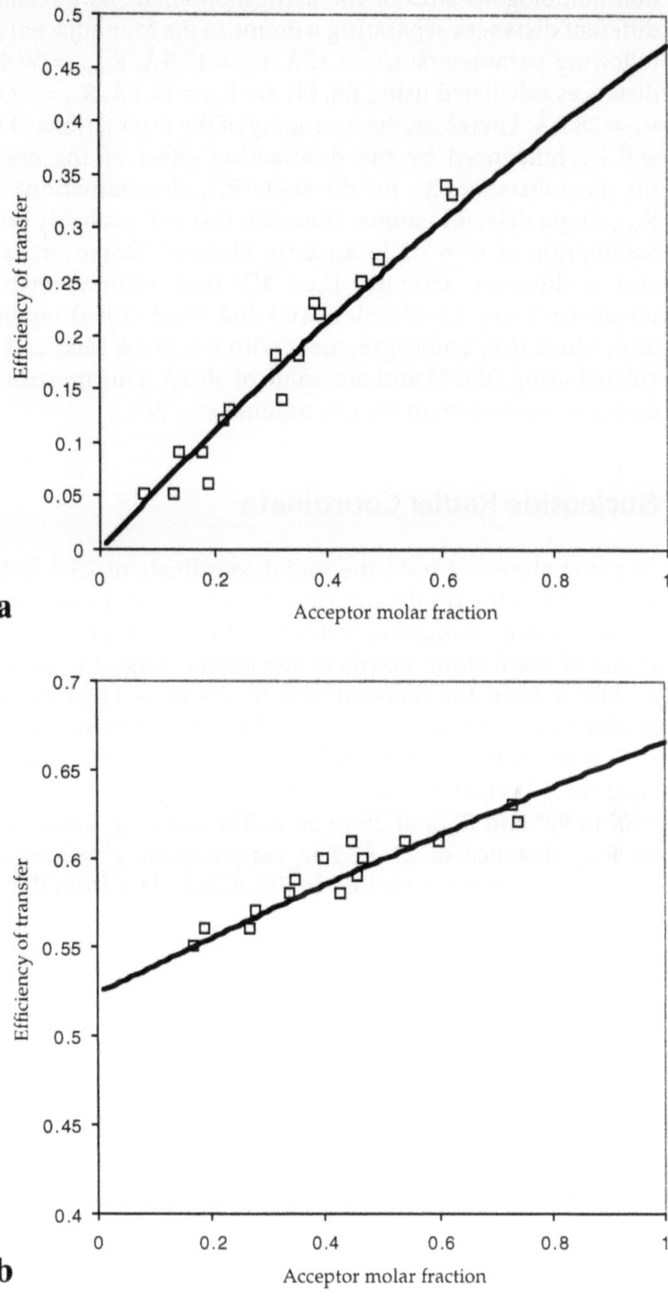

Fig. 7a,b. FRET efficiencies as a function of the acceptor molar fraction for radial coordinates of probes on Cys-374 and nucleotide-binding site of 13.5 Å and 25 Å, respectively. The *solid line* is the theoretical FRET efficiency obtained for an angle of −15°, an R_{intra} of 27 Å and a height of 24 Å. **a** FRET data (*open squares*) obtained for the angle-R_{intra} determination and **b** FRET data (*open squares*) for the R_{intra}-angle determination non-homologous sites in the actin monomers. As a result, there will be four different distances separating a donor to the four adjacent acceptors. Using the following parameters: r_{nuc} = 25 Å, r_{cys} = 13.5 Å, R_{intra} = 30.5 Å and $\phi = -55$, the distances calculated using Eq. (4) are R_{-2} = 42.1 Å, R_{-1} = 32.8 Å, R_{+1} = 61.8 Å and R_{+2} = 78.9 Å. Therefore, the averaging of the orientations of the acceptor dipoles will be influenced by the dominating effect of the energy transfer across the short distances R_{-1} for the angle-R_{intra} determinations and R_{-1}, R_{intra} for the R_{intra}-angle determinations. However, this will probably still not invalidate the assumption of $\kappa^2 = 2/3$ in an actin filament. Moreover, using ε-ATP as donor and a different acceptor (i.e., 4[N-(iodoacetoxy)ethyl-N-methyl]amino-7-nitrobenz-2-oxa-1,3-diazole) Miki and Wahl (1984) obtained a R_{intra} value of 33 Å, which is in good agreement with the 29.2 Å Miki and Mihashi (1978) calculated using DDPM and our value of 30.5 Å. This means that even for a single distance measurement we can assume $\kappa^2 = 2/3$.

Nucleotide Radial Coordinate

In our analysis, we used the radial coordinate of 25 Å determined by Miki et al. (1986) but, the results were only based on four points and they might accommodate a shorter radial coordinate. Also, the C6 of the adenine moiety (which is one of the fixation points of the etheno ring of ε-ATP) is located at 18.7 Å to 20.6 Å from the filament axis in the three Heidelberg models of F-actin (Holmes et al., 1990; Lorenz et al., 1993; Tirion et al., 1995).

Is it possible to fit the FRET data for nucleotide radii of ~20 Å? We calculated the χ^2 values for nucleotide radii ranging from 20 to 25 Å for angles of −90° to 90° and R_{intra} of 25 to 36 Å. The lowest χ^2 values were all obtained for an R_{intra} distance of 27 Å. The variations of χ^2 values with angle for each nucleotide radius are plotted in Fig. 8. It is clear from this figure that negative

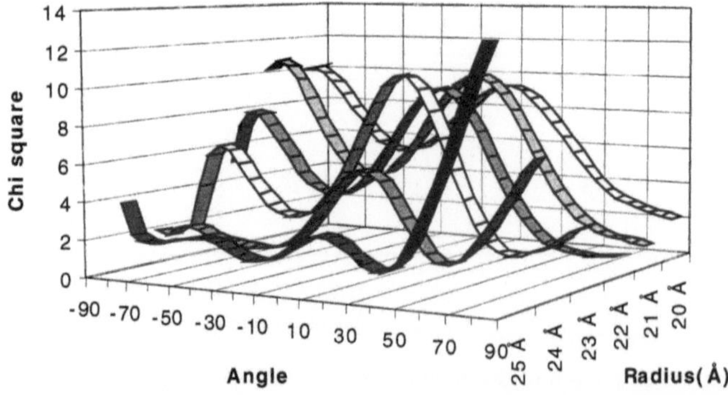

Fig. 8. Variation of χ^2 as a function of the angle with an intramonomer distance of 27 Å for nucleotide radii of 20 to 25 Å. The lowest χ^2 value is obtained for a nucleotide radius of 22 Å

angles between the nucleotide and Cys-374 are unlikely for nucleotide radii smaller than 24 Å. The best fit for all data was obtained for a radius of 22 Å and an angle of 80° to 85°.

Localization of the Probe in the F-Actin Models

ε-ATP differs from ATP by an etheno ring connected to C6 and N1 of the adenine. We used C6 of the nucleotide as a reference to localize the average position of the probes bound to the C-terminus of actin in the filament. The emission dipole of ε-ATP was aligned with C6 and its projection on the filament axis. We then calculated the x, y and z coordinates for the probe in the Lorenz model of F-actin using the parameters for the best fit of the FRET data for the radii 22 Å.

Fig. 9 shows a monomer in the Lorenz et al. (1993) model of F-actin. The green spheres represent two possible positions for the probe. They are localized at the back surface of the monomer, ~15 Å from the S_γ atom of Cys-374. This position is also consistent with the excimer formation reported by Feng et al., where a pyrene on Cys-374 can interact with a pyrene on the loop formed by residues 262-274 between subdomains 3 and 4 of actin (Feng et al., 1997).

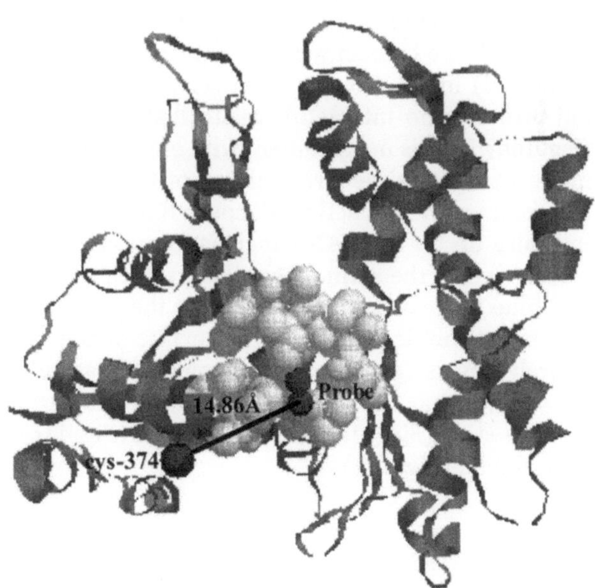

Fig. 9. Ribbon representation of the back side of a monomer from the Lorenz model of F-actin (Lorenz et al. 1993). The probes for the 80° and 85° angles are represented with *small dark gray spheres*. The atoms located within 10 Å of the probes are shown as *light gray spheres*. The *large dark gray sphere* is Sγ of Cys-374

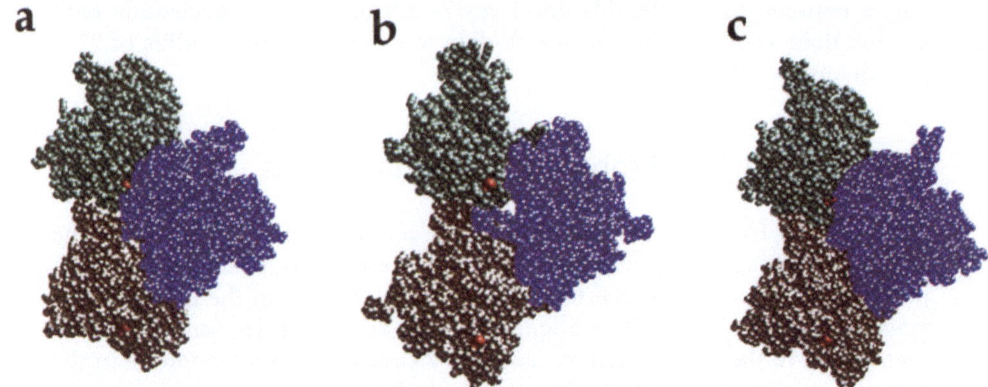

Fig. 10a–c. Illustrates three monomers in the F-actin. Three models of F-actin structure have been proposed: (1) Holmes et al. (1990; Fig. 10a); (2) Lorenz et al. (1993; Fig. 10b); and Tirion et al. (1995; Fig. 10c). The probes are represented by red spheres at a nominal diameter of 5 Å

Figure 10 shows three monomers in the F-actin models from Holmes et al. (1990; Fig. 10a), Lorenz et al. (1993; Fig. 10b) and Tirion et al. (1995; Fig. 10c). The probes are represented by red spheres of 5 Å diameter. The probe localization is consistent with the models of F-actin as it is on the surface of the monomers and not buried deep in the structure. From this figure it is clear that the probe located on the monomer colored in cyan does not interact with the monomer located below on the same strand (white monomer Fig. 10). There could be some contact with the subjacent monomer in the genetic helix (blue monomer) in the Holmes et al. (Fig. 10a) and Tirion et al. (Fig. 10c) models but probably no interactions with that monomer in the Lorenz et al. model (Fig. 10b). This is in agreement with our previous finding that there are no dramatic changes in the environment of the probe between the monomeric and filamentous form of actin (Moens and dos Remedios 1997) as the fluorescence intensity of 1,5-IAEDANS increases by only 7% when actin polymerizes. A closer view of the actin where the residues are colored green for hydrophobic and blue for hydrophilic reveals that the probe (red sphere Fig. 11) is located in a hydrophobic pocket on the surface of the monomer. This localization is entirely consistent with the hydrophobic nature of these probes. This figure also shows that the probe, however, on the surface is less accessible in the Holmes et al. (Fig. 11a) and Tirion et al. (Fig. 11c) models than in the Lorenz et al. model (Fig. 11b).

Do these Results Exclude Any Other Orientation of the Monomer?

The fact that our probe localization is in agreement with the Holmes-Lorenz model of F-actin does not necessarily mean that it is the only possible monomer

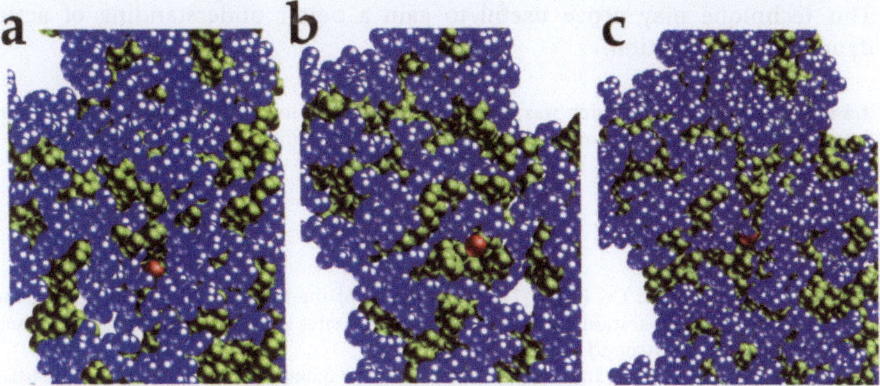

Fig. 11. Closer view of the actin monomer. Residues colored green are hydrophobic, and hydrophilic residues are shown in blue. The probe, coloured red, is located in a hydrophobic pocket on the surface of the monomer. The locations of the probes are illustrated for the Holmes et al. model (Fig. 11a), the Lorenz et al. model (Fig. 11b), and the Tirion et al. model (Fig. 11c)

orientation. We used the best fit to localize the probes but other positions are also possible and refinements of the nucleotide radial coordinate measurements are needed. On the other hand, we do not know whether this position is compatible with a monomer in the orientation proposed by Schutt and colleagues (1997a). Another possibility, which seems to have been forgotten in the discussion over the correct model of F-actin (Egelman et al., 1997; Schutt et al., 1997b; Schutt et al., 1995), is that there might be more than one filament conformation: one might form spontaneously in absence of other proteins and another conformation might be driven by the interaction of actin-binding proteins such as profilin with actin monomers forming a local ribbon structure as an intermediate state.

Conclusions

The functional form of actin, one of the major proteins in eukaryotic cells, is the filament. Filaments have not been crystallized (see Oda et al. this Vol.) so far and the study of their interactions with other proteins and the study of conformational changes in actin often rely on the models of actin filaments published. In this chapter, we used FRET spectroscopy to show that it is possible to determine the position of a probe in space without any prior knowledge of the atomic structure of the actin filament. Our FRET data and the positioning of the probe in F-actin are consistent with the model proposed by Lorenz et al. (1993). This is the first time that FRET has been used to localize a probe in 3-dimensional space in actin filaments. This method can be used not only to test the models of F-actin but also to investigate changes in the position of the probes when actin filaments interact with actin-binding proteins in solution.

This technique may prove useful to gain a better understanding of actin dynamics and function.

Acknowledgments. This research was funded by grants from the National Health and Medical Research Council of Australia and the Australian Research Council.

References

Barden JA, dos Remedios CG (1984) The environment of the high affinity cation binding site on actin and the separation between cation and ATP sites as revealed by proton NMR and fluorescence spectroscopy. J Biochem 96:913–921

Bradford MM (1976) A rapid and sensitive method for the quantitation of microgram quantities of protein utilizing the principle of protein-dye binding. Anal Biochem 72:248–254

Censullo R, Martin JC, Cheung HC (1992) The use of the isotropic orientation factor in fluorescence resonance energy transfer (FRET) studies of the actin filament. J Fluoresc 2:141–155

dos Remedios CG, Moens PDJ (1995a) Fluorescence resonance energy transfer spectroscopy is a reliable "ruler" for measuring structural changes in proteins – dispelling the problem of the unknown orientation factor. J Struct Biol 115:175–185

dos Remedios CG, Moens PDJ (1995b) Actin and the actomyosin interface: a review. Biochim Biophys Acta 1228:99–124

Holmes KC, Popp D, Gebhard W, Kabsch W (1990) Atomic model of the actin filament. Nature 347:44–49

Kasprzak AA, Takashi R, Morales MF (1988) Orientation of the actin monomer in the F-actin filament: radial coordinate of glutamine-41 and effect of myosin subfragment-1 binding on the monomer orientation. Biochemistry 27:4512–4522

Lorenz M, Popp D, Holmes KC (1993) Refinement of the F-actin model against X-ray fiber diffraction data by the use of a directed mutation algorithm. J Mol Biol 234:826–836

Miki M, Mihashi K (1978) Fluorescence energy transfer between ε-ADP at nucleotide binding site and N-(4-dimethyl-amino-3,5-dinitrophenyl)-maleimide at Cys-373 of G-actin. Biochim Biophys Acta 533:163–172

Miki M, Wahl P (1984) Fluorescence energy transfer between points in actosubfragment-1 rigor complex. Biochim Biophys Acta 790:275–283

Miki M, Barden JA, dos Remedios CG (1986a) Fluorescence resonance energy transfer between the nucleotide binding site and Cys-10 in G-actin and F-actin. Biochim Biophys Acta 872:76–82

Miki M, Barden JA, Hambly BD, dos Remedios CG (1986b) Fluorescence energy transfer between Cys-10 residues in actin filaments. Biochem Int 12:725–731

Moens PDJ, Yee D, dos Remedios CG (1994) Determination of the radial coordinate of Cys-374 in F-actin using fluorescence resonance energy transfer spectroscopy: effect of phalloidin on random assembly. Biochemistry 33:13102–13108

Moens PDJ, dos Remedios CG (1997) A conformational change in F-actin when myosin binds: fluorescence resonance energy transfer detects an increase in the radial coordinate of Cys-374. Biochemistry 36:7353–7360

Spudich JA, Watt S (1971) The regulation of rabbit skeletal muscle contraction. I Biochemical studies of the interaction of tropomyosin-troponin complex with actin and the proteolytic fragments of myosin. J Biol Chem 246:4866–4871

Taylor DL, Reidler J, Spudich JA, Stryer L (1981) Detection of actin assembly by fluorescence energy transfer. J Cell Biol 89:362–367

Tirion MM, Benavraham D (1993) Normal Mode Analysis of G-actin. J Mol Biol 230:186–195

Tirion MM, Benavraham D, Lorenz M, Holmes KC (1995) Normal modes as refinement parameters for the F-actin model. Biophys J 68:5–12

Egelman EH, Orlova A, McGough A (1997) Only one F-actin model [letter; comment]. Nature Struct Biol 4:683–684

Feng L, Kim E, Lee WL, Miller CJ, Kuang B, Reisler E, Rubenstein PA (1997) Fluorescence probing of yeast actin subdomain 3/4 hydrophobic loop 262–274. Actin-actin and actin-myosin interactions in actin filaments. J Biol Chem 272:16829–16837

Miki M, Hambly BD, dos Remedios CG (1986) Fluorescence energy transfer between nucleotide binding sites in an F-actin filament. Biochim Biophys Acta 871:137–141

Schutt CE, Kreatsoulas C, Page R, Lindberg U (1997) Plugging into actin's architectonic socket [news] [see comments]. Nature Struct Biol 4:169–172

Schutt CE, Myslik JC, Rozycki MD, Goonesekere NC, Lindberg U (1993) The structure of crystalline profilin-beta-actin. Nature 365:810–816

Schutt CE, Rozycki MD, Myslik JC, Lindberg U (1995) A discourse on modeling F-actin. J Struct Biol 115:186–198

Tirion MM, Ben-Avraham D, Lorenz M, Holmes KC (1995) Normal modes as refinement parameters for the F-actin model. Biophys J 68:5–12

[illegible]

[illegible]

[illegible]

[illegible]

[illegible]

[illegible]

Microscopic Analysis of Polymerization and Fragmentation of Individual Actin Filaments

Shin'ichi Ishiwata[1,2,3,4], Junko Tadashige[1], Ichiro Masui[1], Takayuki Nishizaka[4], and Kazuhiko Kinosita, Jr[4,5]

Introduction

The dynamics of polymerization (and annealing of fragments) and depolymerization (and fragmentation) of actin filaments (F-actin) is a key process in diverse cellular functions, including cell motility. Up to the present, there have been many spectroscopic studies in solution, which have clarified time-dependent but averaged properties, as well as electron microscopic and other studies; but a single filament analysis, which is effective in clarifying not only time-dependent but also nonaveraged properties, has not yet been reported. Thus, direct observation of the polymerization-depolymerization dynamics of individual actin filaments is worthy of investigation.

The polymerization process consists of nucleation and growth phase, and at a steady state, annealing and fragmentation of polymerized filaments occur. The essential elements of these processes have been experimentally clarified and theoretically formulated (Oosawa and Kasai 1962; Oosawa and Asakura 1975). Actin filaments, like microtubules, have a structural polarity, such that the polymerization and depolymerization rates at the two ends of the filaments are different (Oosawa and Asakura 1975; Woodrum et al. 1975; Kondo and Ishiwata 1976; Hayashi and Ip 1976; Pollard and Cooper 1986). The end of the filaments at which the polymerization rate is larger is defined as the barbed (B-) end and the other is called the pointed (P-) end.

After the steady state of polymerization is attained, a treadmill process is expected to occur (Wegner 1976). The B-end of the filament is where filament assembly occurs, while depolymerization occurs at the P-end, such that the length of the filament is maintained nearly constant. The treadmill process has been experimentally demonstrated in solution (cf Wegner 1976; Korn et al. 1987). After actin monomers (G-actin) with ATP are polymerized, the bound

[1] Department of Physics, School of Science and Engineering

[2] Advanced Research Institute for Science and Engineering

[3] Materials Research Laboratory for Bioscience and Photonics, Waseda University, Tokyo 169-8555, Japan

[4] Core Research for Evolutional Science and Technology (CREST) Genetic Programming Team 13, Japan

[5] Department of Physics, Faculty of Science and Technology, Keio University, Yokohama 223-8522, Japan

Results and Problems in Cell Differentiation, Vol. 32
C. dos Remedios (Ed.): Molecular Interactions of Actin

ATP is hydrolyzed and inorganic phosphate (Pi) is released, leaving ADP attached.

When the rate of polymerization is greater than that of ATP hydrolysis, actin molecules with ATP should cap that end of the filaments. Depending on whether each end of the actin filaments is capped by either ATP-bound actin (ATP-cap) or ADP-bound actin (ADP-cap), dynamic instability of the filaments may occur (Korn et al. 1987; Carlier 1989). However, in contrast to microtubules, the dynamic instability has not yet been experimentally proved in actin filaments. Thus, the examination of whether this dynamic process occurs is a challenging problem.

About a decade ago, it became possible to visualize single actin filaments under a fluorescence microscope by labeling the filaments with rhodamine-phalloidin (Rh-Ph; Yanagida et al. 1984) or with fluorescein 5-isothiocyanate (FITC; Honda et al. 1986). In both studies, phallotoxins (Wieland et al. 1975) were added in order to suppress the depolymerization of actin filaments due to extremely low concentration (an order of nM) of actin, lower than the critical concentration for polymerization. Because the filament structure is stabilized, it became possible to visualize single filaments for a sufficiently long time to examine quantitatively not only the polymerization process but also the fragmentation process. Thus, the use of fluorescent-dye conjugated phallotoxins was very useful for stably and clearly visualizing the filament under a conventional fluorescence microscope without disturbance of background fluorescence (cf. Ishiwata 1998). The advantage of this technique is that Rh-Ph does not bind to G-actin and the fluorescence intensity of Rh-Ph increases several-fold upon binding to actin filaments (Harada et al. 1991; Huang et al. 1992).

In this chapter, we describe the properties of the polymerization and fragmentation processes of single actin filaments examined by direct observation through the fluorescence image of actin filaments in the presence of phallotoxins under a conventional fluorescence microscope (Tadashige et al. 1992; Masui et al. 1995).

We ask the reader to bear in mind that although the stabilization by phallotoxins is convenient for visualizing single actin filaments, there are several disadvantages to this method: (1) The greatest disadvantage is that spontaneous depolymerization is suppressed (Estes et al. 1981; Coluccio and Tilney 1984; Sampath and Pollard 1991), (2) The structure of actin filaments may be modified by the binding of phallotoxins (Drubin et al. 1993; Lorenz et al. 1993), so that the results may not represent polymerization properties of pure actin filaments. In this respect, however, this method can, conversely, be considered unique in studying the interaction of phallotoxins with actin filaments.

We have recently succeeded in visualizing polymerization and depolymerization of single actin filaments by using fluorescent dye(rhodamine)-labeled actin without using rhodamine-phallotoxins (phalloidin or phallacidin) under evanescent field illumination (Takahashi and Ishiwata 1998; Fujiwara et al. 1998). Thus it is now possible to visualize at a video rate both polymerization and depolymerization processes for every filament. The results will be published in more detail elsewhere.

How to Image the Polymerization (and Fragmentation) Process of Actin Filaments

To analyze polymerization and fragmentation of individual actin filaments under a fluorescence microscope, it is necessary to observe the same filaments at least for 30 min. Although single filaments are visualized in solution under a conventional fluorescence microscope, it is difficult to trace the polymerization process for each filament floating in solution because of its violent Brownian motion (Yanagida et al. 1984; Isambert et al. 1995).

To overcome this difficulty, we fixed short actin filaments to a glass surface through the cross-linked HMM molecules, as schematically illustrated in Fig. 1 [myosin easily adheres to a collodion(nitrocellulose)-coated glass surface], so that we can keep observing the same filaments. Thus, a short filament adhering to the glass surface was used as a nucleus for observing the polymerization process and as a tool to fix the long filaments so as to observe the fragmentation (severing) process.

To visualize the polymerization process at a constant concentration of G-actin, the actin solution was infused into a flow cell just after the addition

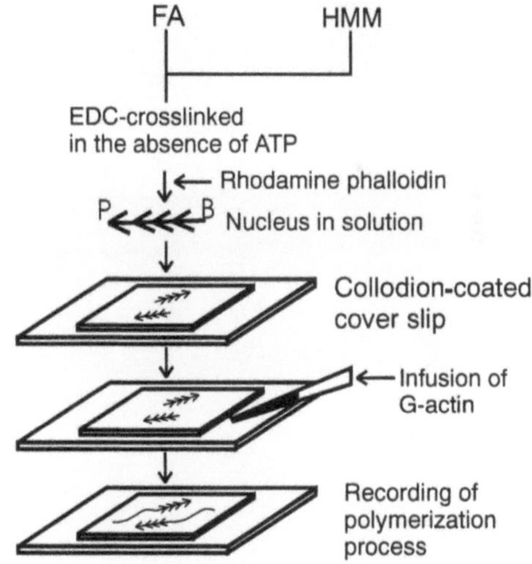

Fig. 1. Scheme showing how to measure the polymerization rate of actin on single actin filaments under a fluorescence microscope. Actin filaments decorated with HMM in the absence of ATP were cross-linked with 1-ethyl-3-(3-dimethyl-aminopropyl)-carbodiimide (EDC) and labeled with Rh-Ph to clearly identify the nuclei (Nishizaka et al. 1993). The EDC cross-linked acto-HMM complexes were mildly sonicated to make filaments short and used as nuclei for polymerization of actin. Immediately after the addition of salt, G-actin solution was infused into the flow cell sandwiched between a pair of coverslips, one of which was coated with nitrocellulose (collodion). The polymerization process was observed under an inverted fluorescence microscope and recorded on a videotape through a SIT camera (C1000; Hamamatsu Photonics). The length of the actin filaments was analyzed using a digital image processor (DIPS-C2000; Hamamatsu Photonics) (Nishizaka et al. 1995)

of salt to the G-actin solution, and the initial process of polymerization was recorded on a videotape at a rate of 30 frames s^{-1}. The length of actin filaments was determined by accumulating and averaging the images for 1 s every 5 (and 6, 7 or 10) min. Thus, the accuracy of the estimation of length change was less than 0.2 μm.

Visualization of the Polymerization Process of Actin Filaments

Figure 2 illustrates a series of fluorescence micrographs showing the polymerization process of individual actin filaments. The concentration of Rh-Ph was reduced to 15 nM, 1/7 that of phalloidin (or phallacidin), which was sufficiently high to image actin filaments. Under this condition, polymerization occurred only at one end of the actin nuclei, corresponding to the B-end.

Judging from the polymerization rate of actin (about 100 nm min^{-1} = 0.6 molecules s^{-1} assuming that the pitch of an actin monomer along the

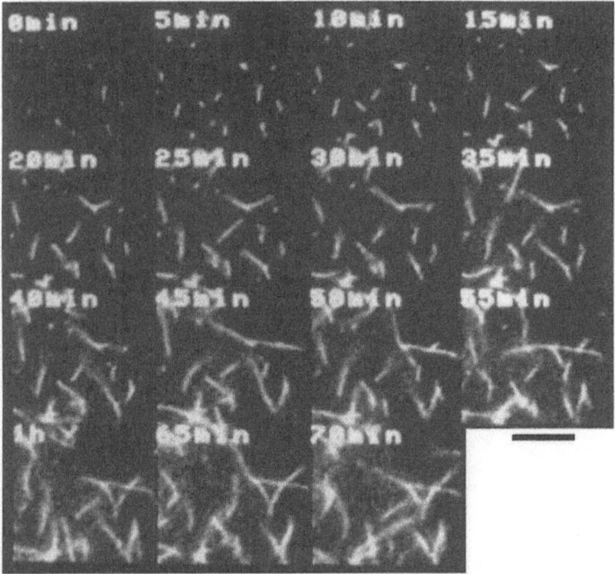

Fig. 2. Fluorescence images showing the time course of the polymerization on single actin filaments. Fluorescence images were recorded for several seconds every 5 min to minimize the photobleaching after the addition of the following actin solution. Condition: 5 μg ml^{-1} G-actin, 30 mM KCl, 1 mM $MgCl_2$, 4 mM ATP, 2 mM MOPS (pH 7), 105 nM phalloidin (Molecular Probes, Inc.), 15 nM Rh-Ph (Molecular Probes, Inc.) and oxygen scavenger system (Harada et al. 1991). The solvent was mixed with G-actin solution containing only ATP and MOPS immediately before the experiments, to minimize the spontaneous polymerization in solution. Temperature 25 °C. Bar, 5 μm

long-pitch helical strand of F-actin to be 5.5 nm) and the concentrations of nuclei and G-actin, the decrease in the concentration of G-actin can be neglected during initial observation (10 min) period. The polymerization rate (shown in Figs. 4 and 5 below) was estimated from the initial phase of elongation for 10 min (cf Fig. 3). We noticed that about one third of the nuclei among those attached to the glass surface did not elongate (see the upper left in each micrograph of Fig. 2). This is probably because G-actin was not accessible to the ends of the nuclei due to an obstacle such as nitrocellulose. Some filaments, which showed bending Brownian motion out of focus, were omitted from the length measurements.

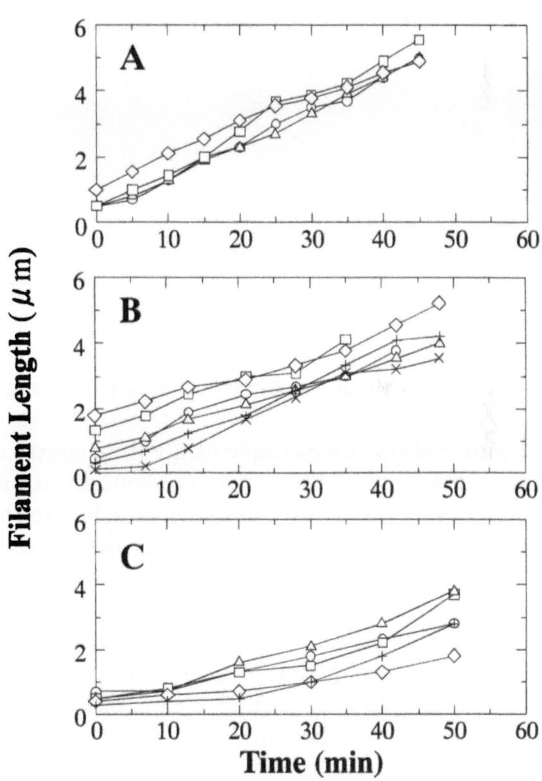

Fig. 3A–C. Time course of polymerization at the B-end of F-actin obtained under a fluorescence microscope at different actin concentrations. Conditions: G-actin, 5 $\mu g\,ml^{-1}$ (**A**, images are shown in Fig. 2), 4 $\mu g\,ml^{-1}$ (**B**) and 3 $\mu g\,ml^{-1}$ (**C**); other solvent conditions, the same as in Fig. 2 (the concentrations of phalloidin and Rh-Ph were maintained at 105 nM and 15 nM, respectively, irrespective of the concentrations of G-actin. Although the total concentration of phalloidin was less than the highest actin concentration examined, we estimated that it was high enough for the binding at the initial stage of polymerization). The filament length at time zero corresponds to the length of nucleus. Temperature 25 °C

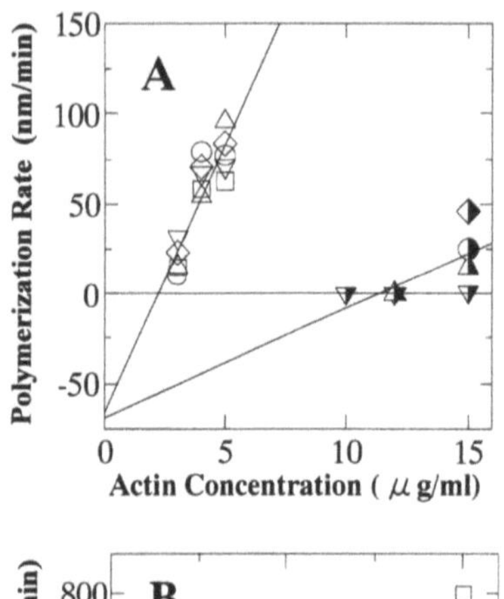

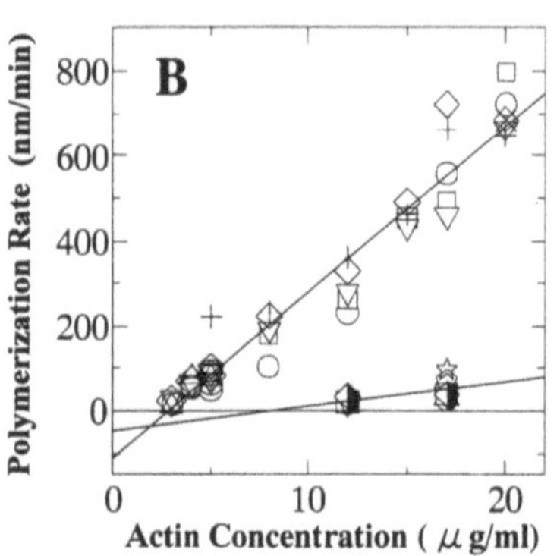

Fig. 4A,B. Initial rate of polymerization vs G-actin concentration in the presence of phalloidin. The polymerization rates at the B- and P-ends of F-actin were examined. When polymerization occurred at both ends of the filament, we decided the end elongated longer to be a B-end and that shorter to be a P-end. **(A)**: polymerization at the B-end in the presence of 1 mM Mg^{2+} (open symbols) or 1 mM Ca^{2+} (half-filled symbols). Conditions as in Fig. 3, except that 1 mM $CaCl_2$ was added in half-filled symbols instead of 1 mM $MgCl_2$. **(B)** Polymerization at the B-end (open symbols and crosses) or the P-end (half-filled symbols, open star and pentagon) in the presence of Mg^{2+}. The data in **A** are also included. Different symbols show different preparations of actin. Temperature, 2 °C

Measurement of the Polymerization Process of Actin Filaments

Figure 3 shows an example of the time course of polymerization of individual actin filaments at three different concentrations of G-actin in the presence of phalloidin. The density of nuclei attached to the glass surface was common to all the experiments. On increasing the concentration of infused G-actin, the average rate of polymerization increased.

The polymerization rate varied from filament to filament (Fig. 3). Even in the same filament, the polymerization rate was not constant over 40–50 min but changed with time beyond the resolution of length determination. The accessibility of G-actin to the ends of the filaments may have fluctuated with time due to the adhesion to the nitrocellulose. Also, these results may be attributable to the nonhomogeneity of the local concentration of G-actin and/or the stochastic and cooperative properties of the polymerization process in the presence of phalloidin.

Depolymerization was suppressed by the presence of phalloidin, but there is a possibility that spontaneous fragmentation occurred. This could be one of the causes of length fluctuations during the polymerization process. In practice, however, we did not notice distinct fragmentation, at least in the middle of the filaments.

The rate of fragmentation in the absence of phalloidin was previously estimated in solution experiments to be $7 \times 10^{-7}\,s^{-1}$ (Kinosian et al. 1993), suggesting that in practice, fragmentation in the presence of phallotoxins does not occur within the period of measurements.

At low concentrations of G-actin, there was a tendency for the polymerization rate to increase with time (Fig. 3). The polymerization curve at $3\,\mu g\,ml^{-1}$ G-actin was convex downward, whereas the curve at $5\,\mu g\,ml^{-1}$ G-actin was nearly straight. This suggests that the polymerization kinetics change with the elongation of the filaments. The shorter the filament, the more the binding of G-actin may be disturbed.

Relation Between Polymerization Rate and Actin Concentration

The average rate of polymerization vs G-actin concentration relation is summarized in Fig. 4. In Fig. 4A, the relation observed at the B-end of F-actin in the presence of Mg^{2+} or Ca^{2+} is compared, while in Fig. 4B, the relation at the B- and P-ends is shown over a comparatively broader range of G-actin concentration.

These experiments were done in the presence of a constant concentration of phalloidin. Contrary to our expectation and differing from previous results obtained in solution (Estes et al. 1981; Coluccio and Tilney 1984; Sampath and Pollard 1991), the critical concentration of polymerization was not zero under all the conditions. This result was not changed even if the polymerization rates were estimated between 40 and 50 min in Fig. 3, where the polymerization rates were larger than the initial ones. However, when the same concentration of phallacidin was used instead of phalloidin, the result was as expected (Fig. 5). In both cases, the same concentration of Rh-Ph, one seventh that of phalloidin (or phallacidin), coexisted in order to visualize single actin filaments.

The relation between the polymerization rate (r) and G-actin concentration (c) can be expressed by:

$$r = k^{+}c - k^{-},$$

where k^{+} and k^{-} are the rate constants for polymerization and depolymerization, respectively. Thus, under the conditions examined in the presence of phalloidin, k^{+} was estimated to be $29.6\,(nm\,min^{-1})/(\mu g\,ml^{-1})$ $(= 7.5 \times 10^{6}\,M^{-1}\,s^{-1})$ and $6.2\,(nm\,min^{-1})/(\mu g\,ml^{-1})$ $(= 1.6 \times 10^{6}\,M^{-1}\,s^{-1})$ for the B-end in the presence of Mg^{2+} and Ca^{2+}, respectively (Fig. 4A), and $39.3\,(nm\,min^{-1})/(\mu g\,ml^{-1})$ $(= 1.0 \times 10^{7}\,M^{-1}\,s^{-1})$ and $5.9\,(nm\,min^{-1})/(\mu g\,ml^{-1})$ $(= 1.5 \times 10^{6}\,M^{-1}\,s^{-1})$ for the B-end and the P-end in the presence of Mg^{2+}, respectively (Fig. 4B). In calculating the values in the parentheses, we assumed that the molecular weight of actin is 42 kDa, and the number density of actin molecules along the filament is 2 molecules/5.5 nm.

On the other hand, in the presence of phallacidin, k^+ at the B-end was estimated to be 34.9 $(nm\,min^{-1})/(\mu g\,ml^{-1})$ (= $8.9 \times 10^6 M^{-1} s^{-1}$) and 5.8 $(nm\,min^{-1})/(\mu g\,ml^{-1})$ (= $1.5 \times 10^6 M^{-1} s^{-1}$) in the presence of Mg^{2+} and Ca^{2+}, respectively (Fig. 5).

The values of k^+ obtained above in the presence of Mg^{2+} (Figs. 4B, 5) were a little larger than, but consistent with, those obtained in solution under the same solvent conditions except for the absence of phallotoxins (the values of k^+ at the B- and P-ends were, respectively, $6.3 \times 10^6 M^{-1} s^{-1}$ and $1.1 \times 10^6 M^{-1} s^{-1}$; Suzuki and Mihashi 1989). This differs, however, from the results that the addition of phallotoxins reduced the value of k^+ by 20 to 50% (Coluccio and Tilney 1984; Wendel and Dancker 1987; Sampath and Pollard 1991). Such an apparent discrepancy may be attributable to the difference in experimental conditions. Note that the value of k^- was almost zero in the presence of phallacidin, consistent with the previous solution experiments (Estes et al. 1981; Coluccio and Tilney 1984; Sampath and Pollard 1991).

We found that the addition of 10 mM Pi in the presence of phallacidin slightly increased the value of k^+ to 39.0 $(nm\,min^{-1})/(\mu g\,ml^{-1})$ (= $9.9 \times 10^6 M^{-1} s^{-1}$) and 6.9 $(nm\,min^{-1})/(\mu g\,ml^{-1})$ (= $1.8 \times 10^6 M^{-1} s^{-1}$) in the presence of Mg^{2+} and Ca^{2+}, respectively, that is, by about 1.1 to 1.2 times (Fig. 5). This demonstrates that Pi does not significantly affect the polymerization rate but stabilizes the filament structure (Nonomura et al. 1975) through the decrease

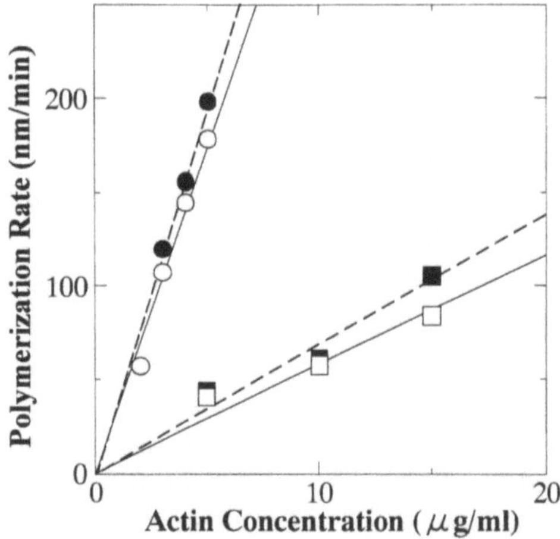

Fig. 5. Initial rate of polymerization at the B-end vs G-actin concentration in the presence of phallacidin. Effects of Pi and divalent cations (Mg^{2+} or Ca^{2+}) were examined. Conditions: various concentrations of G-actin, 30 mM KCl, 2 mM $MgCl_2$ (circles) or $CaCl_2$ (squares), 4 mM ATP, 2 mM MOPS (pH 7), 2 mg ml^{-1} BSA, 1.5 mM NaN_3, with (closed symbols) or without (open symbols) 1.5 mM Pi, 105 nM phallacidin, 15 nM Rh-Ph, 10 mM DTT, 0.3% (w/v) methylcellulose and oxygen scavenger system (Harada et al. 1991). The time course of polymerization of actin filaments was indistinguishable from that in the presence of phalloidin shown in Fig. 3, except that the polymerization curve was slightly convex upward at every G-actin concentration we examined

in the rates of both depolymerization (Rickard and Sheterline 1986, 1988; Funatsu 1986) and fragmentation as described below (Fig. 6).

It is also to be noted that, irrespective of the species of phallotoxins, in other words, independent of whether the critical concentration was zero (Fig. 5 for phallacidin) or not (Fig. 4 for phalloidin), the polymerization rate, k^+, was nearly the same under the same solvent conditions. On the other hand, the value of k^+ for Mg-actin was five-to-six times larger than that for Ca-actin (Figs. 4A, 5). This is in contrast to previous results obtained in solution that k^+ for Mg-actin was, at most, only two times larger than that for Ca-actin (see Estes et al. 1992). This apparent discrepancy may be attributable to the structural changes induced by the binding of phallotoxins and the difference in experimental conditions. Another possibility is that the HMM-cross linked nucleus may have amplified a small difference between the polymerization properties of Mg-actin and Ca-actin.

What is most notable in the above experiments is that the critical concentration for polymerization was practically zero in the presence of 105 nM phallacidin with 15 nM Rh-Ph (Fig. 5) but not zero in the presence of 105 nM phalloidin with 15 nM Rh-Ph (Fig. 4). At first sight, the reason for this difference appears to be attributable to the larger dissociation constant of phalloidin than that of phallacidin with F-actin. In practice, however, the dissociation constants of phalloidin and Rh-Ph with F-actin are, respectively, reported to be 67 and 40 nM (Molecular Probes Data Book; Huang et al. 1992), and those of fluorescent dye-conjugated phallotoxins such as coumarin-, NBD- and bodipy-phallacidin are, respectively, reported to be 24, 18 and 38 nM (Huang et al. 1992). Thus, although the dissociation constant of phallacidin has not yet been reported, there seems to be no large difference between phalloidin and phallacidin. If the dissociation constant itself is practically indistinguishable, the reason for the above difference may be attributable to the different rate constants.

The apparent association rate constant of Rh-Ph for F-actin is reported to be two-to-three orders of magnitude smaller than the rates of polymerization and depolymerization (De la Cruz and Pollard 1994, 1996; $6 \times 10^3 M^{-1} s^{-1}$ at 20 °C; and $5 \times 10^4 M^{-1} s^{-1}$ at 40 °C, Hotta and Ishiwata 1997). If this is also the case for the newly polymerized ends of the filaments, there would be an appreciable probability that the actin molecules once attached to the filament ends are detached, so that the apparent polymerization rate becomes reduced, resulting in the nonzero critical concentration.

This was not the case at least for phallacidin (as observed in Fig. 5), suggesting that the association rate constant of phallotoxins for the ends of the filaments is much larger than that for the bulk of the filaments. In fact, there is a report suggesting that phalloidin binds to the filament ends more rapidly than to the bulk of the filament (Cano et al. 1992). We infer that the rate constant of attachment to the filament ends of phalloidin may be still much slower than that of phallacidin.

Such a subtle difference may be produced by the structural change of actin molecules incorporated into the filament ends. The result shown in Fig. 4A, that the apparent critical concentration for polymerization was larger in the presence of Ca^{2+} than of Mg^{2+}, can be ascribable to this mechanism, and may be related to the difficulty of nucleus formation for Ca-actin. That is, the association rate of phalloidin to the polymerized filament end may be slower for Ca-actin than for Mg-actin. This difference may be overcome in the presence of phallacidin, because the association rate constant of phallacidin at the polymerized ends is large enough to overcome the smaller association rate constant to Ca-actin. These mechanisms should be quantitatively examined in future.

Fragmentation of Actin Filaments: Effect of Anions at High Ionic Strength

Finally, the fragmentation of individual actin filaments was visualized using the same technique as described above. In this case, however, the long actin filaments polymerized beforehand in the presence of Rh-Ph were attached to the glass surface through the short fragments decorated with cross-linked HMM as shown in Fig. 6 (a similar technique has been used before: Bearer 1991; Maciver et al. 1991; Nishizaka et al. 1993). Various salt solutions were then infused into the cell and their effects on the fragmentation were examined under a continuous flow of the solutions.

As observed in a series of micrographs summarized in Fig. 6, shortening of the filaments in the presence of 3 M KSCN occurred mainly by severing with a lifetime of about 2 min (upper micrographs, −Pi) and 10 min (lower micrographs, +Pi), and finally all the filaments disappeared. Gradual depolymerization from the ends of the filaments was not detected. We examined various kinds of salts composed of K^+ as the cation, and Cl^-, I^-, SCN^-, etc as anions.

We found that the severing abilities were in the order of the Hofmeister series (lyotropic number) for anions, ie, the order according to which the hydrophobic interaction is destroyed: SCN^-, I^-, Cl^-, CH_3COO^-, SO_4^{2-}, etc (for the Hofmeister series, see Melander et al. 1984; Cacace et al. 1997). This is in spite of the fact that severing was largely retarded by phalloidin (for the effect of KI in the presence of phalloidin, see Dancker et al. 1975).

The present results are consistent with the previous results obtained by the phase-contrast image analysis of the dissociation process from the P-end of thin filaments in the I-Z-I brush of myofibrils (Funatsu and Ishiwata 1985; Funatsu et al. 1988), which was prepared by mildly etching the P-end of thin filaments with 0.5 M KCl treatment (Ishiwata and Funatsu 1985). The dissociation rate in the presence of 0.5 M salts in the absence of phallotoxins was in the order of SCN^-, I^-, NO_3^-, Cl^- and SO_4^{2-}, being consistent with the Hofmeister series (Funatsu 1986).

These results suggest that a hydrophobic interaction between actin molecules is important for the stabilization of the filament, even after the polymer structure is stabilized by the binding of phalloidin. However, we cannot completely eliminate the possibility that the interaction between phalloidin and

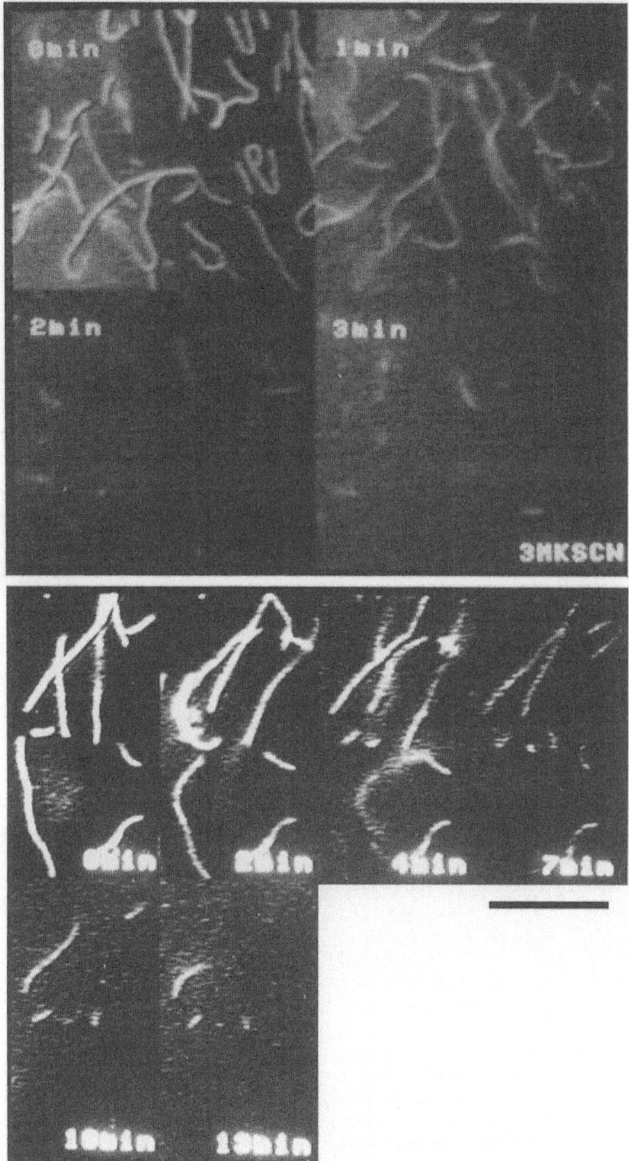

Fig. 6. Fluorescence images showing the time course of fragmentation of actin filaments in the presence of high concentrations of chaotropic anions. To record the time course of fragmentation of actin filaments, the filaments polymerized onto the nuclei of EDC-cross-linked acto-HMM complex in solution were infused into the cell, so that the filaments were attached to the glass surface through the short nuclei portion of the filaments and the polymerized portion of the filaments showed Brownian motion. The assay solvent was then infused. The time (min) after the infusion of the solvent is shown in each micrograph. Conditions as in Fig. 3, except that 3 M KSCN was added instead of 30 mM KCl and in the absence (upper micrographs) or presence (lower micrographs) of 10 mM Pi. Temperature 25 °C. Bars 5 μm

actin is weakened by the anions in the order of the Hofmeister series. It is to be noted that PO_4^{2-} was exceptional because it did not obey the Hofmeister series, ie, in practice, the shortening of the filaments did not occur in the presence of 0.5 M Pi.

The importance of hydrophobic interaction in the polymerization of actin has been recognized for many years (Oosawa and Asakura 1975). Nowadays, this discussion is based on the 3D atomic structure of G-actin (Kabsch et al. 1990) and the molecular model of F-actin (Holmes et al. 1990). These authors suggested that the hydrophobic loop located between subdomains 3 and 4 plays a role in stabilization of F-actin and this suggestion was supported by data from actin mutants or isoforms (Holmes et al. 1990; Allen et al. 1996b; Kuang and Rubenstein 1997).

Although we cannot add any information on the interaction between actin monomers on the atomic level, the results obtained here not only confirmed that the hydrophobic interaction is essential for maintaining the polymer structure but also demonstrated that other kinds of interactions exist between actin monomers, which are stabilized by phalloidin and Pi.

The manner in which phalloidin binds in the 3D structure of F-actin (Drubin et al. 1993; Lorenz et al. 1993; Steinmetz et al. 1998) suggests that phalloidin functions as a glue making connections among actin monomers in the filament. In practice, phalloidin can restore the polymerizability and also the ability of actin to activate myosin ATPase in monomers which have been impaired by chemical modifications of Lys 61 residue with FITC (Miki 1987), with several lysine residues with (m-maleimidobenzoyl)-N-hydroxysuccinimide ester (MBS) (Miki and Hozumi 1991), and by large hydrophobic probes at Cys-374 (Moens et al. 1994). Thus, actin filaments are stabilized by the cooperative nature of several kinds of molecular forces acting between actin and the regulators, ie, phallotoxins and Pi.

It is interesting that the addition of 10 mM Pi substantially lengthened the lifetime of F-actin in the presence of 3 M KSCN (Fig. 6), implying that Pi stabilized the structure of F-actin against the chaotropic effects of anions (for KI, see Dancker and Fischer 1989). For many years, the stabilizing effect of Pi has been well known. For example, with the addition of mM Pi: (1) the fragility of F-actin observed by negative staining in the electron microscope appeared to become stiffer (Nonomura et al. 1975); (2) the depolymerization rate observed in cuvette was slowed down (Rickard and Sheterline 1986, 1988; Carlier 1989); (3) subdomain 2 of the actin monomer in F-actin became oriented so as to interact strongly with the adjacent actin monomers (Orlova and Egelman 1992); and (4) severing activity of gelsolin (Allen et al. 1996a) and actophorin (Maciver et al. 1991) was reduced.

Pi does not greatly accelerate the polymerization rate, as seen in Fig. 5, and suppresses not only the depolymerization but also the fragmentation as observed in Fig. 6, such that Pi shifts the polymerization-depolymerization equilibrium to the polymerization side. The present results strongly suggest that Pi is a stabilizer of the hydrophobic interaction between actin monomers which is essential for the polymerization of actin. On the other hand, phalloidin stabilizes the F-actin-ADP-Pi complex (Dancker and Hess 1990). Thus, phallotoxins and Pi function as allosteric effectors that synergistically stabilize the structure of actin filaments.

Conclusions

The polymerization and fragmentation processes of individual actin filaments can be visualized by imaging the polymerized filaments through binding of the fluorescent dye, Rh-Ph, under a conventional fluorescence microscope. Thus, the polymerization process at the B- and P-ends of actin filaments can be analyzed in the presence of Mg^{2+} or Ca^{2+}. By measuring the rate of polymerization at various G-actin concentrations, we confirmed that the depolymerization process was inhibited by the attachment of phallotoxins to the filaments (phallacidin was more effective than phalloidin), but the polymerization rate itself was not significantly affected, such that the critical concentration of polymerization was essentially zero. Also, the fragmentation process of filaments due to the addition of high concentrations of chaotropic salts such as KSCN and KI (except Pi) can be observed and stabilization of the hydrophobic interactions by Pi was confirmed. The present study demonstrates that the polymerization and fragmentation dynamics on single actin filaments can be examined quantitatively by fluorescence microscopy.

Acknowledgments. The authors would like to thank Dr JMQ Davies of Waseda University for his critical reading of the manuscript. The authors also thank Miss Ikuko Fujiwara for her assistance in preparing figures. This research was supported in part by Grants-in-Aid for Scientific Research, for Scientific Research on Priority Areas and for the HighTech Research Center Project from the Ministry of Education, Science, Sports and Culture of Japan, and Core Research for Evolutional Science and Technology (CREST) from the Japan Science and Technology Corporation (JST), and a Waseda University Special Grant-in-Aid.

References

Allen PG, Laham LE, Way M, Janmey PA (1996a) Binding of phosphate, aluminum fluoride, or beryllium fluoride to F-actin inhibits severing by gelsolin. J Biol Chem 271:4665–4670

Allen PG, Shuster CB, Kas J, Chaponnier C, Janmey PA, Herman IM (1996b) Phalloidin binding and rheological differences among actin isoforms. Biochemistry 35:14062–14069

Bearer EL (1991) Direct observation of actin filament severing by gelsolin and binding by gCap39 and CapZ. J Cell Biol 115:1629–1638

Cacace MG, Landau EM, Ramsden JJ (1997) The Hofmeister series: salt and solvent effects on interfacial phenomena. Q Rev Biophys 30:241–277

Cano ML, Cassimeris L, Joyce M, Zigmond SH (1992) Characterization of tetramethylrhodaminyl-phalloidin binding to cellular F-actin. Cell Motil Cytoskel 21:147–158

Carlier M-F (1989) Role of nucleotide hydrolysis in the dynamics of actin filaments and microtubules. Int Rev Cytol 115:139–170

Coluccio LM, Tilney LG (1984) Phalloidin enhances actin assembly by preventing monomer dissociation. J Cell Biol 99:529–535

Dancker P, Fischer S (1989) Stabilization of actin filaments by ATP and inorganic phosphate. Z Naturforsch C44:698–704

Dancker P, Hess L (1990) Phalloidin reduces the release of inorganic phosphate during actin polymerization. Biochim Biophys Acta 1035:197–200

Dancker P, Löw I, Hasselbach W, Wieland Th (1975) Interaction of actin with phalloidin: Polymerization and stabilization of F-actin. Biochim Biophys Acta 400:407–414

De la Cruz EM, Pollard TD (1994) Transient kinetic analysis of rhodamine phalloidin binding to actin filaments. Biochemistry 33:14387–14392

De la Cruz EM, Pollard TD (1996) Kinetics and thermodynamics of phalloidin binding to actin filaments from three divergent species. Biochemistry 35:14054–14061

Drubin DG, Jones HD, Wertman KF (1993) Actin structure and function: Roles in mitochondrial organization and morphogenesis in budding yeast and identification of the phalloidin-binding site. Mol Biol Cell 4:1277–1294

Estes JE, Selden LA, Gershman LC (1981) Mechanism of action of phalloidin on the polymerization of muscle actin. Biochemistry 20:708–712

Estes JE, Selden LA, Kinosian HJ, Gershman LC (1992) Tightly-bound divalent cation of actin. J Muscle Res Cell Motil 13:272–284

Fujiwara I, Takahashi S, Tadakuma H, Ishiwata S (1998) Visualization of polymerization process of single actin filaments under total reflection microscope. Biophysics (Japanese) 38:S63 (Abstr)

Funatsu T (1986) Structural stability and capping proteins of actin filaments. Doct Diss Waseda University

Funatsu T, Ishiwata S (1985) Characterization of β-actinin: A suppressor of the elongation at the pointed end of thin filaments in skeletal muscle. J Biochem 98:535–544

Funatsu T, Asami Y, Ishiwata S (1988) β-actinin: A capping protein at the pointed end of thin filaments in skeletal muscle. J Biochem 103:61–71

Harada Y, Sakurada K, Aoki T, Thomas DD, Yanagida T (1991) Mechanochemical coupling in actomyosin energy transduction studied by in vitro movement assay. J Mol Biol 216:49–68

Hayashi T, Ip W (1976) Polymerization polarity of actin. J Mechanochem Cell Motil 3:163–169

Holmes KC, Popp D, Gebhard W, Kabsch W (1990) Atomic model of the actin filament. Nature 347:44–49

Honda H, Nagashima H, Asakura S (1986) Directional movement of F-actin in vitro. J Mol Biol 191:131–133

Hotta M, Ishiwata S (1997) Temperature dependence of association and dissociation rate constants of rhodamine phalloidin with actin filaments. J Muscle Res Cell Motil 18:491 (Abstr)

Huang Z, Haugland RP, You W, Haugland RP (1992) Phalloidin and actin binding assay by fluorescence enhancement. Anal Biochem 200:199–204

Isambert H, Venier P, Maggs AC, Fattoum A, Kassab R, Pantaloni D, Carlier M-F (1995) Flexibility of actin filaments derived from thermal fluctuations. Effect of bound nucleotide, phalloidin, and muscle regulatory proteins. J Biol Chem 270:11437–11444

Ishiwata S (1998) The use of fluorescent probes. In: Current Methods in Muscle Physiology (ed by Sugi H) Oxford University Press, Oxford, UK, pp 199–222

Ishiwata S, Funatsu T (1985) Does actin bind to the ends of thin filaments in skeletal muscle? J Cell Biol 100:282–291

Kabsch W, Mannherz HG, Suck D, Pai EF, Holmes KC (1990) Atomic structure of the actin: DNase I complex. Nature 347:37–44

Kinosian HJ, Selden LA, Estes JE, Gershman LC (1993) Actin filament annealing in the presence of ATP and phalloidin. Biochemistry 32:12353–12357

Kondo H, Ishiwata S (1976) Uni-directional growth of F-actin. J Biochem 79:159–171

Korn ED, Carlier M-F, Pantaroni D (1987) Actin polymerization and ATP hydrolysis. Science 238:638–644

Kuang B, Rubenstein PA (1997) Beryllium fluoride and phalloidin restore polymerizability of a mutant yeast actin (V266G, L267G) with severely decreased hydrophobicity in a subdomain 3/4 loop. J Biol Chem 272:1237–1247

Lorenz M, Popp D, Holmes KC (1993) Refinement of the F-actin model against X-ray fiber diffraction data by the use of a directed mutation algorithm. J Mol Biol 234:826–836

Maciver SK, Zot HG, Pollard TD (1991) Characterization of actin filament severing by actophorin from *Acanthamoeba castellani.* J Cell Biol 115:1611–1620

Masui I, Tadashige J, Nishizaka T, Ishiwata S (1995) Microscopic analysis of polymerization process on a single actin filament. Proc Annu Meet Phys Soc Jpn 29:pA3 (Abstr)

Melander WR, Corradini D, Horvath C (1984) Salt-mediated retention of proteins in hydrophobic-interaction chromatography. Application of solvophobic theory. J Chromatogr 317:67–85

Miki M (1987) The recovery of the polymerizability of Lys-61-labelled actin by the addition of phalloidin. Fluorescence polarization and resonance-energy-transfer measurements. Eur J Biochem 164:229–235

Miki M, Hozumi T (1991) Interaction of maleimidobenzoyl actin with myosin subfragment 1 and tropomyosin-troponin. Biochemistry 30:5625–5630

Moens PDJ, Yee D, dos Remedios CG (1994) Determination of the radial coordinate of Cys-374 in F-actin using fluorescence resonance energy transfer spectroscopy: Effect of phalloidin on random assembly. Biochemistry 33:13102–13108

Nishizaka T, Yagi T, Tanaka Y, Ishiwata S (1993) Right-handed rotation of an actin filament in an in vitro motile system. Nature 361:269–271

Nishizaka T, Miyata H, Yoshikawa H, Ishiwata S, Kinosita Jr K (1995) Unbinding force of a single motor molecule of muscle measured using optical tweezers. Nature 377:251–254

Nonomura Y, Katayama E, Ebashi S (1975) Effect of phosphates on the structure of the actin filament. J Biochem 78:1101–1104

Oosawa F, Asakura S (1975) Thermodynamics of the polymerization of proteins. Academic Press, New York

Oosawa F, Kasai M (1962) Theory of linear and helical aggregations of macromolecules. J Mol Biol 4:10–21

Orlova A, Egelman EH (1992) Structural basis for the destabilization of F-actin by phosphate release following ATP hydrolysis. J Mol Biol 227:1043–1053

Pollard TD, Cooper JA (1986) Actin and actin-binding proteins. A critical evaluation of mechanisms and functions. Annu Rev Biochem 55:987–1035

Rickard JE, Sheterline P (1986) Cytoplasmic concentrations of inorganic phosphate affect the critical concentration for assembly of actin in the presence of cytochalasin D or ADP. J Mol Biol 191:273–280

Rickard JE, Sheterline P (1988) Effect of ATP removal and inorganic phosphate on length redistribution of sheared actin filament populations. Evidence for a mechanism of end-to-end annealing. J Mol Biol 201:675–681

Sampath P, Pollard TD (1991) Effects of cytochalasin, phalloidin, and pH on the elongation of actin filaments. Biochemistry 30:1973–1980

Steinmetz MO, Stoffler D, Müller SA, Jahn W, Wolpensinger B, Goldie KN, Engel A, Faulstich H, Aebi U (1998) A correlative analysis of actin filament assembly, structure, and dynamics. J Mol Biol 276:1–6

Suzuki N, Mihashi K (1989) Subunit flow in F-actin under steady-state conditions. Application of a novel method to determination of the rate of subunit exchange of F-actin at the terminals. Biophys Chem 33:177–193

Tadashige J, Nishizaka T, Ishiwata S (1992) Direct observation of the dynamic process of polymerization and depolymerization on single actin filaments. Biophysics (Japanese) 32:S194 (Abstr)

Takahashi S, Ishiwata S (1998) Direct observation of polymerization process of single actin filaments under fluorescence microscope. Biophys J 74:A46 (Abstr)

Wegner A (1976) Head to tail polymerization of actin. J Mol Biol 108:139–150

Wendel H, Dancker P (1987) Influence of phalloidin on both the nucleation and the elongation phase of actin polymerization. Biochim Biophys Acta 915:199–204

Wieland T, de Vries JX, Schäfer AJ, Faulstich H (1975) Spectroscopic evidence for the interaction of phalloidin with actin. FEBS Lett 54:73–75

Woodrum DT, Rich SA, Pollard TD (1975) Evidence for biased bidirectional polymerization of actin filaments using heavy meromyosin prepared by an improved method. J Cell Biol 67:231–237

Yanagida T, Nakase M, Nishiyama K, Oosawa F (1984) Direct observation of motion of single F-actin filaments in the presence of myosin. Nature 307:58–60

Two Conformations of G-Actin Related to Two Conformations of F-Actin

Edward H. Egelman and Albina Orlova[1]

Multiple Conformations of F-Actin

A single high-resolution model for the F-actin filament has been generated by refining the structure of a crystallographically determined skeletal muscle G-actin subunit against X-ray fiber diffraction data (Holmes et al. 1990; Lorenz et al. 1993). However, electron microscopic studies have shown at low-resolution that the actin filament can exist in multiple conformations as a function of the bound metal (Orlova and Egelman 1995), the nucleotide (Orlova and Egelman 1993), the isoform of actin involved (Orlova et al. 1997), and the presence of other proteins bound to actin (Owen and DeRosier 1993; McGough et al. 1997). It is thus informative to ask what is known about different structural states of G-actin.

Different Structural States of G-Actin

Following the initial determination of a structure of the complex of skeletal muscle actin with DNase I (Kabsch et al. 1990), the complexes of a cytoplasmic G-actin with profilin (Schutt et al. 1993) and skeletal muscle G-actin with gelsolin segment I (McLaughlin et al. 1993) were determined. Although differences existed in the G-actin structure among these three studies, the basic conformation of the subunit was quite conserved. The more recent crystallographic determination of a G-actin subunit in a different conformation (Chik et al. 1996) provides an opportunity of asking whether the different G-actin structures can be simply related to differences observed in F-actin structures.

We have taken different G-actin structures and aligned them to the actin subunit in the Lorenz et al. (1993) model for F-actin. Figure 1 shows three different G-actin subunits oriented with respect to the F-actin filament axis as determined for the Lorenz et al. model (Fig. 1b, f). These three G-actin structures were taken from: (1) the complex of skeletal muscle actin with DNase I (Kabsch et al. 1990; Fig. 1a, e); (2) the Lorenz et al. (1993) model (Fig. 1b, f); (3) the complex of cytoplasmic β G-actin and profilin with the nucleotide-binding

[1] Department of Biochemistry & Molecular Genetics University of Virginia Health Sciences Center Charlottesville, Virginia 22908-0733, USA

Results and Problems in Cell Differentiation, Vol. 32
C. dos Remedios (Ed.): Molecular Interactions of Actin

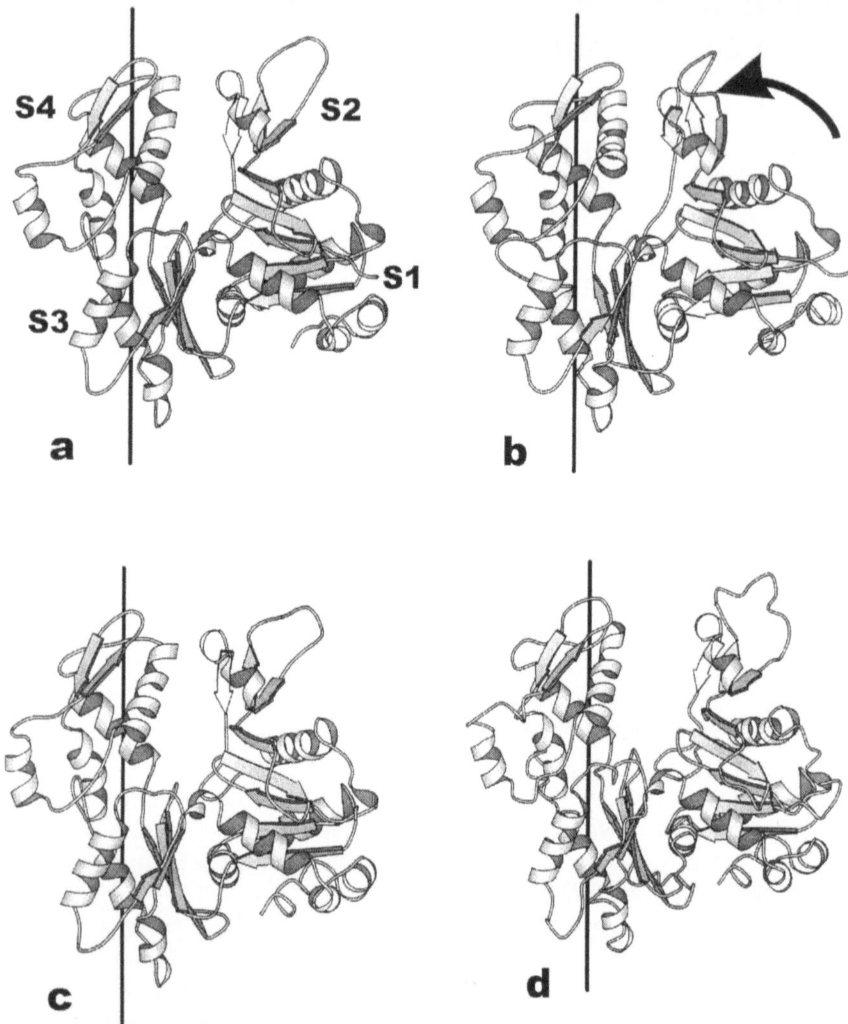

Fig. 1a–h. Ribbon diagrams generated by MOLSCRIPT (Kraulis 1991) for the actin subunit in the actin-DNase I complex (Kabsch et al. 1990) (**a,e**), in the form used to model the F-actin protomer (Lorenz et al. 1993) (**b,f**), in the closed state of the actin-profilin complex (Schutt et al. 1993) (**c,g**), and in the open state of this complex (Chik et al. 1996) (**d,h**). The structures in **e,f,g,h** are rotated by 90° from the structures in **a, b, c, d**, respectively. The subunits in **a,c** and **d** have all been oriented with a least-squares fit to the filament position of the Lorenz et al. model (Lorenz et al. 1993) shown in **b**, using only subdomains 1, 3 and 4. The filament axis is indicated by the vertical line in each panel. The largest difference among these structures involves subdomain 2, which is rotated in towards subdomain 4 when compared to a in the Lorenz et al. model **b**, and rotated out of the plane of the subunit in the open state (**h**).

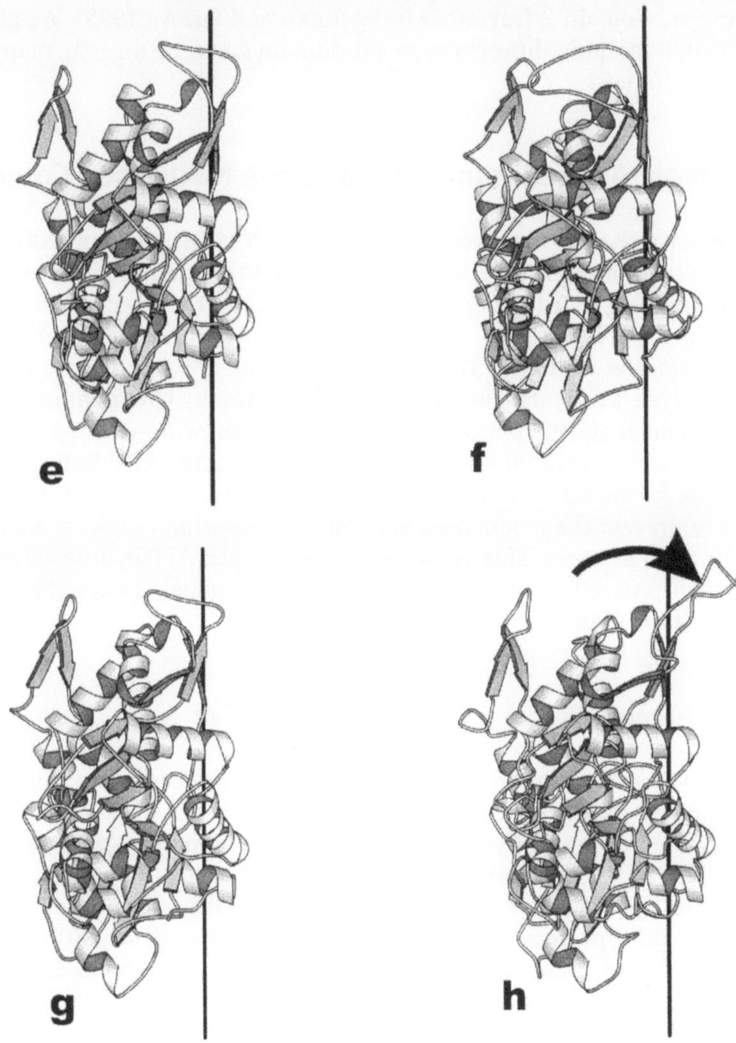

Fig. 1a–h. *Continued*

cleft in a closed state (Schutt et al. 1993; Fig. 1c, g); and (4) the complex of cytoplasmic β G-actin and profilin with the nucleotide-binding cleft in an open state (Chik et al. 1996; Fig. 1d, h).

The main difference between these four structures involves subdomain 2. Interestingly, subdomain 2 is the most highly conserved subdomain within actin (Orlova et al. 1997). We have previously suggested that two of the main differences between distinct conformations of F-actin involved the C-terminus and subdomain 2 (reviewed in Egelman and Orlova 1995). We now ask if the crystallographic differences in subdomain 2 can be directly mapped onto the EM differences.

A Yeast Actin Mutant with an Open Nucleotide Conformation

We have shown that a yeast actin mutant, V159N, exists in a different conformation from wild-type filaments, and that the main difference appears to be an opening of the nucleotide-binding cleft that occurs in the wild-type filaments following Pi release. This opening does not occur in the mutant (Belmont et al. 1999). The mutant filaments are much more stable than the wild-type filaments, but hydrolyze ATP normally (Belmont and Drubin 1998; Belmont et al. 1999). Based upon a comparison with wild-type filaments incubated with beryllium fluoride (generating an actin-ADP-BeF complex, an analogue for the actin-ADP-Pi state), we have shown that the effect of the mutation is to prevent the conformational change following Pi release that destabilizes the actin filament. This mutation thus decouples ATP hydrolysis and Pi release, which proceed normally, from the conformational change that occurs in the wild-type filaments with Pi release.

Wild-Type Actin Has a Closed Nucleotide Cleft

The reconstruction in Fig. 2a is from wild-type yeast actin, and an opening is observed between the two major domains of the actin subunit. The reconstruction in Fig. 2b is from the V159N yeast actin mutant, where we can see a closing of the nucleotide-binding cleft.

The opening of the nucleotide binding cleft has previously been observed in a wild-type yeast actin reconstruction (Orlova et al. 1997), and appears to be modeled quite well by the open state of the G-actin subunit in the β-actin-profilin crystal structure (Fig. 2e). Similarly, the conformation of the closed state of the actin subunit in the β-actin-profilin crystal structure (Fig. 2f) provides a reasonable model for the subunit in the V159N mutant actin (Fig. 2b). This comparison therefore suggests that one component of the variation seen in different F-actin reconstructions can be modeled successfully by observed changes in G-actin structure.

Differences in F-Actin Reconstruction Are Highly Significant

Of course, one needs to establish that the differences observed by electron microscopy and three-dimensional helical reconstruction actually arise from differences in the actin filaments, and not from artifacts due to sampling statistics, differences in resolution, etc. We show in Fig. 2c and d the statistical

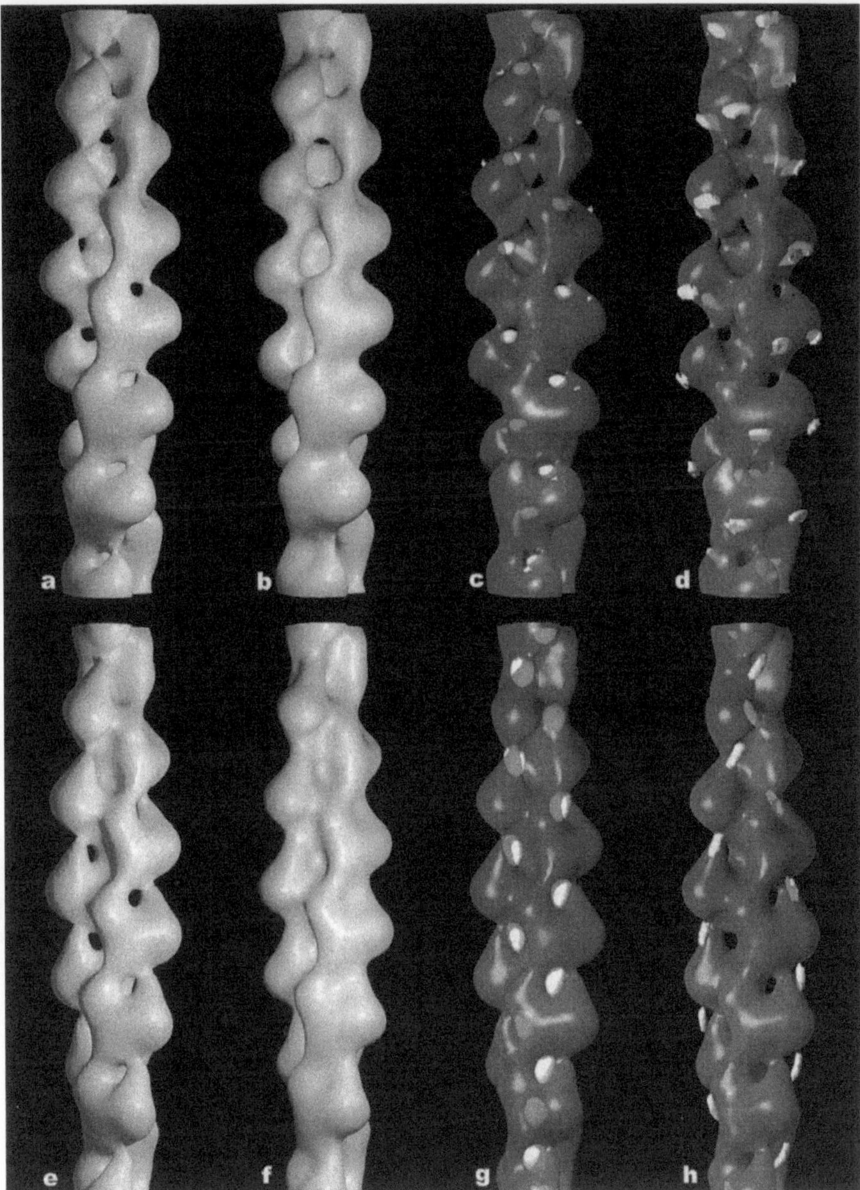

Fig. 2a–h. Three-dimensional reconstructions of yeast wild-type filaments (**a**) and V159N mutants (**b**), are compared with model filaments generated from the β-actin subunit in the open conformation (Schutt et al. 1993) (**e**) and in the closed conformation (Chik et al. 1996) (**f**). The reconstructions in **a** and **b** are each from averages of 42 filaments, or 84 near-far data sets. The statistical difference maps (Egelman and Yu 1989) between the wild-type and V159N mutants are shown in **c** for the positive difference peaks, and in **d** for the negative peaks. The differences between the two crystal models are shown in **g** and **h**. The threshold used for all of the reconstructions corresponds to 100% of the expected molecular volume, assuming a partial specific density of $0.75\,cm^3 g^{-1}$. The threshold for the statistical difference peaks is 3 σ. The difference peaks are shown on a semitransparent surface of the wild-type filament in **c** and **d**, and on a surface of the model in **e** for **g** and **h**.

difference map between the wild-type filaments (Fig. 2a) and the V159N mutants (Fig. 2b). A semitransparent surface of the wild-type filament reconstruction has been superimposed on the positive differences (Fig. 2c) and the negative differences (Fig. 2d). The differences that are shown are those that are greater than three standard deviations, and thus highly significant.

For comparison, difference maps between the two crystal-based filament models (Fig. 2e, f) are shown in Fig. 2g, h. It can be seen that the positive difference map (where the wild-type density is significantly greater than that of the V159N mutant) is quite similar between the actual filaments and the models. However, the negative difference maps are not similar, since the crystal difference map only has a negative peak on the outside of subdomain 2 (where it would be expected, if the conformational change mainly resulted from a rotation of subdomain 2), while the actual filaments show differences on the bottom of subdomains 1 and 3. This suggests that there may be some reorientation of the subunit or change in subdomains 1 and 3 as a result of the opening of the nucleotide-binding cleft in the actual filaments, but more work is needed to clarify this. (ed: These observations are relevant to the data reported in Kekic et al. (this Vol.) who report conformational changes in subdomain 2 of skeletal α-actin when cofilin binds to subdomain 1).

Summary

In summary, a number of different conformational states of F-actin have been described by several different laboratories. Crystal structures have revealed that an opening of the nucleotide-binding cleft, produced by a large rotation of subdomain 2, can occur in G-actin. We have shown that two crystal states of β-actin, in an open and closed form, can provide a very good model for the conformational difference in F-actin between yeast the wild-type and a V159N mutant. This suggests that some of the dynamics associated with G-actin may provide insights into dynamic processes within the F-actin filament.

Acknowledgments. This work was supported by NIH AR42023.

References

Belmont LD, Drubin DG (1998) The yeast v159n actin mutant reveals roles for actin dynamics in vivo. J Cell Biol 142:1289–1299

Belmont LD, Orlova A, Drubin DG, Egelman EH (1999) A change in actin conformation associated with filament instability after Pi release. ProcNatAcadSci, USA 96:29–34

Chik JK, Lindberg U, Schutt CE (1996) The structure of an open state of beta-actin at 265 Å resolution. J Mol Biol 263:607–623

Egelman EH, Orlova A (1995) New insights into actin filament dynamics. Cur Opin Struct Biol 5:172–180

Egelman EH, Yu X (1989) The location of DNA in RecA-DNA helical filaments. Science 245:404–407

Holmes KC, Popp D, Gebhard W, Kabsch W (1990) Atomic model of the actin filament. Nature 347:44–49

Kabsch W, Mannherz HG, Suck D, Pai EF, Holmes KC (1990) Atomic structure of the actin: DNase I complex. Nature 347:37–44

Kraulis PJ (1991) MOLSCRIPT: A program to produce both detailed schematic plots of protein structures. J Appl Crystallogr 24:946–950

Lorenz M, Popp D, Holmes KC (1993) Refinement of the F-actin model against X-ray fiber diffraction data by the use of a directed mutation algorithm. J Mol Biol 234:826–836

McGough A, Pope B, Chiu W, Weeds A (1997) Cofilin changes the twist of F-actin: Implications for actin filament dynamics and cellular function. J Cell Biol 138:771–781

McLaughlin PJ, Gooch JT, Mannherz HG, Weeds AG (1993) Structure of gelsolin segment 1-actin complex and the mechanism of filament severing. Nature 364:685–692

Orlova A, Egelman EH (1993) A conformational change in the actin subunit can change the flexibility of the actin filament. J Mol Biol 232:334–341

Orlova A, Egelman EH (1995) Structural dynamics of F-actin I Changes in the C-terminus. J Mol Biol 245:582–597

Orlova A, Chen X, Rubenstein PA, Egelman EH (1997) Modulation of yeast F-actin structure by a mutation in the nucleotide-binding cleft. J Mol Biol 271:235–243

Owen C, DeRosier D (1993) A 13 Å map of the actin-scruin filament from the *Limulus* acrosomal process. J Cell Biol 123:337–344

Schutt CE, Myslik JC, Rozycki MD, Goonesekere NCW, Lindberg U (1993) The structure of crystalline profilin β-actin. Nature 365:810–816

Actin Structure Function Relationships Revealed by Yeast Molecular Genetics

Lisa D. Belmont and David G. Drubin[1]

Introduction

The yeast *Saccharomyces cerevisiae*, has proven to be a powerful model organism in which to study actin function and structure-function relationships. This yeast has a single actin gene, a feature that greatly simplifies molecular-genetic and biochemical analysis of actin. Yeast actin is 88% identical to mammalian actin and its biochemical properties are similar to all other actins studied (Nefsky and Bretscher 1992). The ease of molecular genetics in yeast has allowed the generation of numerous site-specific actin mutants. Furthermore, the nonlethal actin mutants have been expressed in yeast as the sole source of actin (Shortle et al. 1984), allowing elucidation of biological importance via phenotypic analysis, and purification and biochemical characterization of mutant actins.

The ability to biochemically characterize mutant actins expressed in yeast has been especially useful since functional actin has not been expressed and purified successfully from bacteria. In Table 1, features of the published yeast actin mutants are summarized. These various mutants have provided insights into the mechanisms of polymerization and depolymerization, binding sites of various actin-binding proteins and drugs, regulation of actin by nucleotide, and the in vivo roles of actin in yeast.

Probing Actin Structural Changes and Mechanism of Polymerization and Depolymerization

Rapid turnover of actin filaments is required for diverse processes including cytokinesis, endocytosis and cell motility. Therefore, rapid polymerization and depolymerization are crucial for most actin-dependent processes. A number of studies have been carried out to elucidate the mechanisms of actin polymerization and depolymerization.

[1] Department of Molecular and Cell Biology, 401 Barker Hall, University of California, Berkeley, California 94720-3202, USA

Results and Problems in Cell Differentiation, Vol. 32
C. dos Remedios (Ed.): Molecular Interactions of Actin

Table 1. Yeast actin mutations. It should be noted that all mutants were not tested for the same phenotypes, this is a report of what has been published. For example, many actin mutants exhibit random budding, but only a few were tested. Generally, wild type means that the yeast expressing this mutant actin as their sole source of actin were not temperature– or cold–sensitive. Dominant lethal phenotypes were generally inferred from the inability to recover transformants of the mutant actin in the presence of a Wt actin gene, except in the case of D11K and D11Q where dominant lethality was independently determined. Abbreviations: Wt wild-type, Ts temperature–sensitive, Cs cold–sensitive

Substitution	In vivo Phenotype	Biochemical effects	Reference
DSE2-4 deletion	Slight < in secretion	<S1-ATPase	Cook et al. (1992); Miller et al. (1999b)
D2N/E4Q	Slight < in secretion	<S1-ATPase; <Caldesmon binding	Cook et al. (1992); Crosbie et al. (1994); Miller et al. (1996b)
D2V	Wt		Johannes and Gallwitz (1991)
D2A	Cs, Ts, actin bars, delocalized patches, clumped mitochondria		Wertman et al. (1992); Drubin et al. (1993)
D2V/E4V	Wt		Johannes and Gallwitz (1991)
SEV3-5 EDEV	Wt	>S1-ATPase	Cook et al. (1993)
E4A	Wt		Wertman et al. (1992)
E4V	Wt		Johannes and Gallwitz (1991)
D11A	Recessive lethal		Wertman et al. (1992)
D11K	Dominant lethal		Johannes and Gallwitz (1991)
D11Q	Dominant lethal		Johannes and Gallwitz (1991)
D11E	Wt		Cook and Rubenstein (1989)
D11N	No viable spores		Cook and Rubenstein (1989)
S14A	Ts, depolarized patches, loss of cables, actin bars	<Assembly stimulated ATPase	Chen and Rubenstein (1995)
S14G	Recessive lethal		Chen and Rubenstein (1995)
S14C	Recessive lethal		Chen and Rubenstein (1995)
S14T	Wt		Chen and Rubenstein (1995)
M16I	Ts		Karpova et al. (1995)
D24A/D25A	Recessive Cs, Ts, delocalized patches, faint cables, clumped mitochondria	<Weak binding to myosin	Wertman et al. (1992); Drubin et al. (1993); Miller et al. (1996b)
P32L	Ts		Shortle et al. (1984)

R37A/R39A	Recessive Cs, Ts, clumped patches, partially depolarized actin, actin bars, clumped mitochondria		Wertman et al. (1992); Drubin et al. (1993)
K50A/D51A	Recessive Cs, Ts, delocalized patches, faint cables, clumped mitochondria		Wertman et al. (1992); Drubin et al. (1993)
D56A	Ts		Adams and Botstein (1989)
D56A/E57A	Recessive Ts, depolarized patches, loss of cables		Wertman et al. (1992); Drubin et al. (1993)
A58T	Ts		Shortle et al. (1984)
K61N	Ts		Adams and Botstein (1989)
K61A/R62A	Partial dominant lethal		Wertman et al. (1992)
R68A/E72A	Wt		Wertman et al. (1992)
D80A/D81A	Recessive Ts, Cs		Wertman et al. (1992)
E83A/K84A	Recessive Ts, Cs, depolarized patches, actin bars		Wertman et al. (1992); Drubin et al. (1993)
H88Y	Ts		Adams and Botstein (1989)
T89I	Wt		Adams and Botstein (1989)
E93A/R95A	Recessive lethal		Wertman et al. (1992)
E99A/E100A	Recessive Ts, depolarized patches, loss of cables, clumped mitochondria	<Weak binding to myosin	Wertman et al. (1992); Drubin et al. (1993); Miller et al. (1996b)
R116A/E117A/K118A	Recessive Ts, depolarized patches, loss of cables		Wertman et al. (1992); Drubin et al. (1993)
D154A/D157A	Recessive lethal		Wertman et al. (1992)
D157E	Wt	>ATP exchange	Belmont et al. (1999b)
G158A	Ts, depolarized actin	<ATP exchange	Belmont et al. (1999b)
V159N	Recessive weak Ts, extra F-actin	Slow filament depolymerization	Belmont and Drubin (1998); Belmont et al. (1999a)
R177A/D179A	Recessive Ts, depolarized actin, loss of cables, clumped actin patches	Slow polymerization, no phalloidin binding	Wertman et al. (1992); Buzan and Frieden (1996); Drubin et al. (1993)
R183A/D184A	Wt		Wertman et al. (1992)
D187A/K191A	Wt		Wertman et al. (1992)

Table 1. *Continued*

Substitution	In vivo Phenotype	Biochemical effects	Reference
K191M	Wt		Johannes and Gallwitz (1991)
K191M/C374A	Wt		Johannes and Gallwitz (1991)
E195A/R196A	Wt		Wertman et al. (1992)
E205/R206/E207A	Dominant lethal		Wertman et al. (1992)
R210A/D211A	Recessive weak Ts, loss of cables		Wertman et al. (1992)
K213A/E214A/K215A	Recessive Cs, Ts		Wertman et al. (1992)
D222A/E224A/E226A	Recessive Ts, depolarized patches, loss of cables		Wertman et al. (1992); Drubin et al. (1993)
E237A/K238A	Recessive lethal		Wertman et al. (1992)
E241A/D244A	Recessive lethal		Wertman et al. (1992)
E253A/R254A	Recessive lethal		Wertman et al. (1992)
R256A/E259A	Recessive Ts, Cs, slightly depolarized patches, reduced staining of actin structures, clumped Mitochondria		Wertman et al. (1992); Drubin et al. (1993)
E259V	Recessive Ts		Dunn and Shortle (1990)
S265C	Wt	Fast nucleation	Feng et al. (1997)
S265C/C374A	Wt	Fast nucleation	Feng et al. (1997)
V266G/L267G	Slight Ts, Cs, multi-nucleate, depolarized actin, actin bars	<ATPase of G-actin, <Tm, <ATP exchange, no polymerization	Kuang and Rubenstein (1997a,b)
V266G	Random budding	Wt	Kuang and Rubenstein (1997a)
V266D	Slight Ts	Slow nucleation and polymerization at 4°C	Kuang and Rubenstein (1997a)
V266F	Wt	Wt	Kuang and Rubenstein (1997a)
L267G	Slight Ts	Wt	Kuang and Rubenstein (1997a)
L267D	Cs	Cs for polymerization	Chen et al. (1993)
L269D	Wt	Wt	Kuang and Rubenstein (1997a)
L269K	Wt	Wt	Kuang and Rubenstein (1997a)
E270A/D275A	Recessive lethal		Wertman et al. (1992)
D268A/D288A	Recessive lethal		Wertman et al. (1992)

R290A/K291A/E292A	Recessive lethal		Wertman et al. (1992)
E292K	Ts		Karpova et al. (1995)
E311A/R312A	Recessive Ts Cs		Wertman et al. (1992)
K315A/E316A	defects in pseudo-hyphal growth		Wertman et al. (1992); Cali et al. (1998)
K326A/K328A	Dominant lethal		Wertman et al. (1992)
E334A/R335A/K336A	Recessive lethal		Wertman et al. (1992)
K336M	Wt		Johannes and Gallwitz (1991)
K336Q	Wt		Johannes and Gallwitz (1991)
I341A	Wt	<Strong binding to myosin	Miller et al. (1996a)
I341F	Dominant lethal		Miller et al. (1996a)
I345A	Wt		Miller et al. (1996a)
I345F	Recessive lethal		Miller et al. (1996a)
I341A/I345A	Recessive lethal		Miller et al. (1996a)
I341K/I345K	Dominant lethal		Miller et al. (1996a)
W356Y	Wt		Johannes and Gallwitz (1991)
W356H	Wt		Johannes and Gallwitz (1991)
K359A/E361A	Wt		Wertman et al. (1992)
D363A/E364A	Recessive Ts, depolarized patches, loss of cables, clumped mitochondria		Wertman et al. (1992)
K373M	Wt		Johannes and Gallwitz (1991)
K373 Term	Recessive lethal		Johannes and Gallwitz (1991)
C374A	Wt		Johannes and Gallwitz (1991)
C374A/K191M	Wt		Johannes and Gallwitz (1991)
C374 Term	Ts, depolarized actin		Johannes and Gallwitz (1991)
F375Term	Ts, depolarized actin		Johannes and Gallwitz (1991)

Probing the Nucleotide-Binding Cleft

The hydrolysis of ATP and subsequent release of Pi provides a switch between conformations that favor assembly or disassembly and imparts upon the actin filament a functional polarity so that subunits may treadmill through filaments. Several mutagenesis studies have sought to elucidate the mechanism by which ATP hydrolysis and Pi release regulate actin dynamics. When the actin/DNase I cocrystal structure was solved, the amino acid residues that contact the bound nucleotide were identified (Kabsch et al. 1990).

The amide of Ser14 makes hydrogen bonds with the γ-phosphate and β-phosphate of the bound ATP or ADP. Chen and coworkers mutated the Ser14 to Cys, Thr, Gly, and Ala (Chen and Rubenstein 1995). S14C and S14G were lethal mutations, while strains expressing S14A as their sole actin source were temperature-sensitive. Strains expressing S14T actin appeared to be wild type. Purified S14A actin assembled more rapidly than wild-type actin, with no apparent lag phase. The critical concentration was unaffected, but the steady-state ATP hydrolysis rate was dramatically reduced compared to wild-type yeast actin. ATP hydrolysis rates at steady state are an indication of the rate of actin filament treadmilling. Further studies (Chen et al. 1995) revealed a decrease in the affinity of S14A actin for ATP and in the intrinsic ATPase rate, as well as a decrease in the T_m. The temperature sensitivity and low denaturation temperature may be due to loss of a polar bridge formed between the hydroxyl of S14 and the amide nitrogen of Gly74. This possibility is supported by the observation that the S14T mutation did not have a deleterious effect. Three-dimensional reconstructions of S14A and wild-type yeast actin filaments revealed that wild-type yeast actin has a more open nucleotide-binding cleft than that of rabbit muscle actin, while S14A had a closed nucleotide-binding cleft, more similar to that of rabbit muscle actin (Orlova et al. 1997).

In another study, we mutagenized Val159 to Asn. V159 is in the nucleotide binding cleft and the amide of V159 makes a hydrogen bond with the γ-phosphate of bound ATP (Kabsch et al. 1990). The V159N mutation has an interesting in vivo phenotype of increased F-actin structures, despite normal total actin levels as detected by Western blotting (Belmont and Drubin 1998). Studies on purified V159N actin revealed slow filament depolymerization and a decreased critical concentration. The rates of Pi release during initial assembly are the same as for wild-type actin, but at steady state the rate of Pi release is dramatically reduced, suggesting slow filament treadmilling.

Three-dimensional EM reconstructions revealed that V159N actin filaments as well as wild-type yeast actin polymerized with ADP-BeF_3^- (which mimics ADP-Pi) have a more closed nucleotide binding cleft (Belmont et al. 1999a). This strongly suggests that the slow filament dynamics of V159N actin are due to the failure to make the conformational change that normally accompanies Pi release, and that this conformational change consists primarily of an opening of the nucleotide-binding cleft. It is interesting that S14A actin filaments have a conformation similar to that of V159N and ADP-BeF_3^- filaments. While these filaments also appeared to exhibit slow steady-state ATP hydrolysis in vitro, consistent with increased filament stability, they did not appear to be hyperstabilized in vivo. This could be due to the greater temperature sensitivity of S14A actin. Thus, the general instability of S14A actin, rather than decreased rates of actin filament turnover, might have the greatest effect in vivo. Such an observation might explain why the S14A mutant showed a loss of cables and depolarization of patches in vivo, phenotypes that are typical of many partial loss-of-function actin mutants.

Testing the Hydrophobic Plug Model

Critical to understanding how filaments assemble and disassemble is elucidation of the determinants of filament stability. The hydrophobic plug model predicts that the hydrophobic loop encompassing residues 262–272 can insert into a hydrophobic "pocket" formed by the interface of two actin subunits on opposing strands, thus stabilizing the filament (Holmes et al. 1990). This interaction requires that the hydrophobic plug undergo a large movement away from the surface of the actin monomer where it is "parked" in the X-ray crystal structure.

Two actin mutagenesis studies designed to test this model have been performed in yeast. In the first study, Leu267 (referred to as L266 in the paper) was changed to Asp (Chen et al. 1993). This residue corresponds to I267 in the rabbit muscle actin sequence. Yeast expressing the L267D mutation grew at wild-type rates and exhibited only a slight decrease in growth rate at decreased temperatures. The purified L267D actin was severely inhibited in nucleation at low temperatures. Because hydrophobic interactions are cold-sensitive, this result was in agreement with predictions from the hydrophobic plug model. However, the T_m of this actin mutant was lowered, suggesting that a more global conformational change resulting from the L267D mutation might have contributed to the nucleation defect.

The next set of studies (Kuang and Rubenstein 1997a,b) involved a more extensive mutagenesis of the proposed hydrophobic plug residues. Kuang and Rubenstein made the following mutations: V266G, V266D, V266F, L267G, L269D, L269K, and V266G/L267G (referred to as GG-actin). Purified V266G, V266F, L267G, L269D, and L269K actins polymerized normally at temperatures ranging from 4 to 25 °C. V266D actin exhibited cold-sensitive defects in nucleation and elongation, although the extent of polymerization was almost the same as that of wild-type actin. The most dramatic effects were observed with GG-actin. At all temperatures tested, this actin failed to polymerize. This polymerization defect could not be rescued by the addition of stabilized actin seeds. Polymerization could, however, be rescued by addition of wild-type yeast actin. This wild-type actin facilitated assembly of some of the GG-actin into filaments. The extent to which GG–actin assembled into copolymers was sensitive to low temperature.

The CD spectrum of GG-actin is nearly identical to that of wild-type yeast actin, suggesting that there are no large structural changes in this mutant actin. However, the T_m is decreased significantly. GG–actin will also polymerize in the presence of phalloidin, and these filaments show an increased susceptibility to protease compared to wild-type filaments. This observation is consistent with the hypothesis that the mutant lowers cross-strand stabilization.

The Holmes model also predicts that the strongest hydrophobic interactions should involve the most N-terminal region of the hydrophobic plug. However, the L267D mutation caused more severe effects than the V266D mutation. These results are more consistent with the model proposed by Tirion et al. (1995), in which the hydrophobic plug can adapt to its environment. While all of these results are generally consistent with hydrophobic plug models in that they implicate the proposed plug region in intermonomer stabilization, they do not address the region with which this hydrophobic loop interacts. A rigorous test of the model would be to isolate compensating mutations in the residues of the proposed hydrophobic pocket that restore normal polymerization properties to hydrophobic plug mutants.

Introduction of Chemical Probes at Unique Sites

Mutagenesis can also be used to engineer sites that can be specifically labeled with chemical probes. Traditionally, Cys374 is labeled with pyrene to monitor actin polymerization (see Sheterline et al. 1996 for a more complete discussion of chemical modifications of actin). Actin polymerization places this moiety in a more hydrophobic environment, causing an increase in fluorescence.

Feng and coworkers reasoned that by changing the position of the pyrene label in yeast actin they might be able to test features of molecular models for the actin filament. To allow pyrene labeling in the loop between actin subdomains 3 and 4, Feng et al. (1997) constructed two yeast actin mutants: S265C and S265C/C374A. With these and the wild type it was then possible to label actin at positions 374 (wild-type), 265 (S265C/C374A), or both 265 and 374 (S265C).

These mutations caused no detectable defects when expressed in yeast, and the purified actin had a critical concentration, T_m, intrinsic ATPase, and protease digestion patterns that were essentially the same as those of wild-type yeast actin. The only difference observed was a faster rate of nucleation by both S265C and S265C/C374A actin. Pyrene-labeled S265C/C374A actin showed a decrease in pyrene fluorescence upon polymerization, indicating that pyrene attached to position 265 shifts into a more hydrophilic environment after polymerization. This is consistent with predictions from the Holmes filament model (Holmes et al. 1990) that there is a space between the two strands of the helix in an actin filament.

Residue 265 lies just outside the hydrophobic plug and serves as a pivot point for the rotation required for insertion of the plug into the hydrophobic pocket. This movement of the plug away from its "parked" position against the surface of actin may place residue 265 in a more hydrophilic environment. The Feng et al. study also revealed that pyrene-labeled S265C actin exhibited intermolecular quenching between pyrenes at positions 374 and 265 on separate actin subunits in a filament, creating a new excimer peak between 450 and 550 nm. This indicates that the two residues to which the pyrenes are attached are within 18 Å of each other. This is in better agreement with the Holmes et al. model (Holmes et al. 1990), that predicts this distance to be approximately 24 Å, than with the ribbon model, that predicts a distance of at least 40 Å (Schutt et al. 1989, 1993).

Engineering specific sites where one can attach chemical probes is a very powerful approach with the potential to provide many more insights into the conformational changes that occur upon polymerization, nucleotide hydrolysis, or Pi release.

Identification of Binding Sites of Proteins and Drugs

Elucidation of the intrinsic polymerization and depolymerization properties of actin is essential to our understanding of actin-related processes. However, in vivo, actin dynamics are regulated by a variety of actin-binding proteins. The large array of yeast actin mutants listed in Table 1 have proven to be quite useful in functionally mapping the binding sites of various actin-binding proteins and drugs on the surface of actin. The largest collection of actin mutations was generated in the "clustered charged to alanine scan" of yeast actin (Wertman et al. 1992). This study was based on the observation that clusters of charged residues tend to be on the surface of a protein (Bass et al. 1991; Bennett et al. 1991; Gibbs and Zoller 1991). Groups of two or more charged amino acids that appeared within windows of five amino acids were systematically mutated to alanines, generating a collection of 36 actin alleles. Although the mutagenesis was begun before the actin structure had been solved, analysis of the crystal structure of actin (Kabsch et al. 1990) revealed that indeed, nearly all of these mutations were on the surface of actin. As a result, many of these mutants have subsequently been found to specifically disrupt the interaction of actin–binding proteins and/or drugs with actin.

We will discuss how these and other actin mutants have been used to map binding sites of actin-binding proteins and drugs on the surface of actin. See the chapter by Doyle and Reisler, this volume, for a discussion of how yeast actin mutants have been used to study the interactions between actin and myosin.

Mapping the Binding Sites of Actin Binding Proteins on the Surface of Actin

Fimbrin (Sac6p) was identified as an actin-binding protein in yeast both by affinity chromatography (Drubin et al. 1988) and as a suppressor of an actin mutation (Adams et al. 1989). The fimbrin-binding site on yeast actin was genetically mapped by selecting for actin mutants that suppress the temperature sensitivity of *sac6* mutants (Honts et al. 1994). The rationale behind this approach was that *sac6* mutations impairing binding of fimbrin (Sac6p) to actin could be compensated for by a mutation in the fimbrin-binding site on actin. Indeed, actin alleles that suppressed (or were suppressed by) *sac6* mutations, mapped to subdomain 2 and adjacent regions of subdomain 1 (Figure 1A). As further confirmation that these mutations directly affect actin residues involved in the interaction with fimbrin, the relevant mutant actins were purified and shown to have defects in binding to fimbrin.

An alternative genetic approach was taken by Holtzman et al. (1994). Their approach was based on the assumption that an actin mutant that specifically disrupted interactions with Sac6p would have similar phenotypes and genetic interactions to a sac6 null mutant. Of 22 actin alleles analyzed, two met the criteria for being defective in Sac6p binding. They were synthetically lethal with null mutations of *ABP1* and *SLA2*, but not *SLA1*. The mutations identified were E99A/E100A and P32L, two of the eight mutations identified by Honts and coworkers. In support of these genetic results, electron cryomicroscopy of actin-binding domains of α-actinin and fimbrin bound to actin filaments showed that they bind to a concave surface formed between actin subdomains 1 and 2 (Hanein et al. 1997). In addition, Hanein et al. found that subdomain 1 of actin undergoes a conformational shift upon binding to the amino-terminal actin-binding domain of fimbrin.

In another study to map the fimbrin-binding site using actin mutations, a fusion protein of green fluorescent protein (GFP) and Sac6p (GFP-Sac6p) was introduced into strains containing the actin alanine scan mutations. The cells were then observed for GFP localization to actin patches. This study identified three actin alleles that had been implicated in fimbrin-binding previously, as well as one new allele (D222A, E224A, E226A) located on subdomain 4, away from the fimbrin binding site (T. Doyle, pers. comm.). It is possible that this latter mutation disrupts the conformational change that occurs upon fimbrin binding to the actin filament. These studies demonstrate the power of genetic approaches for mapping protein-binding sites. They also show that caution must be used when interpreting the results of such studies.

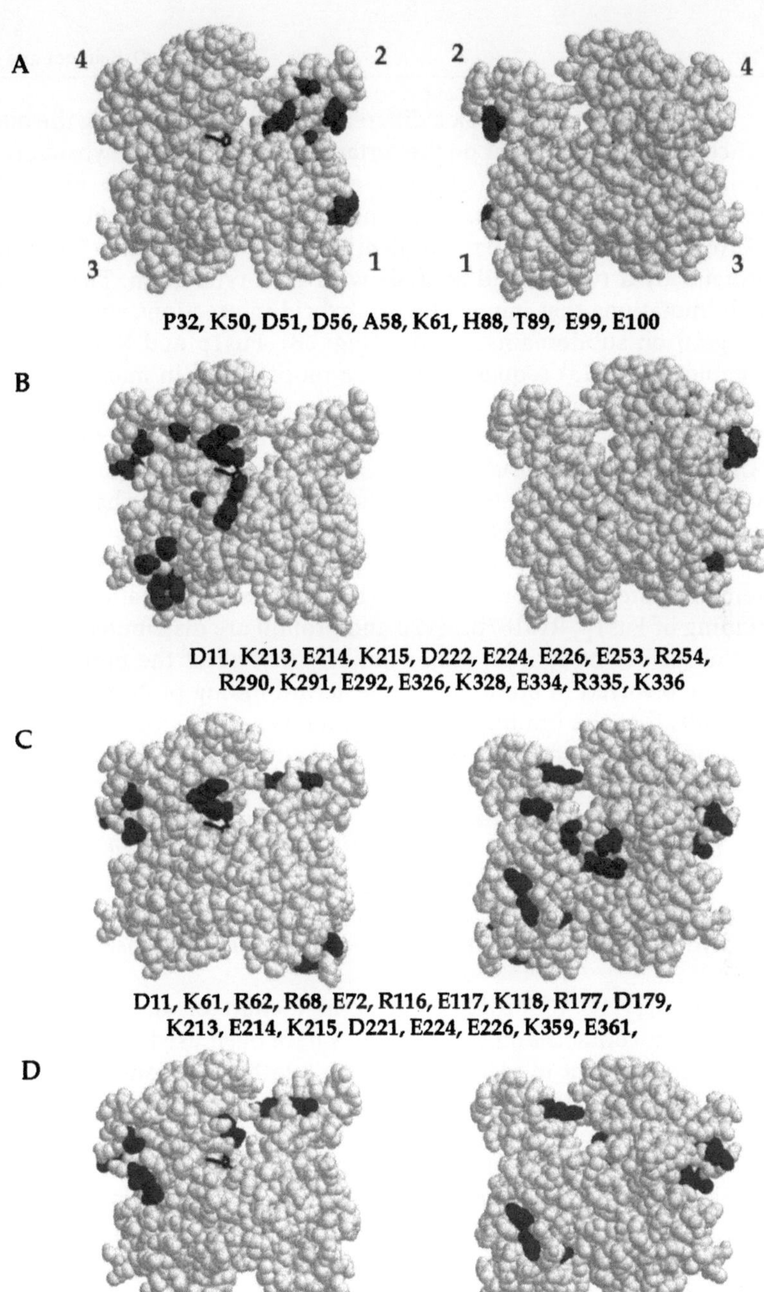

Fig. 1A–D. The binding sites of several actin-binding proteins have been mapped onto the surface of actin using actin mutants. Front and back views of the surface of rabbit muscle actin (Kabsch et al. 1990), the four actin subdomains are labeled. **A** The location of mutated residues in actin alleles that suppress or are suppressed by fimbrin (*sac6*) mutations (Honts et al. 1994). P32, E99 and E100 were also identified genetically as residues required for fimbrin binding by Holtzman et al. (1994), and K50, D51, E83, K84, E99, and E100 were identified as mutations that disrupt the localization of GFP–Sac6p (T. Doyle, pers. comm.). The location of actin mutations that disrupt two-hybrid interactions between actin and **B** Aip1p, **C** both profilin and Rvs167p, and **D** Srv2p. There is, in addition, a significant overlap between the proposed binding site of Fus1p and that of profilin and Rvs167p (Amberg et al. 1995). Fus1p and Rvs167p both contain SH3 domains

Amberg et al. (1995) took a different approach to mapping the binding sites of actin-binding proteins on the surface of actin. The two-hybrid protein interaction reporter system (Chien et al. 1991) was used to map interactions between the alanine scan actin mutants and actin, profilin, Srv2p, Aip1p, Rvs167p, and Fus1p. Aip1p was identified in this study as an "actin-interacting protein" by a two-hybrid analysis with wild-type actin. The location of the actin mutations that impair the two hybrid interactions with Aip1p cluster to a region on subdomains 3 and 4 (Fig. 1B). Fus1p and Rvs167p contain Src-homology 3 (SH3) sequence motifs, a motif found in many cytoskeletal proteins (Pawson and Gish 1992). The pattern of actin mutations that disrupt two-hybrid interactions are similar between Fus1p, Rvs167p, and profilin (Fig. 1C). However, these results are not consistent with the actin/profilin cocrystal, in which the contacts between these two proteins span the bases of subdomains 1 and 3 (Schutt et al. 1993).

The proposed Srv2p-binding site partially overlaps with that of the two SH3 domain proteins and profilin (Fig. 1D). Because the mutations that disrupt the binding of Fus1p, Rvs167p, Srv2p and profilin are distributed over a large part of the surface of actin, it seems unlikely that all of the mutated residues are directly involved in binding of these actin-binding proteins. It is possible, for example, that the binding of these proteins requires (or is enhanced by) the binding of other proteins (see Kekic et al., this Vol.) and that is why the region mapped by two-hybrid interactions is so large.

Amberg et al. (1995) also identified regions on the surface of actin that are required for actin-actin interactions in the two-hybrid system. These interactions might represent dimer formation, the first step in actin polymerization. Many of these mutations mapped to regions expected to form the hydrophobic pocket, in support of the model proposed by Holmes et al. (1990).

The cocrystal of profilin and actin (Schutt et al. 1993) revealed novel inter-actin contacts and these contacts have been used to propose an alternative actin filament model. However, the two-hybrid results show no evidence that these contacts are utilized in this reporter system. One caveat to using the two-hybrid system to map binding sites is that the mapped interactions may not be direct. Because the interactions are measured in vivo, they could be mediated through one or several other proteins. This system therefore provides a very powerful initial screen for mutations that disrupt protein-protein interactions, but the results should be followed up with biochemistry on purified components.

Mapping the Binding Sites of Drugs on the Surface of Actin

In addition to mapping the binding sites of proteins, the actin alanine scan collection has been used to map the binding site of the actin-sequestering drugs, latrunculin A and tolytoxin, and the actin filament binding toxin,

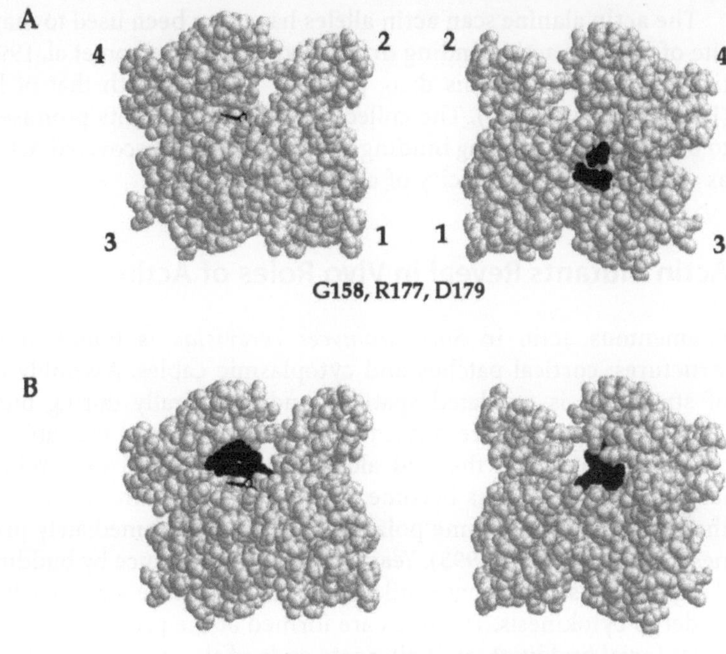

Fig. 2A,B. The location of the phalloidin and latrunculin A binding sites have been mapped onto the surface of actin using actin mutants. **A** Amino acid residues, when mutated to alanine, that disrupt rhodamine-phalloidin staining of actin filaments in fixed cells (Drubin et al. 1993; L. D. Belmont et al. 1999b). **B** The location of mutations that cause resistance to latrunculin A (Ayscough et al. 1997; L. D. Belmont et al. 1999b)

phalloidin. The phalloidin-binding site (Fig. 2A) was identified from the observation that the mutations R177A/D179A and G158A caused failure to stain with rhodamine phalloidin (Drubin et al. 1993; Belmont et al. 1999b). The placement of the phalloidin–binding site is consistent with the phalloidin-binding site proposed by Lorenz et al. based on structure analysis (Lorenz et al. 1993).

To map the latrunculin A binding site, actin alleles were tested for relative sensitivity to latrunculin A. Four alleles were found to be completely resistant. The residues implicated by this analysis map to a region near the actin nucleotide–binding cleft (Fig. 2B). In support of the conclusion that these residues are part of the binding site for this compound, latrunculin A was shown to inhibit nucleotide exchange on purified yeast actin, but not on two resistant alleles (Ayscough et al. 1997; Belmont et al. 1999b). These results not only identified the latrunculin A binding site, but also demonstrated that the toxicity of latrunculin A is a direct result of its binding to actin. This is because a mutation in actin is sufficient to cause resistance to the compound.

The actin alanine scan actin alleles have also been used to map the binding site of another actin-binding drug, tolytoxin (Patterson et al. 1993). The putative binding site for this drug partially overlaps with that of latrunculin A (Belmont et al. 1999b). The collection of actin mutants promises to continue to be useful for mapping binding sites of as yet undiscovered natural products, as well as testing specificity of actin binding toxins.

Actin Mutants Reveal in Vivo Roles of Actin

Filamentous actin in *Saccharomyces cerevisiae* is found in two types of structures: cortical patches and cytoplasmic cables. Assembly of both types of structures is regulated spatially and temporally during bud emergence. The cortical patches are concentrated in the bud, and the cables run between the mother cell and the bud along the mother-bud axis. When the bud is large, the actin patches become depolarized for isotropic growth, and late in the cell cycle they become polarized at the neck immediately prior to cytokinesis (Lew and Reed 1995). Yeasts normally reproduce by budding to produce daughter cells that grow until they reach a mature size, at which time they undergo cytokinesis. The buds are formed at the previous bud site in haploid yeast (axial budding), or at alternate ends of the yeast cell in diploids (bipolar budding).

The collection of actin mutations has been extremely useful in the identification of various cellular processes for which actin is required. Cytochalasins, while inhibiting the assembly of yeast actin in vitro, appear not to be taken up into yeast cells, so until the discovery that latrunculin A was an effective actin–sequestering drug in yeast, actin mutants were the only way with which to test the in vivo roles of actin. In addition, the actin mutant collection provides information about the actin surface and interactions required for different functions because some actin mutants impair a subset of actin-dependent functions by disrupting interactions between actin and one or a small number of actin-binding proteins. Another advantage of having a large actin mutant collection is that mutations disrupting a single one of actin's interactions rather than the entire actin cytoskeleton may reveal actin functions that are masked when the entire actin cytoskeleton is disrupted by drugs or by actin mutants with pleiotropic effects.

Studies on the first two actin mutants revealed roles for actin in localized deposition of chitin on the cell surface, secretion of the periplasmic protein, invertase, and bud formation (Novick and Botstein 1985). Actin mutants were also found to have increased osmotic sensitivity and an accumulation of secretory vesicles. In addition, actin is required for the extension of mating projections (Read et al. 1992). Subsequent studies demonstrated that actin mutants have defects in orientation of the mitotic spindle (Palmer et al. 1992) and positioning nuclei properly in pheromone-treated cells (Read et al. 1992). These two observations suggest cross-talk between the actin and microtubule cytoskeletons.

A more extensive study on 13 of the alanine scan actin mutants revealed additional roles for actin. Actin mutants had defects in bud site selection, bud morphology (sometimes resulting in elongated buds), and cell size. Mitochondria were seen to line up along actin cables in wild-type yeast but not in actin mutants that had mutations in the "myosin footprint" (Drubin et al. 1993). Additional studies revealed that mitochondrial movements in meiosis (Smith et al. 1995) and mitosis (Simon et al. 1995) are dependent on actin. Kübler and Riezman (1993) demonstrated that actin is required for the first step of receptor-mediated internalization of α–factor. Subsequent studies on cofilin and actin mutants suggest that rapid actin turnover may be required for fluid phase endocytosis (Lappalainen and Drubin 1997; Belmont and Drubin 1998).

In all, actin mutants have revealed roles for actin in endocytosis, secretion, polarized growth, morphogenesis, bud site selection, spindle orientation, nuclear migration, and mitochondrial organization and movement.

The studies described above were performed on yeast form (YF) cells, which grow by axial or bipolar budding to produce ovoid daughters. However, when *Saccharomyces cerevisiae* are starved for nitrogen they switch to a pseudohyphal (PH), mode of growth (Gimeno et al. 1992). PH yeast cells are elongated, bud in a unipolar manner, and invade the agar on which they are grown. In addition, the actin patches are polarized to the distal tip of the of the daughter cell throughout bud growth (Kron et al. 1994). This contrasts with YF cells in which there is a distinct period during late bud growth when the patches become distributed throughout the bud (Lew and Reed. 1995).

A recent study has examined the effects of 12 alanine scan actin mutants on various aspects of PH growth (Cali et al. 1998). In this study various defects in agar invasion, cell elongation, and unipolar bud site selection were found for the actin mutants. There was also a correlation between actin patch depolarization and defects in cell elongation, suggesting that the extended period of patch polarization is required for cell elongation. In addition, some of the actin-related defects in PH growth were shown to be due to the failure of the mutant actin to bind to fimbrin. Finally, the *act1–104* (K315/E316) mutant, which has no known phenotype for YF growth, disrupts filamentation. This actin mutant may, therefore, provide a useful tool with which to identify an actin binding protein that has a specific function in PH growth.

Concluding Remarks

The ease of molecular genetics in *Saccharomyces cerevisiae* allows the rapid and efficient generation of mutants. The approach of generating large collections of mutations to functionally map the surfaces of a protein has been quite informative with actin as well as with other cytoskeletal proteins, including tubulin (Huffaker et al. 1988) and cofilin (Lappalainen et al. 1997).

Studies of actin mutants have allowed researchers to test a number of predictions from biophysical studies, such as the hydrophobic plug model, the actin filament model, and to describe conformational shifts after ATP hydrolysis and Pi release. Studies with actin mutants generally support the hydrophobic plug model (Holmes et al. 1990) and the Lorenz (Lorenz et al. 1993) filament model. In addition, mutagenesis studies support biochemical results indicating a shift in subdomain 2 after Pi release (Strzelecka-Golaszewska et al. 1993).

The ability to mutate residues to allow the placement of chemical probes at specific sites in actin provides a powerful approach for testing these models in even greater molecular detail.

The actin mutants have allowed the sites of actin-binding proteins to be mapped onto the surface of actin. The molecular-genetic approaches complement structural studies such as those employing X-ray crystallography of actin and actin-binding proteins, and electron cryomicroscopy of actin decorated with actin-binding proteins. Site-directed mutagenesis provides a powerful approach with which to test the contributions of specific amino acids to molecular interactions. Furthermore, the molecular-genetic approaches provide in vivo identification and verification of binding sites.

Actin mutants have also been useful in mapping binding sites of drugs onto the surface of actin. Furthermore, actins that are resistant to various actin-binding compounds can demonstrate the specificity of actin-binding drugs and have the potential to be useful research tools. For example, inducing expression of wild-type actin into a strain expressing a mutant actin that does not bind phalloidin could potentially allow one to distinguish new and old populations of actin filaments by comparing anti-actin immunofluorescence patterns to rhodamine phalloidin staining patterns.

Mutational analysis of yeast actin has also identified numerous functions for actin in vivo. Moreover, well-characterized actin mutants can shed light on the mechanisms of actin-dependent processes. For example, the V159N actin mutant, which slows down actin turnover, has only minor effects on the development of a polarized yeast actin cytoskeleton (Belmont and Drubin 1998). This suggests that yeast polarize their actin by some method other than selective depolymerization and polymerization. Future studies of the numerous actin mutants that fail to polarize their actin cytoskeletons should help to reveal the mechanisms of establishment and maintenance of actin polarity in yeast.

In conclusion, molecular-genetic analysis of yeast actin has contributed greatly to our mechanistic understanding of actin dynamics as well as its regulation and cellular organization. Actin is highly conserved, and conclusions from these studies in yeast are as a result proving to be generally applicable to other eukaryotes.

References

Adams AE, Botstein D (1989) Dominant suppressors of yeast actin mutations that are reciprocally suppressed. Genetics 121:675–683

Adams AE, Botstein D, Drubin DG (1989) A yeast actin-binding protein is encoded by SAC6, a gene found by suppression of an actin mutation. Science 243:231–233

Amberg DC, Basart E, Botstein D (1995) Defining protein interactions with yeast actin in vivo. Nat Struct Biol 2:28–35

Ayscough KR, Stryker J, Pokala N, Sanders M, Crews P, Drubin DG (1997) High rates of actin filament turnover in budding yeast and roles for actin in establishment and maintenance of cell polarity revealed using the actin inhibitor latrunculin-A. J Cell Biol 137:399–416

Bass SH, Mulkerrin MG, Wells JA (1991) A systematic mutational analysis of hormone-binding determinants in the human growth hormone receptor. Proc Natl Acad Sci USA 88:4498–502

Belmont LD, Drubin DG (1998) The Yeast V159N actin mutant reveals roles for actin dynamics in vivo. J Cell Biol 142:1289–1299

Belmont LD, Orlova A, Drubin DG, Egelman EH (1999a) A change in actin conformation associated with filament instability and Pi release. Proc Natl Acad Sci USA 96:29–34

Belmont LD, Patterson GML, Drubin DG (1999b) New actin mutants allow further characterization of the nucleotide binding cleft and drug binding sites. J Cell Sci 112(9):1325–1336

Bennett WF, Paoni NF, Keyt BA, Botstein D, Jones AJ, Presta L, Wurm FM, Zoller MJ (1991) High resolution analysis of functional determinants on human tissue-type plasminogen activator. J Biol Chem 266:5191–5201

Buzan JM, Frieden C (1996) Yeast actin: polymerization kinetic studies of wild type and a poorly polymerizing mutant. Proc Natl Acad Sci USA 93:91–95

Cali BM, Doyle TC, Botstein D, Fink GR (1998) Multiple functions for actin during filamentous growth of *Saccharomyces cerevisiae*. Mol Biol Cell 9:1873–1889

Chen X, Cook RK, Rubenstein PA (1993) Yeast actin with a mutation in the "hydrophobic plug" between subdomains 3 and 4 (L266D) displays a cold-sensitive polymerization defect. J Cell Biol 123:1185–1195

Chen X, Peng J, Pedram M, Swenson CA, Rubenstein PA (1995) The effect of the S14A mutation on the conformation and thermostability of *Saccharomyces cerevisiae* G-actin and its interaction with adenine nucleotides. J Biol Chem 270:11415–11423

Chen X, Rubenstein PA (1995) A mutation in an ATP-binding loop of *Saccharomyces cerevisiae* actin (S14A) causes a temperature-sensitive phenotype in vivo and in vitro. J Biol Chem 270:11406–11414

Chien CT, Bartel PL, Sternglanz R, Fields S (1991) The two-hybrid system: a method to identify and clone genes for proteins that interact with a protein of interest. Proc Natl Acad Sci USA 88:9578–9582

Cook RK, Rubenstein PA (1989) The effects of actin asp10 mutants on yeast viability. J Cell Biol 109:271a

Cook RK, Blake WT, Rubenstein PA (1992) Removal of the amino-terminal acidic residues of yeast actin. Studies in vitro and in vivo [published erratum appears in J Biol Chem 267:13780]. J Biol Chem 267:9430–9436

Cook RK, Root D, Miller C, Reisler E, Rubenstein PA (1993) Enhanced stimulation of myosin subfragment 1 ATPase activity by addition of negatively charged residues to the yeast actin NH2 terminus. J Biol Chem 268:2410–2415

Crosbie RH, Miller C, Chalovich J, Rubenstein P, Reisler E (1994) Caldesmon, N-terminal yeast actin mutants, and the regulation of actomyosin interactions. Biochemistry 33:3210–3216

Drubin DG, Miller KG, Botstein D (1988) Yeast actin-binding proteins: Evidence for a role in morphogenesis. J Cell Biol 107:2551–2561

Drubin DG, Jones HD, Wertman KF (1993) Actin structure and function: roles in mitochondrial organization and morphogenesis in budding yeast and identification of the phalloidin-binding site. Mol Biol Cell 4:1277–1294

Dunn TM, Shortle D (1990) Null alleles of SAC7 suppress temperature-sensitive actin mutations in *Saccharomyces cerevisiae*. Mol Cell Biol 10:2308–2314

Feng L, Kim E, Lee WL, Miller CJ, Kuang B, Reisler E, Rubenstein PA (1997) Fluorescence probing of yeast actin subdomain 3/4 hydrophobic loop 262–274. Actin-actin and actin-myosin interactions in actin filaments. J Biol Chem 272:16829–16837

Gibbs CS, Zoller MJ (1991) Rational scanning mutagenesis of a protein kinase identifies functional regions involved in catalysis and substrate interactions. J Biol Chem 266:8923–8931

Gimeno CJ, Ljungdahl PO, Styles CA, Fink GR (1992) Unipolar cell divisions in the yeast *S. cerevisiae* lead to filamentous growth: Regulation by starvation and RAS. Cell 68:1077–1090

Hanein D, Matsudaira P, DeRosier DJ (1997) Evidence for a conformational change in actin induced by fimbrin (N375) binding. J Cell Biol 139:387–396

Holmes KC, Popp D, Gebhard W, Kabsch W (1990) Atomic model of the actin filament. Nature 347:44–49

Holtzman DA, Wertman KF, Drubin DG (1994) Mapping actin surfaces required for functional interactions in vivo. J Cell Biol 126:423–432

Honts JE, Sandrock TS, Brower SM, O'Dell JL, Adams AE (1994) Actin mutations that show suppression with fimbrin mutations identify a likely fimbrin-binding site on actin. J Cell Biol 126:413–422

Huffaker TC, Thomas JH, Botstein D (1988) Diverse effects of beta-tubulin mutations on microtubule formation and function. J Cell Biol 106:1997–2010

Johannes FJ, Gallwitz D (1991) Site-directed mutagenesis of the yeast actin gene: a test for actin function in vivo. Embo J 10:3951–3958

Kabsch W, Mannherz HG, Suck D, Pai EF, Holmes KC (1990) Atomic structure of the actin:DNase I complex. Nature 347:37–44

Karpova TS, Tatchell K, Cooper JA (1995) Actin filaments in yeast are unstable in the absence of capping protein or fimbrin. J Cell Biol 131:1483–1493

Kron SJ, Styles CA, Fink GR (1994) Symmetric cell division in pseudohyphae of the yeast *Saccharomyces cerevisiae*. Mol Biol Cell 5:1003–1022

Kuang B, Rubenstein PA (1997a) Beryllium fluoride and phalloidin restore polymerizability of a mutant yeast actin (V266G, L267G) with severely decreased hydrophobicity in a subdomain 3/4 loop. J Biol Chem 272:1237–1247

Kuang B, Rubenstein PA (1997b) The effects of severely decreased hydrophobicity in a subdomain 3/4 loop on the dynamics and stability of yeast G-actin. J Biol Chem 272:4412–4418

Kübler E, Riezman H (1993) Actin and fimbrin are required for the internalization step of endocytosis in yeast. EMBO J 12:2855–2862

Lappalainen P, Drubin DG (1997) Cofilin promotes rapid actin filament turnover in vivo. Nature 388:78–82

Lappalainen P, Fedorov EV, Fedorov AA, Almo SC, Drubin DG (1997) Essential functions and actin-binding surfaces of yeast cofilin revealed by systematic mutagenesis. EMBO J 16:5520–5530

Lew DJ, Reed SI (1995) A cell cycle checkpoint monitors cell morphogenesis in budding yeast. J Cell Biol 129:739–749

Lorenz M, Popp D, Holmes KC (1993) Refinement of the F-actin model against X-ray fiber diffraction data by the use of a directed mutation algorithm. J Mol Biol 234:826–836

Miller CJ, Doyle TC, Bobkova E, Botstein D, Reisler E (1996a) Mutational analysis of the role of hydrophobic residues in the 338–348 helix on actin in actomyosin interactions. Biochemistry 35:3670–3676

Miller C, Wong W, Bobkova E, Rubenstein P, Reisler E (1996b) Mutational Analysis of the role of the N terminus of actin in actomyosin interactions. Comparison with other mutant actins and implications for the cross-bridge cycle. Biochemistry 35:16557–16565

Nefsky B, Bretscher A (1992) Yeast actin is relatively well behaved. Eur J Biochem 206:949–955

Novick P, Botstein D (1985) Phenotypic analysis of temperature-sensitive yeast actin mutants. Cell 40:405–416

Orlova A, Chen X, Rubenstein PA, Egelman EH (1997) Modulation of yeast F-actin structure by a mutation in the nucleotide-binding cleft. J Mol Biol 271:235–243

Palmer RE, Sullivan DS, Huffaker T, Koshland D (1992) Role of astral microtubules and actin in spindle orientation and migration in the budding yeast, *Saccharomyces cerevisiae*. J Cell Biol 119:583–593

Patterson GM, Smith CD, Kimura LH, Britton BA, Carmeli S (1993) Action of tolytoxin on cell morphology, cytoskeletal organization, and actin polymerization. Cell Motil Cytoskel 24:39–48

Pawson T, Gish GD (1992) SH2 and SH3 domains: from structure to function. Cell 71:359–362

Read EB, Okamura HH, Drubin DG (1992) Actin- and tubulin-dependent functions during *Saccharomyces cerevisiae* mating projection formation. Mol Biol Cell 3:429–444

Schutt CE, Lindberg U, Myslik J, Strauss N (1989) Molecular packing in profilin: actin crystals and its implications. J Mol Biol 209:735–746

Schutt CE, Myslik JC, Rozycki MD, Goonesekere NC, Lindberg U (1993) The structure of crystalline profilin-beta-actin. Nature 365:810–816

Sheterline P, Clayton J, Sparrow J (1996) Protein profile: actins. Academic Press, London

Shortle D, Novick P, Botstein D (1984) Construction and genetic characterization of temperature-sensitive mutant alleles of the yeast actin gene. Proc Natl Acad Sci USA 81:4889–4893

Simon VR, Swayne TC, Pon LA (1995) Actin-dependent mitochondrial motility in mitotic yeast and cell-free systems: identification of a motor activity on the mitochondrial surface. J Cell Biol 130:345–354

Smith MG, Simon VR, O'Sullivan H, Pon LA (1995) Organelle-cytoskeletal interactions: actin mutations inhibit meiosis-dependent mitochondrial rearrangement in the budding yeast *Saccharomyces cerevisiae*. Mol Biol Cell 6:1381–1396

Strzelecka-Golaszewska H, Moraczewska J, Khaitlina SY, Mossakowska M (1993) Localization of the tightly bound divalent-cation-dependent and nucleotide-dependent conformation changes in G-actin using limited proteolytic digestion. Eur J Biochem 211:731–742

Tirion MM, ben-Avraham D, Lorenz M, Holmes KC (1995) Normal modes as refinement parameters for the F-actin model. Biophys J 68:5–12

Wertman K, Drubin D, Botstein D (1992) Systematic mutational analysis of the yeast ACT1 gene. Genetics 132:337–350

Actin-Binding Proteins: An Overview

Enrique M. De La Cruz[1]

Types of Actin Structures

Actin filaments in cells are temporally and spatially arranged into a variety of highly ordered assemblages (Fig. 1). Some of these structures are very stable, but others are extremely dynamic and turn over in a few seconds. Continuous reorganization of actin filaments is required for many types of cellular movements. It is believed that the assembly of actin filaments provides some of the force necessary for extension of the leading edge in motile cells (Fig. 2). Rapid depolymerization is required to replenish the monomer pool. The precise timing and location of filament assembly, elongation, turnover and network formation are determined by the interactions of actin with a number of accessory proteins collectively referred to as actin-binding proteins.

Classification of ABPs

Actin-binding proteins have traditionally been classified according to their biochemical properties or binding mechanisms (Fig. 3). Some interact with free actin monomers. Others bind two or more monomer subunits incorporated in filaments. In many cases, a given actin-binding protein binds both monomers and filaments but the interaction with one of the two forms usually dominates. Binding can vary with changes in pH, nucleotide and ionic composition of the medium. Several classes of monomer and filament-binding proteins are defined, each with multiple members. Myosins bind along the sides of filaments and couple the free energy of ATP hydrolysis with changes in structure to perform mechanical work. Cross-linking proteins like alpha-actinin and filamin possess two actin-binding sites and assemble filaments into bundles or meshed networks. The specific organization of the networks and bundles depends on the number, affinity, spacing and orientation of filament-binding sites on the actin-binding protein, as well as filament length and the concentrations of actin and cross-linking protein.

[1] University of Pennsylvania School of Medicine, Department of Physiology, Pennsylvania Muscle Institute, Philadelphia, Pennsylvania 19104, USA. The author is a Burroughs Wellcome Fund Fellow of the Life Sciences Research Foundation

Results and Problems in Cell Differentiation, Vol. 32
C. dos Remedios (Ed.): Molecular Interactions of Actin

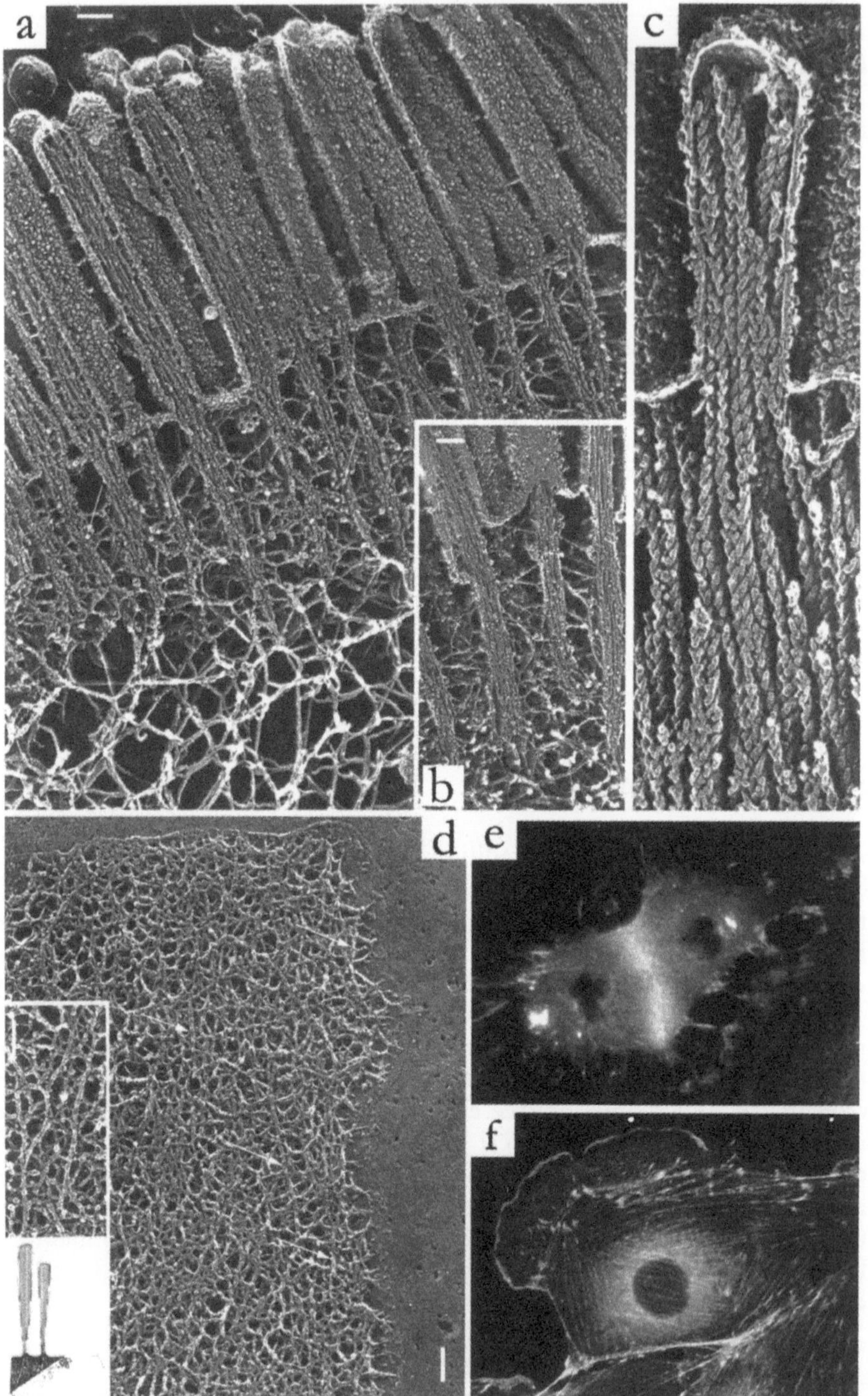

Fig. 1a–f. Organization of actin filaments in cells. **a–c** The brush border of mouse intestinal epithelia; *bar* 0.1 μM. In **c** actin filaments are decorated with myosin subfragment 1. **d** The macula of chick inner ear hair cells; *bar* 0.1 μM. **e–f** Localization of α-actinin in PtK2 cells during mitosis and interphase. **a–c** reproduced with permission from Hirokawa et al. (1982) J. Cell Biol. 94:425–443. **d** with permission from Hirokawa, N., Tilney, LG (1982) J. Cell Biol. **95**, 249–261

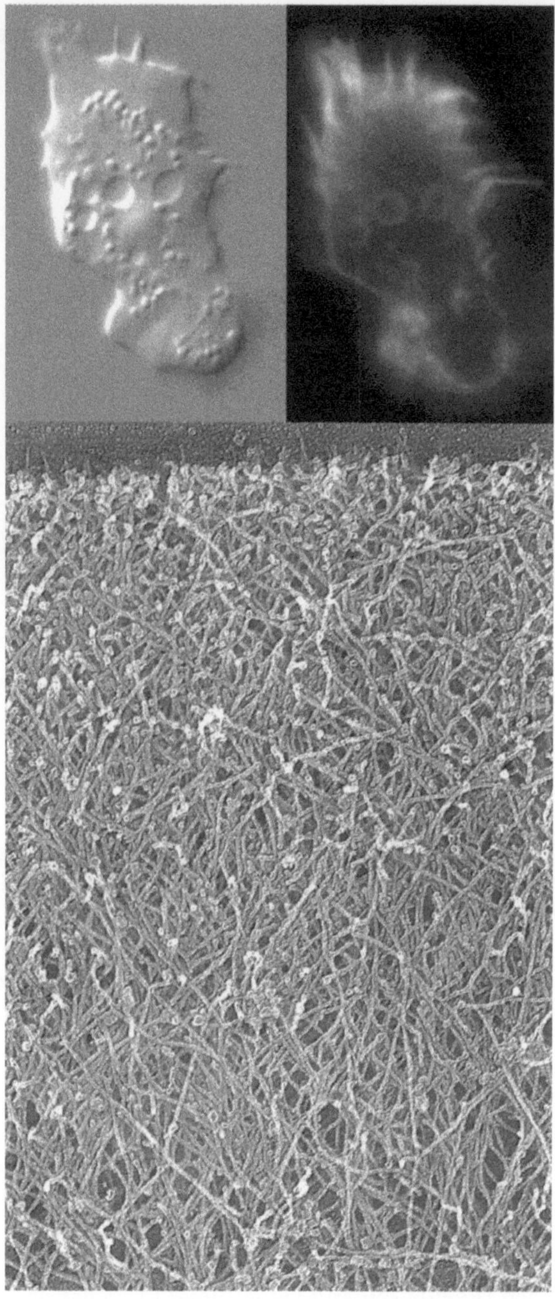

Fig. 2. The leading edge of motile cells. *Top* DIC image (*left*) and fluorescent localization of actin filaments with rhodamine phalloidin (*right*) of a migrating *Acanthamoeba*. Note the filopodia extending from the F-actin-rich cortex. *Bottom* The actin network in lamellipodium of migrating *Xenopus* keratocyte

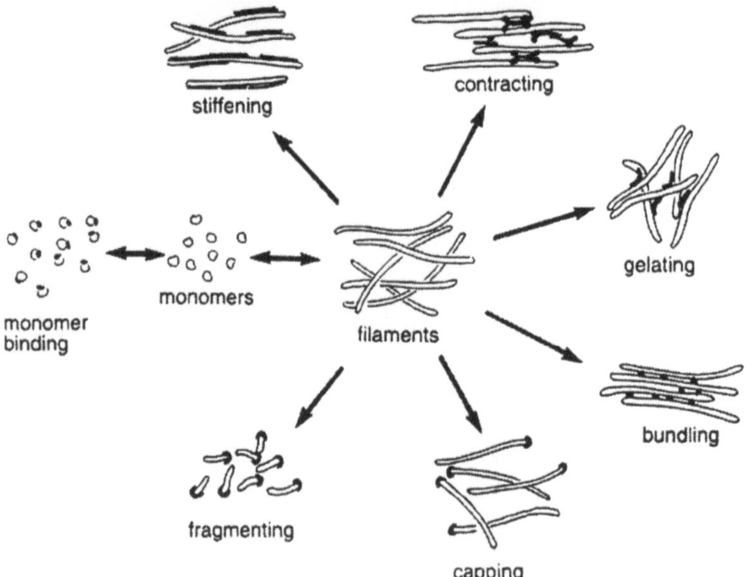

Fig. 3. Schematic diagram summarizing the interaction of actin-binding proteins with actin. For simplicity, only proteins that directly affect the assembly of actin are shown. Reproduced with permission by Garland Publishing (New York, NY)

Actin-binding proteins that bind terminal subunits of filaments influence the rates of monomer addition and dissociation at that particular end. Severing proteins such as gelsolin and cofilin disrupt the lateral and longitudinal contacts between monomers in filaments, reducing the mean equilibrium filament length and contributing two new filament ends to the population per severing event. Gelsolin remains bound to the newly formed barbed end and inhibits elongation from this end. Monomer-binding proteins are classified according to whether they maintain an unpolymerized monomer pool or have effects on polymerization other than simple sequestration.

The Big Picture and How the Details Define it

Our collective goal as researchers in the actin field is to answer the following five general questions concerning actin-binding proteins: (1) What are the identities of all actin-binding proteins (Components)? (2) How do they associate with actin and each other (Interactions)? (3) How are their activities modified (Regulation and Dynamics)? (4) What do they look like (Structure)? and (5) What roles do they play in living systems (Cells and Organisms)? Although these questions do not cover all aspects of actin-binding proteins, they provide a rigorous framework for understanding their mechanisms of action at the molecular level and in living systems. Until all of these questions are answered completely and accurately, actin researchers will continue to be challenged with important issues.

The following overview discusses the significance of these questions in developing our views of the mechanisms and functions of actin and actin-binding proteins. To provide a broad perspective, reference is made to a number of different proteins rather than focusing on one or two specific examples even though there are particular actin-binding protein families, such as myosin, which have been characterized in greater detail than others. To keep the number of citations to a minimum, I will refer to a chapter in this book or a recent review of a particular subject, with the understanding that earlier work will be referenced therein.

Components

Over 75 classes of actin-binding proteins have been identified, often with multiple isoforms (Pollard 1993). Fifteen are members of the myosin family of molecular motors, over a dozen are cross-linking proteins and at least six are monomer-binding proteins. New actin-binding proteins with unique or modified chemical and physical properties are discovered every year. Identification of homologues and variants within different biological kingdoms suggests that many activities have been highly conserved throughout the course of evolution. Many cells have multiple isoforms of a given class, each with distinct cellular functions, indicative of strong evolutionary pressure for specialization within a given class.

Actin-binding proteins may be identified from their amino acid sequences (McCann and Craig 1997; Lappalainen et al. 1998). The number of potential actin-binding proteins will certainly explode as genomic sequencing provides us with the sequences of all proteins in a given organism. However, identification by sequence alone does not verify actin-binding activity, and eventually binding to actin monomers, filaments or both must be demonstrated biochemically with purified components. Whether actin-binding proteins function as monomers, dimers or other oligomers also needs to be addressed.

Interactions

Actin-binding proteins induce multiple effects on the structure, conformation and assembly of actin. Monomer-binding proteins decrease the rates of spontaneous nucleation. Elongation is suppressed at both barbed and pointed ends of filaments by thymosins. Profilin-actin monomers elongate from the barbed ends but not the pointed ends of filaments, thus inducing a unidirectional assembly of filaments. Nucleotide exchange from actin monomers is accelerated by proteins that open the nucleotide-binding cleft (Chik et al. 1996) and inhibited by proteins that keep the cleft closed (see Egelman and Orlova, this Vol.).

Filament binding can alter the interactions between monomer subunits and change the affinity of monomers for filament ends, thereby inducing depolymerization (Carlier et al. 1997) or assembly of filaments (Freeman et al. 1998). Capping proteins bind filament ends and consequently reduce the incorporation of monomers at the capped ends. ATP hydrolysis at the barbed ends of elongating filaments may be accelerated by profilin (Pantaloni and Carlier 1993) and actophorin, a member of the cofilin/ADF family, increases the rate of phosphate release from filaments (Blanchoin and Pollard 1999). Hydrolysis and phosphate release destabilize the polymer structure and may facilitate the rapid turnover of these filament populations *in vivo*.

Quantitative measurements of actin binding are essential for describing the interactions of actin-binding proteins with actin. The equilibrium and rate constants are most informative. They can assess the validity of a binding mechanism and, since they are directly related to the energetics of the interaction, provide a direct measurement of the thermodynamics of binding. Expanding the free energy of binding into its enthalpic and entropic components gives an indication to the nature of the bonds and interactions that contribute to the stability of the associated complex.

Many proteins bind actin weakly with affinities in the low micromolar range. In most cases this is due to a rapid dissociation rate constant (~1–10 s^{-1}) that permits rapid reorganization on a time scale necessary for many cellular movements (Pollard 1993). Though most associations approach the diffusion limit of bimolecular protein reactions (~10^6–$10^7 M^{-1} s^{-1}$), some ligands bind actin filaments 2 to 3 orders of magnitude more slowly (De La Cruz and Pollard 1996; Ressad et al. 1998). Presumably this occurs because the binding sites are not accessible until a conformational change on the actin exposes the vacant site. Binding mechanisms that distinguish between different populations of actin according to their structure and dynamics may regulate the interactions of some proteins with particular classes of actin filaments in cells.

Most, if not all, actin-binding proteins bind other ligands in addition to actin. Ligands include other proteins, membranes, lipids, ions, nucleotides (ATP, GTP) and nucleic acids (DNA, aa-tRNA). Binding sites for various ligands can overlap, favoring interaction with one or more components depending on their affinities and concentrations. Alternatively, binding of one ligand can introduce a conformational change that modifies interaction for a second (see Kekic et al., this Vol.). Multiple ligands can bind simultaneously to form, in the simplest case, a ternary complex, or in other cases, a multicomponent complex.

The focal adhesion contains over 20 different proteins, including actin, actin-binding proteins, adhesion molecules and signaling kinases (Burridge and Chrzanowska-Wodnicka 1996). The result is a macromolecular assemblage with the ability to respond specifically to a variety of signals depending on their intensity, source and duration. Other examples of multi-component complexes containing actin and actin-binding proteins include adherens junctions (Stevenson and Keon 1998), cleavage furrows (Lippincott and Li 1998), I-Z-I bodies of developing myofibrils (Holtzer et al. 1997), and actin filament-elongation complexes (Zigmond et al. 1998).

Regulation and Dynamics

Cells respond to external cues such as growth factors, the extracellular matrix, neighboring cells, chemoattractants and repellents by rapidly reorganizing their actin cytoskeleton (for more details see Beck et al. this Vol.). These signals initiate the polarized assembly of filaments allowing cells to migrate in a specific direction, divide or adhere to the underlying substrate. Accordingly, the series of reactions responsible for generating these responses must be turned on and off as well as possess some form of modulation for the system.

Cells have adapted many regulatory mechanisms to accomplish these tasks and in many cases target only one or two key reaction steps even in very complicated processes. If a single step is affected, it tends to be the rate limiting step because it alone determines the overall rate of the reaction. Consequently, regulation results directly from a change in the equilibrium constant of the reaction.

Consider the dissociation of gelsolin from the barbed ends of filament (see Burtnick et al., this Vol.). The complex is very stable (K_d in the low pM range) due to a very slow dissociation rate. As a result, gelsolin competes effectively with free monomers for filament ends and barbed end elongation is inhibited even at very high monomer concentrations. Inositol phospholipids dissociate gelsolin and reduce the affinity of gelsolin for filament ends. Once gelsolin dissociates, the uncapped filaments elongate at rates up to 500 subunits s^{-1} until capping suppresses further elongation. A similar mechanism applies to the regulation of filament elongation by capping protein (Schafer et al. 1996).

Alternatively, association and dissociation rates can both be regulated, generating little or no change in binding affinity while changing the overall reaction rate. For example, profilin binding increases the dissociation and association rates of MgATP binding to actin monomers, but has a minimal effect on the affinity of actin for ATP which is needed to prevent denaturation of the monomer (Vinson et al. 1998). Profilin causes actin to equilibrate with the free nucleotide in the cell (MgATP) and replenishes ATP-actin monomers for rapid elongation more quickly. By increasing both rates profilin makes the system more dynamic and allows it to change on a time scale in accord with many cellular events.

The strategy of subcellular localization is used frequently to promote the interactions of specific cellular components. Localization can be generated through cooperative binding, ligand clustering or preferential binding. Cooperative binding, if strong enough, ensures that sparsely distributed target sites are recognized and occupied by even a few molecules. Such a binding mechanism may explain how members of the ADF/cofilin family rapidly depolymerize specific populations of actin filaments (see McGough et al., this Vol.) at the leading edge of motile cells. In many cases very weak interactions can be favored due to high local concentrations of reacting components (i.e., ligand clustering; see Kang et al. 1997 for discussion on *Listeria* tail formation) by increasing the encounter frequency which could otherwise be rate-limiting if relying entirely on diffusion (Bray 1998). Localization to specific sites in cells such as the plasma membrane can be favored with posttranslational modifications, preferential binding to phospholipids or association with components localized in these regions.

Structure

The determinants for molecular recognition, specificity and affinity are dictated by the shape, structure and chemical surfaces of the interacting components. Establishing the relationship between structure and activity is fundamental for understanding the molecular basis of the interactions that contribute to function. Structures determined at atomic resolution by X-ray crystallography provide the most information, but lower resolution techniques such as electron microscopy (McGough 1998), multidimensional NMR (Hatanaka et al. 1996), fluorescence (see Moens and dos Remedios, this Vol.) hydrodynamic (Mullins et al. 1997), scattering (Patkowski et al. 1990; see Krueger and Trewhella, Vol. 2) and rheological (see chapter by Janmey et al. in this volume) measurements provide valuable information about the structure and organization of actin and actin-binding proteins unavailable by crystallography. Molecular modeling of high-resolution structures yields insight into the interactions and dynamics of the system.

Knowing the structures of actin-binding proteins can help us generate hypotheses about their mechanisms of action that can be tested with the rational design of mutants. Residues of yeast cofilin that affect binding to actin monomers or filaments when mutated map to distinct sites on the crystal structure of cofilin (Lappalainen et al. 1997). The structures of plasma gelsolin and gelsolin segment 1 bound to actin give insight about the molecular mechanisms of filament severing, capping and nucleation (see Kekic et al., and Burtnick et al., both in this Vol.). Ca^{2+} binding opens up the gelsolin molecule, exposing the cryptic actin-binding site on segment 2. This facilitates association of segment 1 with the filament. Binding induces a structural change in the rest of the gelsolin molecule that severs the filament then caps the newly formed barbed end.

Several actin-binding proteins possess similar structural folds despite sharing little or no sequence homology. Gelsolin segment 1 and members of the ADF/cofilin family all have a six-stranded β sheet structure surrounded by four α helices, two on each face of the β-sheet core. The structural homologies probably reflect similar or overlapping actin-binding sites (Wriggers et al. 1998). Common folds have been found in proteins that do not bind actin.

Fascin possesses the β-trefoil fold of interleukin-1 and fibroblast growth factor (Fedorov et al. 1998).

Several conserved actin-binding domains are used by a large number of different actin-binding proteins (Lappalainen et al. 1998). The diversity of their actions relies on the organization, spacing and modification of these domains, or "modules", on the actin-binding protein (see Puius et al. 1998). For example, filament cross-linking proteins are characterized by a conserved 27 kDa actin-binding domain (ABD) composed of two predominantly α-helical calponin homology (CH) domains. The geometric organization of the ABDs determines what type of filament network is formed: ABDs arranged in tandem (i.e., fimbrin) produce tightly packed parallel bundles; ABDs separated by a variable length α-helical (α-actinin, spectrin) or β-sheet (ABP-120, ABP-280, filamin) segment generate loose bundles or networks.

Cells and Organisms

To understand the roles of actin-binding proteins in living systems, we must know their cellular and tissue expression, distribution, concentrations, and dynamic localization of the components. These are essential for modeling and relating intracellular processes and activities with *in vitro* measurements.

Localization with specific antibodies provides valuable information concerning the intracellular and tissue distribution of actin-binding proteins (Fig. 1d, e). Localization in fixed samples offers less temporal resolution about dynamic localization and molecular interactions than observation of fluorescently labelled proteins and green fluorescent protein (GFP – Ludin and Matus 1998) in live cells.

Yeast-capping protein Cap2p-GFP localizes to highly motile cortical actin patches (Waddle et al. 1996) and coronin accumulates at the leading edge of migrating *Dictyostelium* cells exposed to chemoattractants (Gerisch et al. 1995). GFP-Arp2/3 complex and GFP-capping protein colocalize with actin filaments at the leading edge and motile spots of fibroblasts (Schafer et al. 1998). With GFP-α-actinin the entire process of myofibrillogenesis in developing muscle cells was monitored in a single living cell (Dabiri et al. 1998).

The organization of actin filaments in cells gives insight into the in vivo functions of actin-binding proteins and the mechanisms of cell crawling and extension. Fish epidermal keratocytes are an excellent model cell because they are highly motile and display a distinct polarization in their shape and actin organization (see Svitkina et al. 1998). Actin filaments at the leading edge are arranged in orthogonal networks with the majority of their barbed ends free and oriented towards the plasma membrane. Filament density is highest at the leading edge and decreases gradually rearwards towards the nucleus. The pointed ends of filaments at the leading edge are not free but terminate at the sides of other filaments in T-junctions. Taken together, the morphology and organization of the actin network suggests that nucleation and barbed-end elongation occur adjacent to the plasma membrane at the leading edge of the cell. Nucleation must occur at the sides of filaments or newly formed nuclei associate with filaments to produce the ordered orthogonal network (Fig. 2c). Since filament branching and cross-linking rarely occur with pure actin, it is reasonable to conclude that one or more actin-binding proteins are responsible for this formation of filaments. Two potential candidates for performing these functions are filamin and the Arp2/3 complex (see Machesky and May, this Vol.). Evidence implicating the Arp2/3 complex is accumulating rapidly (Mullins et al. 1997) and several putative cellular activators have been identified (Ma et al. 1998; Machesky et al. 1999).

The physiological functions of actin-binding proteins are most easily identified using organisms with deleted or disrupted genes. Knockouts in yeast, *Drosophila*, *C. elegans*, *Dictyostelium* and mice have been successful and informative (Ayscough 1998). Deletion of profilin arrests *S. pombe* at cytokinesis and is lethal in *Drosophila* and mice embryos. Deletion or disruption of cofilin/ADF family members kills *S. cerevisiae*, *Dictyostelium*, *Drosophila* and *C. elegans*. Gelsolin, capping protein and a few filament cross-linking proteins are not essential for viability, but mutants have defects in actin-based cellular processes and in some cases the females are sterile.

Interpretation of in vivo function with genetically disrupted organisms is complicated by several factors including functional or genetic redundancy from multiple isoforms and cellular adaptation. As a result, quite often no obvious phenotypic alterations are observed until multiple genes are disrupted. Furthermore, deleting one component may induce the expression of a different protein or sets of proteins with similar, overlapping functions even though in the native organism these compensatory proteins would be involved in distinct cellular processes (Golla et al. 1997).

I favor the notion that equally important information about the functions of actin-binding proteins can be obtained using evolutionary diverse organisms, from plants and protists to vertebrates, for it is the similarities and differences in these systems that will inform us of the engineering, design and function of the actin cytoskeleton that have evolved for the particular activities and demands of each organism. The acrosome reactions of echinoderm sperm (Tilney et al. 1973) provided the earliest evidence that profilin sequesters actin monomers and that the profilactin complex elongates from free barbed but not pointed ends of filaments in activated cells (Tilney et al. 1983). Pathogenic bacteria (Dramsi and Cossart 1998) and protozoan parasites (Tilney and Tilney 1996) are also interesting models that can help us understand the mechanisms and functions of actin assembly and actin-binding proteins in all living systems.

Acknowledgments. I thank Tom Pollard, Henry Shuman, Joe and Jean Sanger, and Scott Appoldt for reading the manuscript and offering suggestions. High five to my colleagues that contributed figures: L. Tilney (Fig. 1a, b, c), J. and J. Sanger (Fig. 1d, e), E.M. Ostap (Fig. 2 top), T. Svitkina (Fig 2 bottom) and D. Bray (Fig. 3). I thank the Burroughs Wellcome Fund Fellow and the Life Sciences Research Foundation for their support.

References

Ayscough KR (1998) In vivo functions of actin-binding proteins. Curr Opin Cell Biol 10:102–111

Bray D (1998) Signaling complexes: biophysical constraints on intracellular communication. Annu Rev Biophys Biomol Struct 27:59–75

Blanchoin L, Pollard TD (1999) Mechanism of interaction of Acanthamoeba actophorin (ADF/cofilin) with actin filaments. J Biol Chem 274:15538–15546

Burridge K, Chrzanowska-Wodnicka M (1996) Focal adhesions, contractility, and signaling. Annu Rev Cell Dev Biol 12:463–519

Carlier M-F, Laurent V, Santolini J, Melki R, Didry D, Xia G-X, Hong Y, Chua N-H, Pantaloni D (1997) Actin depolymerizing factor (ADF/cofilin) enhances the rate of filament turnover: Implications in actin-based motility. J Cell Biol 136:1307–1323

Chik JK, Lindberg U, Schutt C (1996) The structure of an open state of actin at 265 Å resolution. J Mol Biol 263:607–623

Dabiri GA, Turnacioglu KK, Ayoob JC, Sanger JM, Sanger JW (1998) Transfections of primary muscle cell cultures with plasmids coding for GFP linked to full-length and truncated muscle proteins. In: Sullivan KF, Kay SA (eds) Methods in cell biology: green fluorescent proteins Academic Press, San Diego, pp. 240–259

De La Cruz EM, Pollard TD (1996) Kinetics and thermodynamics of phalloidin binding to actin filaments from three divergent species. Biochemistry 35:14054–14061

Dramsi S, Cossart P (1998) Intracellular pathogens and the actin cytoskeleton. Annu Rev Cell Dev Biol 14:137–166

Fedorov AA, Fedorov L, Ono S, Matsumara F, Almo SC (1998) The crystal structure of human fascin. Mol Biol Cell 9:384a

Freeman JL, De La Cruz EM, Pollard TD, Lefkowitz RJ, Pitcher JA (1998) Regulation of G protein-coupled receptor kinase 5 (GRK5) by actin. J Biol Chem 273:20653–20657

Gerisch G, Albrecht R, Heizer C, Hodgkinson S, Maniak M (1995) Chemoattractant-controlled accumulation of coronin at the leading edge of *Dictyostelium* cells monitored using a green fluorescent protein-coronin fusion protein. Curr Biol 5:1280–1285

Golla R, Philp N, Safer D, Chintapalli J, Hoffman R, Collins L, Nachmias VT (1997) Co-ordinate regulation of the cytoskeleton in 3T3 cells overexpressing thymosin-β4. Cell Motil Cytoskel 38:187–200

Hatanaka H, Ogura K, Moriyama K, Ichikawa S, Yahara I, Inagaki F (1996) Tertiary structure of destrin and structural similarity between two actin-regulating protein families. Cell 85:1047–1055

Hirokawa N, Tilney LG (1982) Interactions between actin filaments and between actin filaments and membranes in quick-frozen and deeply etched hair cells of the chick ear. J Cell Biol 95:249–261

Hirokawa N, Tilney LG, Fujiwara K, Heuser JE (1982) Organization of actin, myosin, and intermediate filaments in the brush border of intestinal epithelial cells. J Cell Biol 94:425–443

Holtzer H, Hijikata T, Lin ZX, Zhang ZQ, Holtzer S, Protasi F, Franzini-Armstrong C, Sweeney HL (1997) Independent assembly of 16-microns-long bipolar MHC filaments and I-Z-I bodies. Cell Struct Function 22:83–93

Kang F, Laine RO, Bubb MR, Southwick FS, Purich DL (1997) Profilin interacts with the Gly-Pro-Pro-Pro-Pro-Pro sequences of vasodilator-stimulated phosphoprotein (VASP): implications for actin-based *Listeria* motility. Biochemistry 36:8384–8392

Korenbaum E, Nordberg P, Bjorkegren-Sjogren C, Schutt CE, Lindberg U, Karlsson R (1998) The role of profilin in actin polymerization and nucleotide exchange. Biochemistry 37:9274–9283

Lappalainen P, Fedorov EV, Fedorov AA, Almo SC, Drubin DG (1997) Essential functions and actin-binding surfaces of yeast cofilin revealed by systematic mutagenesis. EMBO J 16:5520–5530

Lappalainen P, Kessels MM, Cope JTV, Drubin DG (1998) The ADF homology (ADF-H) domain: a highly exploited actin-binding module. Mol Biol Cell 9:1951–1959

Lippincott J, Li R (1998) Sequential assembly of myosin II, an IQGAP-like protein, and filamentous actin to a ring structure involved in budding yeast cytokinesis. J Cell Biol 140:355–366

Ludin B, Matus A (1998) GFP illuminates the cytoskeleton. Trends Cell Biol 8:72–77

Ma L, Rohatgi R, Kirschner MW (1998) The Arp2/3 complex mediates actin polymerization by the small GTP-binding protein cdc42. Proc Natl Acad Sci 95:15362–15367

Machesky LM, Mullins RD, Higgs HN, Kaiser DA, Blanchoin L, May RC, Hall ME, Pollard TD (1999) Scar, a WASp-related protein, activates clendritic nucleation of actin filaments by the Arp2/3 complex. Proc Natl Acad Sci 96:3739–3744

McCann RO, Craig SW (1997) The I/LWEQ module:a conserved sequence that signifies F-actin binding in functionally diverse proteins from yeast to mammals. Proc Natl Acad Sci 94:5679–5684

McGough A (1998) F-actin binding proteins. Curr Opin Struct Biol 8:166–176

Mullins RD, Stafford WF, Pollard TD (1997) Structure, subunit topology, and actin-binding activity of the Arp2/3 complex from *Acanthamoeba.* J Cell Biol 136:331–343

Pantaloni D, Carlier M-F (1993) How profilin promotes actin filament assembly in the presence of thymosin β4. Cell 75:1007–1014

Patkowski A, Seils J, Hinssen H, Dorfmuller T (1990) Size, shape parameters and calcium-induced conformational change of the gelsolin molecule: a dynamic light-scattering study. Biopolymers 30:427–434

Pollard TD (1993) Actin and actin binding proteins In: Kreis T, Vale R (eds) Guidebook to the cytoskeletal and motor proteins. Oxford University Press, Oxford, pp 3–11

Puius YA, Mahoney NM, Almo SC (1998) The modular structure of actin-regulatory proteins. Curr Opin Cell Biol 10:23–34

Ressad F, Didry D, Xia G-X, Hong Y, Chua N-H, Pantaloni D, Carlier M-F (1998) Kinetic analysis of the interaction of actin-depolymerizing factor (ADF)/Cofilin with G- and F-actins. J Biol Chem 273:20894–20902

Schafer DA, Jennings PB, Cooper JA (1996) Dynamics of capping protein and actin assembly in vitro: capping barbed ends by polyphosphoinositides. J Cell Biol 135:169–179

Schafer DA, Welch MD, Machesky LM, Bridgman PC, Meyer SM, Cooper JA (1998) Visualization and molecular analysis of actin assembly in living cells. J Cell Biol 143:1919–1930

Stevenson BR, Keon BH (1998) The tight junction: morphology to molecules. Annu Rev Cell Dev Biol 14:89–109

Svitkina T, Verkhovsky AB, McQuade KM, Borisy GG (1998) Analysis of the actin-myosin II system in fish epidermal keratocytes: Mechanism of cell body translocation. J Cell Biol 139:397–415

Tilney LG, Tilney MS (1996) The cytoskeleton of protozoan parasites. Curr Opin Cell Biol 8:43–48

Tilney LG, Hatano S, Ishikawa H, Mooseker MS (1973) The polymerization of actin: its role in the generation of the acrosomal process of certain echinoderm sperm. J Cell Biol 591:99–126

Tilney LG, Bonder EM, Coluccio LM, Mooseker MS (1983) Actin from *Thyone* sperm assembles on only one end of an actin filament: a behavior regulated by profilin. J Cell Biol 97:112–124

Vinson VK, De La Cruz EM, Higgs HN, Pollard TD (1998) Interactions of *Acanthamoeba* profilin with actin and nucleotides bound to actin. Biochemistry 37:10871–10880

Waddle JA, Karpova TS, Waterston RH, Cooper JA (1996) Movement of cortical actin patches in yeast. J Cell Biol 132:861–870

Wriggers W, Tang JX, Azuma T, Marks PW, Janmey PA (1998) Cofilin and gelsolin segment-1: molecular dynamics simulation and biochemical analysis predict a similar actin binding mode. J Mol Biol 282:921–932

Zigmond SH, Joyce M, Yang C, Brown K, Huang M, Pring M (1998) Mechanism of cdc42-induced actin polymerization in neutrophil extracts. J Cell Biol 142:1001–1012

The ADF/Cofilin Family: Accelerators of Actin Reorganization

Amy McGough[1], Brian Pope[2], and Alan Weeds[2]

Introduction

Actin polymerization and cytoskeletal reorganization play an essential role in cell locomotion and many forms of motility, including phagocytosis and cytokinesis. The rate of assembly of actin filaments in vitro is virtually diffusion controlled (Drenckhahn and Pollard 1986), but depolymerization rates are too slow to regenerate the monomer pool required for cells to advance at rates of up to 30 $\mu m/min^{-1}$ (Zigmond 1993). Actin Depolymerizing Factors (ADF/cofilin) are ideal candidates to aid in this process. They are localized together with actin in motile regions of cells (Fig. 1A) and are essential for cell viability in yeast, *C. elegans*, *Drosophila* and *Dictyostelium*. Here we describe the structure and properties of these proteins and highlight recent work on their interactions with actin.

ADF/Cofilin: Distribution in Organisms and Tissues

Two well-characterized classes of actin monomer-binding proteins have been identified that are present across the complete spectrum of eukaryotic organisms: the profilins (12–15 kDa) and the ADF/cofilin family (15–19 kDa). ADF was first identified as an actin-depolymerizing factor in chick embryonic brain (Bamburg et al. 1980) where its level is 0.44% of total protein (i.e. >30% that of actin) (Bamburg and Bray 1987). Accumulation in growth cones and at the leading edge of ruffling membranes (Fig. 1A) suggested a role in cell motility. Later work identified two closely related proteins in pig brain, destrin and cofilin, that reduced the viscosity of F-actin (Maekawa et al. 1984). Cofilin was so named because it bound to filaments whereas destrin induced their depolymerization (Nishida et al. 1984). Brain cofilin differed from destrin in its sensitivity to pH (Yonezawa et al. 1985), but more recent work has shown there is a similar pH sensitivity for both proteins. They bind filaments at $pH < 7$ and induce depolymerization at $pH > 7.5$ (Hawkins et al. 1993; Hayden et al. 1993). Because the sequences of pig

[1] Department of Biological Sciences, Purdue University, West Lafayette, Indiana 47907-1392, USA

[2] MRC Laboratory of Molecular Biology, Cambridge, CB2 2QH, England

Results and Problems in Cell Differentiation, Vol. 32
C. dos Remedios (Ed.): Molecular Interactions of Actin

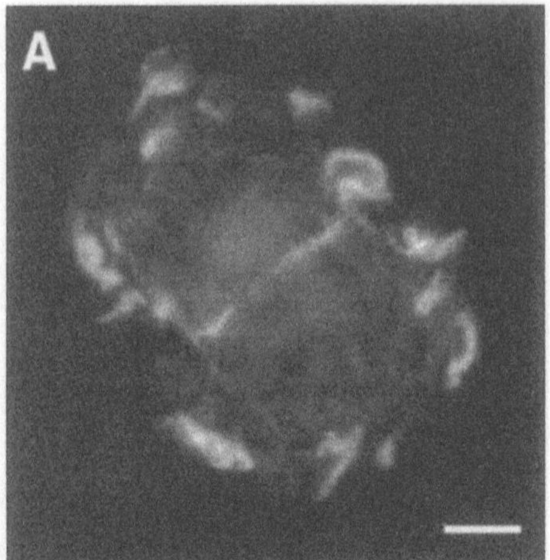

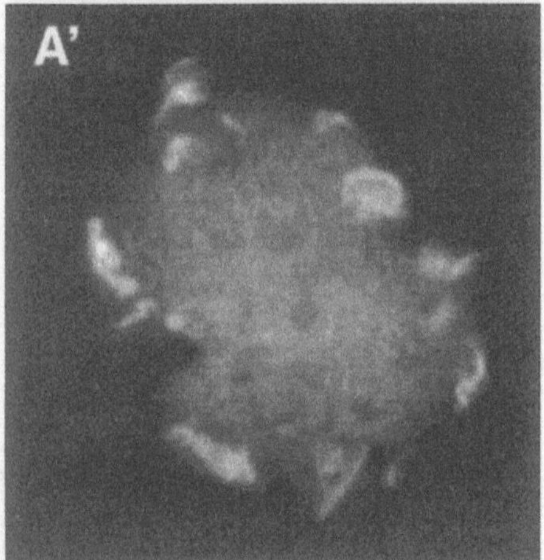

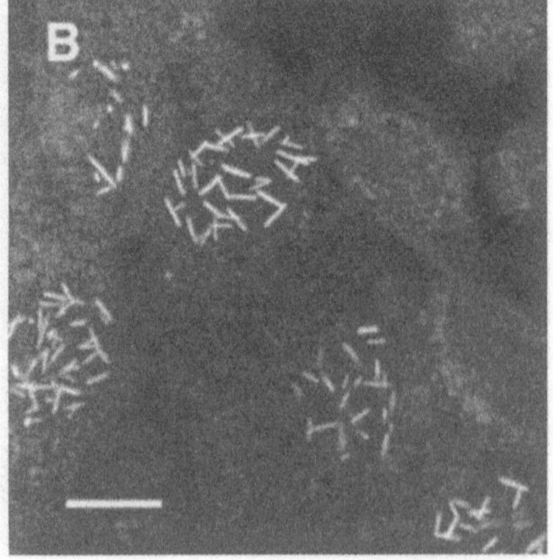

Fig. 1A,A′,B. Immunocytochemical localization of cofilin in A431 cells. Immunofluorescent images of A431 (human epidermoid carcinoma) cells following fixation with formaldehyde and staining with a monoclonal anti-β-actin or a rabbit polyclonal anti-cofilin antibody. Cells stimulated with EGF (200 ng ml^{-1}) show enrichment of actin (**A**) and cofilin (**A′**) in the same regions of cortex and ruffling membranes. Treatment with 10% DMSO produces translocation of each to the nucleus to form colocalized cofilin/actin rods (confocal microscope image, **B**). (Reproduced with permission Dr. Sabine Gonsior). *Bars* 20 μm (**A, A′**), 17 μm (**B**)

destrin and human ADF are identical, the earlier term ADF is used here for this protein.

Not unexpectedly, the sequences of vertebrate ADF and cofilin (Matsuzaki et al. 1988; Abe et al. 1990; Adams et al. 1990; Moriyama et al. 1990) show extensive homology. A phylogenetic analysis of about 30 sequences has revealed two distinct classes: vertebrate and plant (Fig. 2). Yeast, *Dictyostelium* cofilins and *Acanthamoeba* actophorin fit on the edge of the plant classification. *Drosophila* and *Caenorhabditis elegans* (Unc60A and B) cofilins form yet another small sub group. Sequences from echinoderm (depactin) and another *Dictyostelium* homologue (coactosin) are the most divergent, but despite their size relationship to the ADF/cofilins, they may yet be classified into related groups such as the Abp1s (Lila and Drubin 1998).

Vertebrates contain both ADF and cofilin in the same cells. There is also a muscle-specific cofilin in addition to the nonmuscle variant in both humans (Gillett et al. 1996) and *C. elegans* (Ono and Benian 1998). Multiple forms also exist in other species: for example, *Zea mays* has three ADFs (Lopez et al. 1996). The need for different isoforms and the possible roles for each are not clear. The fact that the two proteins are expressed differently in myoblasts

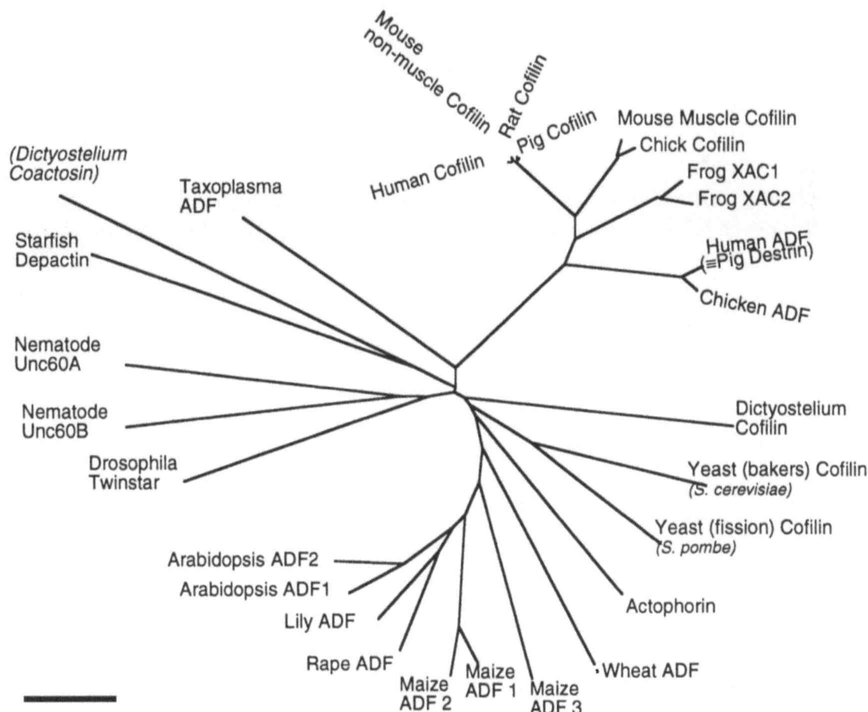

Fig. 2. An unrooted phylogenetic tree of ADF/cofilins. Sequences aligned and bootstrapped using ClustalW (Thompson et al. 1994) and analyzed in PHYLIP. *Bar* 10% divergence

transfected with a mutant of β-actin suggests that they may have distinct roles in cells (Minamide et al. 1997). Moreover, the high degree of functional equivalence between the two proteins is indicated by the fact that either mammalian ADF or cofilin is able to rescue deletion mutations of cofilin in yeast (Iida et al. 1993; Moon et al. 1993).

Detailed analysis of sequences has revealed that ADF/cofilin comprises one branch of a larger superfamily of proteins based on the ADF homology (ADF-H) domain (Lappalainen et al. 1998). The other two branches of this superfamily correspond to the twinfilins and the drebrin/Abp1s. Twinfilins have tandem repeats of the ADF/cofilin sequence (Goode et al. 1998) while drebrin/Abp1s contain a variable insertion and a C-terminal SH3 domain (Lappalainen et al. 1998). Apart from size, these groups differ from the true ADF/cofilins in that twinfilins bind only to monomeric actin and Abp1s only to filamentous actin, whereas most ADF/cofilins bind both F- and G-actins.

Sequence Analysis

Sequence alignments (Fig. 3) highlight the differences between the various branches of the phylogenetic tree. Vertebrate ADF/cofilins (165–168 amino acids) are larger than most of the other ADF/cofilins (137–152 amino acids) with the exception of nematode UNC 60A which also contains 165 amino acids. The smallest members of the family are the plant proteins (~139 residues). Human ADF and cofilin are 72% identical and contain 10% conservative substitutions: 11 of the 31 residues that differ are located within the C-terminal 24 residues.

In vertebrate ADF/cofilins, a stretch of basic residues $_{30}$KKRKK$_{34}$ is absolutely conserved and bears a striking resemblance to the archetypal sequence (KKKRK) that is responsible for nuclear localization of the SV40 large tumor antigen (Karderon et al. 1988). As with SV40, this sequence is also preceded by a β-turn. A second basic region (NLS1 in Fig. 3) suggests a bipartite NLS configuration when compared to the SWI5 transcription factor of *S. cerevisiae* (Jans et al. 1995). DMSO treatment induces translocation of ADF and cofilin to the nucleus in mammalian cells (Nishida et al. 1987). This effect is shown in Fig. 1B. Mutation of KKRKK to KTLKK inhibits the nuclear translocation of cofilin confirming the activity of this region as a nuclear localization signal (Iida et al. 1992). Nuclear localization has also been demonstrated both in *Dictyostelium* and in the root cells of *Z. mays* (in response to treatment with cytochalasin D). The equivalent sequences are KLGRK (Aizawa et al. 1995) and KRLHR (Jiang et al. 1997), respectively.

Atomic Models of ADF, Cofilin and Actophorin

The first atomic model for a member of the ADF/cofilin family (pig ADF) was obtained by nuclear magnetic resonance (NMR) spectroscopy (Fig. 4a;

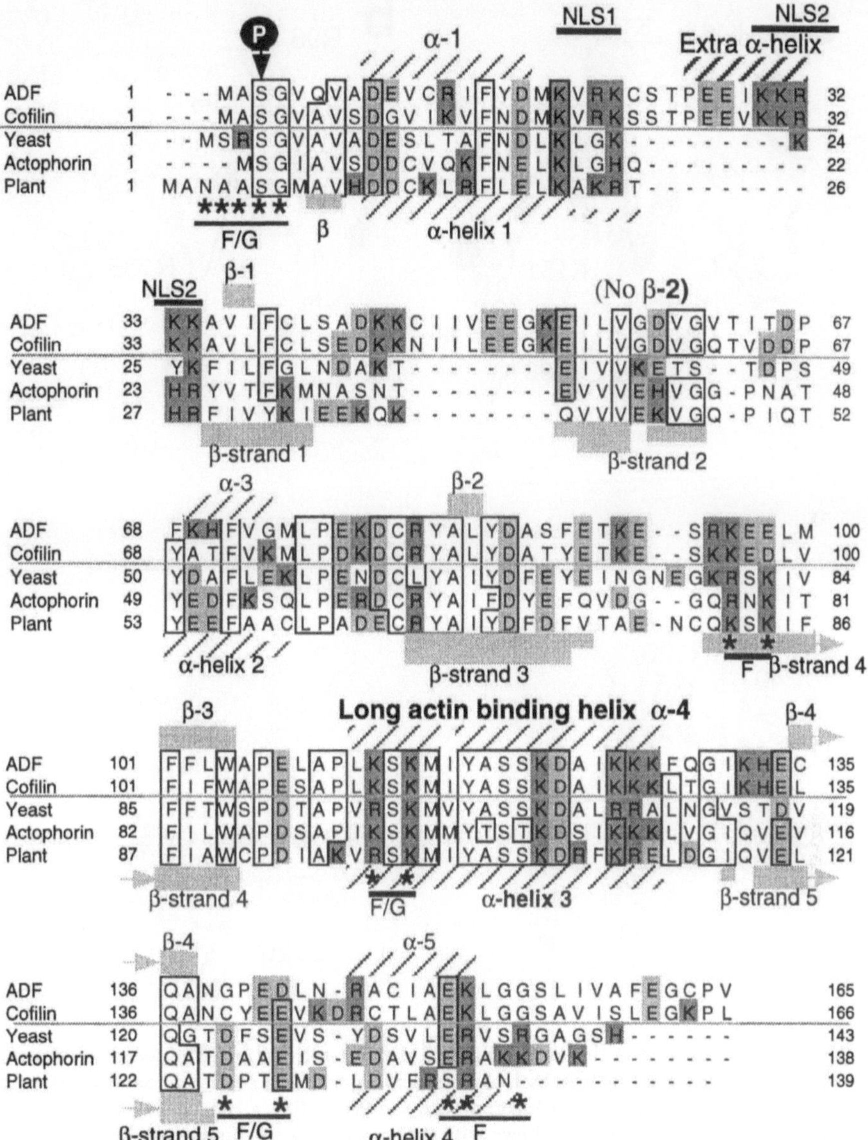

Fig. 3. Sequence comparisons and features. Sequences from human ADF and cofilin (COF1), yeast (*S. cerevisiae*), *Acanthamoeba* (actophorin) and plant (*A. thaliana* ADF1) aligned in ClustalW (Thompson et al. 1994) showing basic (*dark shading*) and acidic (*light shading*) residues with those implicated in binding to actin (*), stretches of β-strand (*dot matrix*), α-helix (*hatched*) and nuclear localization signals (*NLS*). Helices and β-strands from human ADF are indicated *above the sequence* and from yeast and actophorin *below the panels*. Phosphorylation occurs on Ser_3 (P). A small β-strand in yeast at A_7V_8 has not been numbered

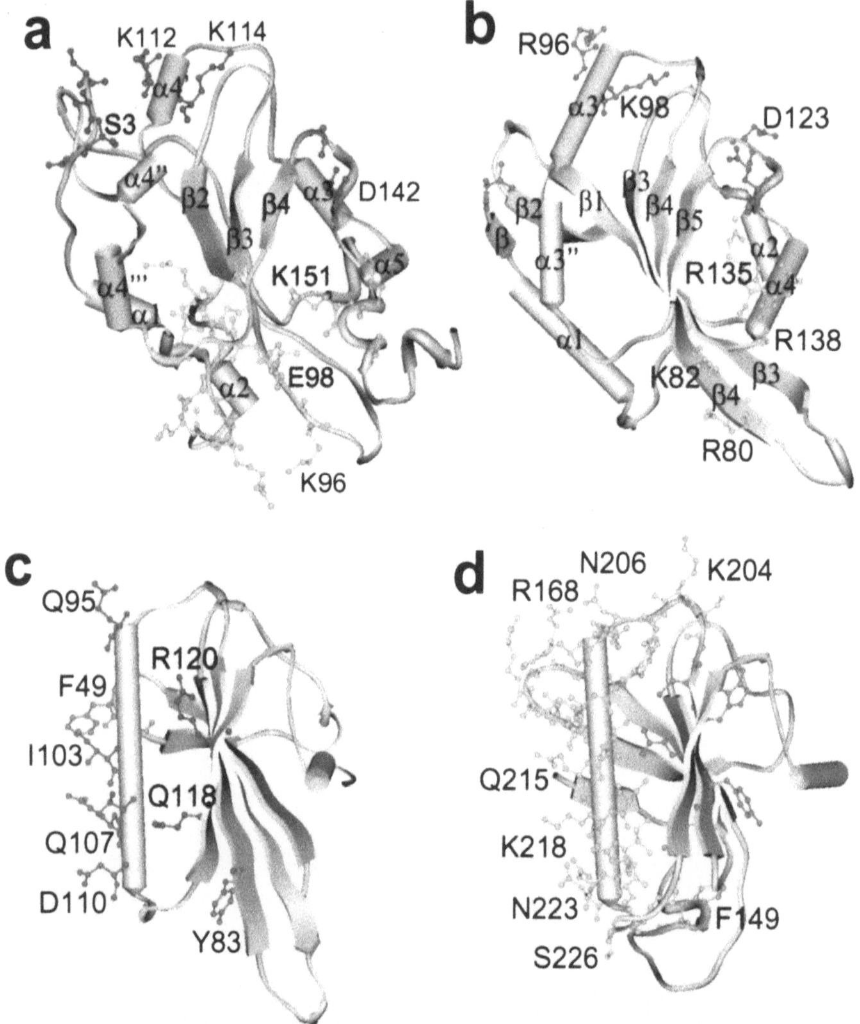

Fig. 4a–d. Comparison of the ADF/cofilin fold with the gelsolin fold. The secondary structural elements were assigned by a Kabsch and Sanders algorithm (Kabsch and Sander 1983) using InsightII (Molecular Simulations Incorporated, San Diego, CA). **a** Human ADF structure determined by Hatanaka et al. (1996) shown with the equivalent actin-binding residues highlighted. The nuclear localization signals, NLS1 and NLS2, are highlighted in *green*. The long helix ($\alpha 4$), which is kinked in the NMR structure, is shown as three shorter α-helices ($\alpha 4'$, $\alpha 4''$ and $\alpha 4'''$). The numbering for the α-helices and β-strands corresponds to those used in Fig. 3. **b** Yeast cofilin (Fedorov et al. 1997) structure with residues required for actin binding (*red*) or specifically in F-actin binding (*yellow*) (Lappalainen et al. 1997). The long kinked helix ($\alpha 3$) is depicted as two shorter α-helices ($\alpha 3'$ and $\alpha 3''$). **c** Gelsolin segment-1 (Pro_{39}–Tyr_{133}; (Burtnick et al. 1997)) structure with residues involved in G-actin binding (McLaughlin et al. 1993) highlighted in red. **d** Gelsolin segment-2 (Gly_{137}–Leu_{247}) structure with residues implicated in F-actin binding (Sun et al. 1994; VanTroys et al. 1996) highlighted in *yellow* (accessible) or *orange* (buried). Phe_{149}, which is involved in G-actin contacts (McLaughlin et al. 1993), is shown in pink

Hatanaka et al. 1996). Subsequently, the structures of yeast cofilin (Fig. 4b; Fedorov et al. 1997) and *Acanthamoeba* actophorin (Leonard et al. 1997) were determined by X-ray crystallography. The ADF/cofilin structure closely resembles the prototypical fold found in the gelsolin family of severing proteins (Figs. 4c, d; McLaughlin et al. 1993; Markus et al. 1994; Schnuchel et al. 1995; Burtnick et al. 1997) despite low sequence homology (18%) between the two protein families (Leonard et al. 1997).

The ADF/cofilin fold consists of a central mixed β-sheet flanked by α-helices. Although the lengths and numbers of strands and helices are variable, their overall arrangement and connectivity are conserved. The mixed β-sheet forms a tightly packed hydrophobic core centered around β2 and β3 of human ADF (corresponding to residues 63 to 73 and 78 to 89 in yeast cofilin, Fig. 3). A number of apolar residues in the interior are conserved, indicating their importance in maintaining this core. The central β-sheet is generally flanked by two pairs of α-helices: the longest helix, 16 to 18 residues, has been implicated in actin binding in both the gelsolin and the ADF/cofilin families. In actophorin this helix is disrupted by a type-I β-turn (at residues $_{97}$YTST$_{100}$) yielding two shorter α-helices (α3, 91–98; α4, 101–107). In yeast cofilin and human ADF this helix is kinked at the equivalent position, but in each of gelsolin's actin-binding domains it is straight (Burtnick et al. 1997; see Fig. 4 for examples).

Comparing the sequences of various members of the ADF/cofilin family in the context of the available three dimensional structures suggests that plant cofilins were progenitors of mammalian forms. There are several insertions in mammalian ADF/cofilin (Fig. 3); the first of these ($_{23}$CSTPEEIKKR$_{32}$) is α-helical and contains part of the NLS sequence (green residues in Fig. 4a). In human ADF these residues fall primarily on the surface opposite the long, helix leaving the NLS exposed. This would be required if cofilin is actively translocated into the nucleus with an actin cargo as is suggested by the appearance of nuclear cofilin-actin rods (Fig. 1B). The second insertion ($_{46}$CIIVEEGK$_{53}$) forms a loop in the ADF structure that replaces the β2 strand in yeast cofilin (Fig. 4b) and lies behind Ser$_3$ and other residues implicated in actin binding (Fig. 4a). This insertion sequence is highly conserved in vertebrates ranging from *Xenopus* to man. Finally, vertebrate ADF/cofilins have an additional 12 to 14 residues at their carboxy termini extending from the fourth α-helix away from the putative actin-binding surface.

Cofilin/Actin Interactions

The Binding Site for Actin on Cofilin

Mutational studies, chemical cross linking and peptide mapping have been used to define the sites on cofilin that may interact with actin (Fig. 4a, b). Much of this work has suggested that some or all of the long helix is involved in actin binding. Competitive binding of peptides implicated residues 104–115 of the vertebrate sequence in the interacton of ADF/cofilin to monomeric but not to filamentous actin (Yonezawa et al. 1991). Mutations of K_{112} and K_{114} abolished all actin binding (Moriyama et al. 1992) as did the corresponding residues in yeast (R_{96} and K_{98}) (Lappalainen and Drubin 1997). A sequence adjacent to this region ($_{122}$DAIKKKL$_{128}$) has also been implicated in actin binding by virtue of homology to a region near the N-terminus of tropomyosin ($_{2}$DAIKKK$_{7}$).

Tropomyosin and cofilin compete for binding to actin (Nishida et al. 1984) and high concentrations of the DAIKKK peptide compete with cofilin for F-actin (Yonezawa et al. 1991). The inference that the complete helix from residues L_{111} to L_{128} (human ADF) is involved in actin binding was further substantiated using peptide mimetics (VanTroys et al. 1997). Hatanaka et al. (1996) proposed a model for ADF binding to G-actin based on its structural similarity to gelsolin segment-1 (Fig. 4c). A key feature of this model (and a subsequent model obtained by molecular dynamics simulation refinement (Wriggers et al. 1998)) is the involvement of the long helix in actin interaction. Similarly, the long helix in gelsolin segment-2 (Fig. 4d) has also been implicated in actin binding (VanTroys et al. 1996) providing further support for this concept. However, mutational studies in yeast cofilin found no defects in vivo when the long helix residues $K_{105,}$ $D_{106,}$ R_{109} or R_{110} were replaced with alanines (Lappalainen et al. 1997). These same studies identified regions elsewhere on the molecule that are required for actin interaction, some of which are specific for F-actin binding. Taken together, these results suggest that there may be two distinct sites on cofilin for actin, one of which interacts with both G-actin and F-actin, while the other facilitates F-actin binding.

The Binding Site for Cofilin on Actin

One fascinating property of the ADF/cofilins is their ability to bind either monomeric or filamentous actin. Unlike gelsolin, which accomplishes this by partitioning G- or F-actin specificity into different domains, cofilin employs a single actin-binding module to accomplish the task. In many cases, the switch from G- to F-actin binding can be demonstrated by lowering the pH of the medium (below pH 7). This process is fully reversible and does not appear to require conformational changes in the cofilin. This suggests that the sites for cofilin on G-actin monomers and F-actin subunits are similar.

Electron cryomicroscopy has provided the first view for how cofilin binds to actin by determining the cofilin/F-actin structure to 27 Å resolution (McGough et al. 1997). Cofilin makes extensive contacts with two longitudinally-associated actin subunits when it binds. Because actin is a relatively asymmetric object at low resolution, it can be readily fitted into electron microscopy (EM) reconstructions (Rayment et al. 1993; Schroder et al. 1993; McGough et al. 1994) thus providing higher-resolution information about binding sites than would otherwise be available from electron microscopy alone.

Figure 5a shows the molecular footprint for cofilin on actin which is obtained after fitting the actin coordinates into the EM reconstruction. The cofilin-binding site involves subdomains 1 and 3 of the upper actin subunit and subdomains 1 and 2 of the lower. Comparison of cofilin's footprint with those made by other F-actin–binding proteins (McGough 1998) reveals that it is most similar to that of gelsolin segment-2, the F-actin-binding domain (Fig. 5b). The segment-2 footprint was determined from a 45 Å resolution EM reconstruction of F-actin decorated with gelsolin G2-6 (McGough et al. 1998).

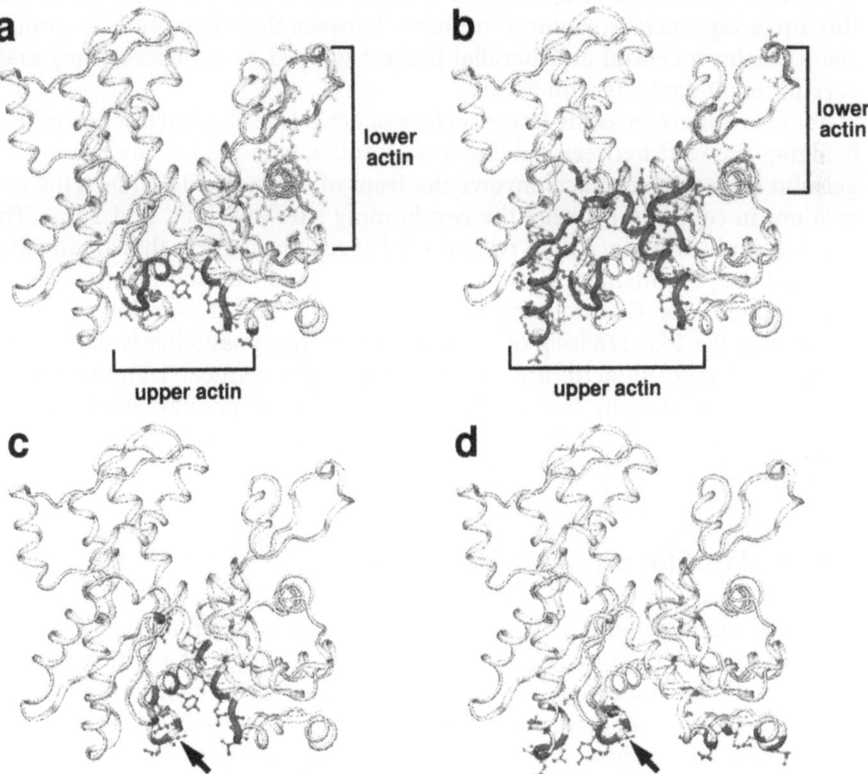

Fig. 5a–d. Molecular footprints for actin-binding proteins. **a** Actin residues that fall at or near the cofilin-binding site as identified by electron cryomicroscopy of cofilin:F-actin to 27 Å resolution (McGough et al. 1997) shown on the atomic model for actin (Kabsch et al. 1990). Cofilin binds two longitudinally associated actin subunits in the filament. Residues from subdomains 1 and 3 of the upper actin subunit form one contact (*black*); residues in subdomains 1 and 2 of the lower actin (gray) form the second contact. **b** Actin residues at or near the gelsolin segment-2 binding site determined by electron cryomicroscopy of gelsolin G2-6:F-actin to 45 Å resolution (McGough et al. 1998). Gelsolin also binds two longitudinally associated actin subunits. **c** Actin residues involved in gelsolin segment-1 binding determined by X-ray crystallography to 2.5 Å resolution (McLaughlin et al. 1993). **d** Actin residues involved in binding profilin determined by X-ray crystallography to 2.55 Å resolution. (Schutt et al. 1993)

Although the lower resolution of this reconstruction means that the binding site is not as well defined (hence the footprint is larger), it shows that contacts with the upper subunit are centered at the cleft between subdomains 1 and 3, just as they are with cofilin. Likewise, contacts with the lower actin subunit involve subdomains 1 and 2.

At present, there is no crystal structure of cofilin bound to G-actin; however, binding studies have shown that cofilin competes with gelsolin segment-1 (A. Weeds, unpub. work) and profilin (Blanchoin and Pollard 1998; Didry et al. 1998) for monomer binding. Both segment-1 (McLaughlin et al. 1993) and profilin (Schutt et al. 1993) have been shown by X-ray crystallography to bind G-actin at the barbed end, between subdomains 1 and 3 (Fig. 5c, d). Despite this and a superficial structural similarity between these two proteins (profilin also contains a central anti-parallel β-sheet flanked by α-helices), they make very different contacts with actin.

Profilin binds across the lower surface of actin (see also Burtnick, this Vol.), bridging the cleft between subdomains 1 and 3, whereas the majority of the gelsolin segment-1 contacts involve the front of the molecule. In fact, the only residues in common between the two binding sites are Glu_{167} and Tyr_{169}. This may be significant, since both residues fall at regions in actin that are involved in stabilizing the filament (Lorenz et al. 1993). Because cofilin competes with both proteins for G-actin binding, it is possible that it, too, interacts with residues in the 162–176 loop of actin. An alternative possibility is that conformational changes in actin are induced by gelsolin segment-1 and/or profilin that prevent binding by cofilin. Although no such changes have been reported for gelsolin segment-1, profilin has been shown to alter actin structure under some conditions (Chik et al. 1996).

Effect of Cofilin on Actin Structure

Cofilin is unique among F-actin-binding proteins in one remarkable respect: electron cryomicroscopy has shown it reduces the angle of rotation between subunits along the short-pitched left-handed helix (the so-called genetic helix). However, the rise per actin subunit is unaffected. As a result, the length of the actin cross-over is dramatically reduced from about 370 to ~270 Å. However, the overall filament length remains unchanged (Fig. 6; McGough et al. 1997).

The effects on filament twist occur at both low and high pH and with different isoforms of actin. All cofilins tested to date have shown this effect, but the absolute degree of twist change reflects the source of the cofilin (A. McGough, unpub. observ.).

At low resolution (27 Å), the conformation of the actin subunit itself appears largely unaltered by the binding of cofilin (for evidence of conformational changes in actin see Kekic et al., this Vol.). Thus, the twist change appears to involve rigid body rotations of actin subunits. In addition, examination of the actin filament model built using the cofilin-induced twist reveals no major

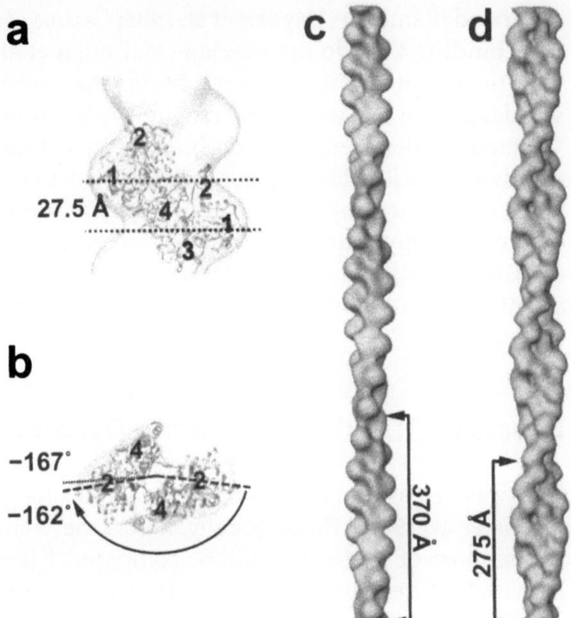

Fig. 6a–d. Effect of cofilin on actin filament twist. (adapted from Chiu et al. 1999) **a** Side and **b** top views of two actin subunits in a filament are shown as ribbon diagrams. The four subdomains of actin are labeled. In standard actin filament geometry, one subunit is related to the next by a rise of 27.5 Å and an average rotation of −167°. When cofilin is bound the rotation per subunit is reduced by about 5°, but the rise per subunit does not change. **c** The twisting nature of the actin filament is evident by the appearance of crossovers that typically average 370 Å in length. **d** Reconstruction of cofilin:F-actin showing the effect of cofilin on filament twist. In the twisted cofilin:actin filament the actin cross-overs are reduced in length by about 25%. Because the rise per subunit is constant, the overall length of the filament does not change. Images were generated using Ribbons 2.65 (Carson and Bugg 1986) and Iris Explorer (Numerical Algorithms Group Ltd., Downers Grove, Illinois)

intersubunit clashes. Nevertheless, imposing the twist change as a rigid body rotation on the actin filament model produces certain clashes between subdomains in adjacent subunits. For example, the side chains of two residues in the upper actin subunit (Lys_{291} and Glu_{167}) appear to clash with residues in subdomain 2 of the lower. These residues are significant in that each is involved in the intermolecular contacts that stabilize F-actin (Lorenz et al. 1993).

There are also disruptions in the lateral contacts that are caused by cofilin's effects on filament twist. Analysis of the cofilin/F-actin reconstruction at high contour levels reveals that the strongest features in the filament are the actin-cofilin-actin contacts running along the long-pitched two-start helix. In addition, phalloidin binding, which involves a site of lateral contacts between three actin subunits (Lorenz et al. 1993; Steinmetz et al. 1998) is significantly altered by the change in twist. As a result, phalloidin is unable to bind cofilin–

decorated filaments (Hayden et al. 1993; Carlier et al. 1997) despite the fact that their binding sites do not overlap (McGough et al. 1997). Thus, it appears that both the longitudinal and lateral bonds that contribute to filament stability in standard actin are weakened, raising the possibility that the actin filament is "primed" to disassemble upon binding by cofilin even at low pH where little depolymerization takes place. Direct evidence for the disruption of actin filament contacts by cofilin birding has recently been provided by electron cryomicroscopy studies of F-actin decorated with a variety of ADF/cofilins (McGough and Chiu 1999). These studies show the two long-pitch filament strands separated over distances ranging from a few to over 20 actin subunits in length.

Effects of ADF/Cofilin on Actin Dynamics in Vitro

ADF was discovered at the same time as plasma gelsolin (Bamburg et al. 1980; Harris et al. 1980). Consequently, it was inevitable that the properties of the two families of proteins would be compared. Kinetic evidence for the severing activities of gelsolin domains came from fluorescence changes when these domains bind labeled actin (NBD–actin or PI–actin) (Way et al. 1989). The absence of similar fluorescence changes when vertebrate cofilins bind labelled actins made it difficult to analyze the mechanism of ADF interaction (Hawkins et al. 1993; Hayden et al. 1993).

Light microscopy showed clear evidence for fragmentation of filaments (Maciver et al. 1991; Hawkins et al. 1993) and these observations formed the basis for interpreting kinetic experiments. However, recent work has shown that this is an oversimplification. Some believe that there is no severing activity whatsoever (Carlier et al. 1997; Didry et al. 1998) even though rapid shortening of filaments evidenced by electron microscopy strongly supports fragmentation as one component in the depolymerization process (Maciver et al. 1998). Elsewhere, anomalies in the kinetic behavior of yeast actin during polymerization in the presence of yeast cofilin have been interpreted on the basis of filament fragmentation (Du and Frieden 1998).

ADF binds to G-actin, but its affinity for actin-ADP is over 60-fold greater than for actin-ATP under physiological conditions. Their K_d values for human ADF binding are 0.1 and 6 μM respectively (Ressad et al. 1998). ADF is clearly not a sequestering protein in comparison to thymosin β4 (Carlier et al. 1997; Blanchoin and Pollard 1998). There is disagreement about the extent of depolymerization of F-actin by ADF at pH 8. Ressad et al. (1998) found the maximum concentration of ADF-actin-ADP to be ~8 μM, while our own experiments have indicated a much higher value, ~20 μM (S. Yeoh and A. Weeds, unpub.).

ADF and cofilin bind cooperatively to actin filaments as evidenced both from sigmoidal binding curves and electron cryomicroscopy (McGough et al. 1997). The extent of cooperativity differs with the different forms of ADF, as does the pH sensitivity of its binding to actin (Du and Frieden 1998; Maciver et al. 1998). For example, plant ADF and actophorin bind F-actin equally well at pH 8 and pH 6.5, while human ADF and cofilin bind more extensively at pH 6.5 than at pH 8 (Maciver et al. 1998).

A unique feature of the effects of ADF/cofilin on F-actin is the acceleration of the dissocation rate constant of subunits from the pointed ends of filaments (Fig. 7). There is at least a 30-fold activation, giving an off-rate of ~9 s^{-1} (Ressad et al. 1998). Using an ATPase assay, the rate declines at very high concentrations of ADF, presumably due to a buildup of ADF-actin-ADP complexes (Carlier et al. 1997). Although the dissociation of these complexes is in rapid equilibrium, with a rate constant at 4°C ~15 s^{-1} (Ressad et al. 1998), the slow exchange of ADP for ATP on actin (~0.007 s^{-1}) (Teubner and Wegner 1998) results in a buildup of complexes which severely limits the depolymerizing effect of ADF.

This may explain the observations that overexpression of ADF in *Dictyostelium* leads to an increase in the level of F-actin (Aizawa et al. 1996). However, the recent demonstration that profilin catalyzes the exchange of nucleotide in these binary complexes (Blanchoin and Pollard 1998; Didry et al. 1998) as well as in G-actin, means that the rate-limiting step in the kinetic cycle, nucleotide exchange, (which is even slower in the binary complexes than on actin itself) is bypassed and G–actin–ATP is thereby made rapidly available for filament elongation (Fig. 7). When profilin is present, nucleotide exchange in the ADF-actin-ADP complexes is accelerated and the complexes rapidly dissociate, thereby increasing the rate of treadmilling up to 125-fold (Didry et al. 1998).

Regulation of Cofilin Activity

Biochemical experiments have shown differential effects of ADF/cofilin on actin with pH, suggesting that actin disassembly may be regulated by changes in intracellular pH (Hawkins et al. 1993; Hayden et al. 1993; Meberg et al. 1998). Studies using Swiss 3T3 cells have shown that pH also affects ADF/cofilin activity in vivo (B. Bernstein and J. Bamburg, pers, comm.). Interestingly, phosphorylation of ADF/cofilin appears to increase with pH-induced activation of depolymerizing activity, consistent with a compensating homeostatic mechanism. Phosphoinositides were shown to inhibit the depolymerizing activity of cofilin as they do also the capping and severing activities of gelsolin (Yonezawa et al. 1990). Subsequently, it became clear that vertebrate cofilins are inhibited by phosphorylation (Morgan et al. 1993), and there is a considerable body of evidence that dephosphorylation is associated with increased motility in a variety of cell types (reviewed by Moon and Drubin 1995).

Dephosphorylation of cofilin accompanies cytokinesis in oocytes (Abe et al. 1996) and recent work has started to elucidate the signaling pathways involved when growth factors such as NGF stimulate growth cone activity in neurons (Meberg et al. 1998). Cofilin dephosphorylation occurs at the end of a number

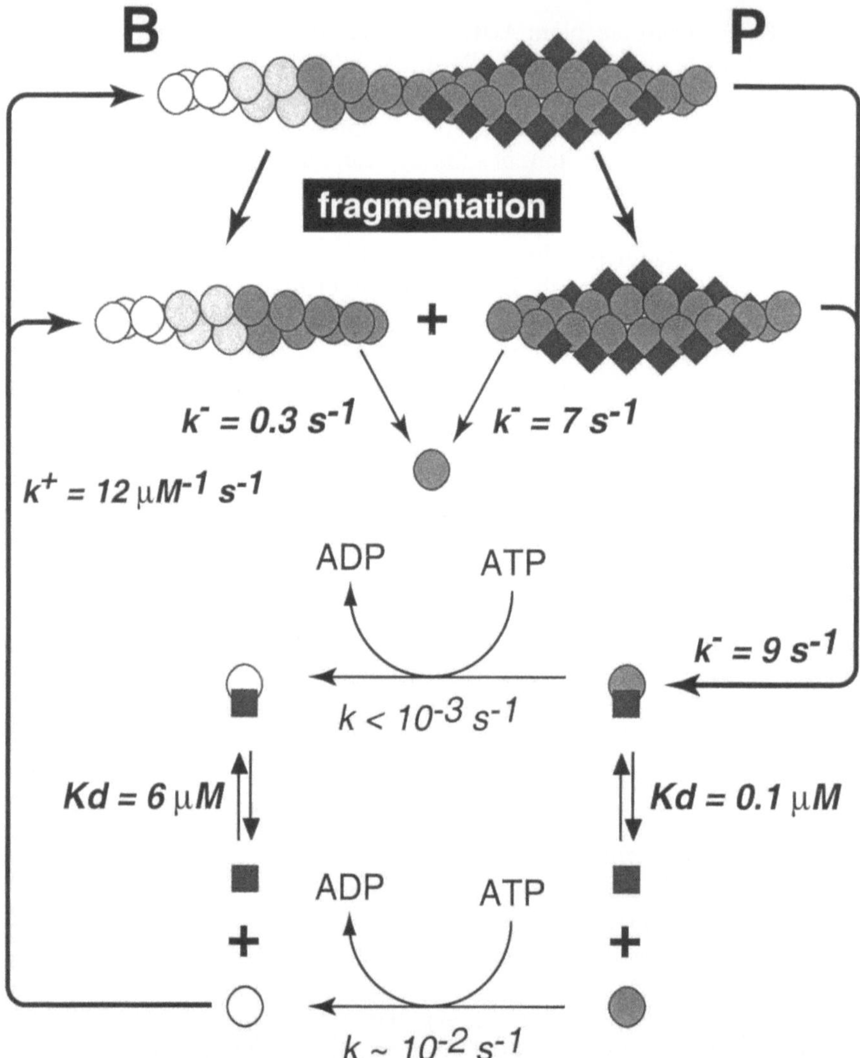

Fig. 7. Schematic representation of effects of ADF/cofilin on actin treadmilling. ADF/cofilin have been shown to fragment filaments and accelerate disassembly at the pointed ends of filaments. Actin filament (*dark gray spheres* ADP-actin; *light gray spheres* ADP.Pi-actin; *white spheres* ATP-actin) partially decorated with ADF/cofilin (*diamonds* and *squares* to represent F-actin and G-actin bound forms). Rate constants are shown for association of ATP-actin onto the barbed ends, dissociation of ADP-actin from barbed and pointed ends (Pollard 1986), dissociation of complexes from the pointed ends (Carlier and Pantaloni 1997; Ressad et al. 1998), and for ADP exchange on actin monomers and complexes. These latter associations are greatly accelerated by profilin (Blanchoin and Pollard 1998; Didry et al. 1998)). Dissociation constants are also shown for the complexes (Carlier et al. 1997; Ressad et al. 1998)

of important signaling pathways and much effort is therefore being directed at identifying the regulating enzymes involved.

Although cofilin is dephosphoylated in vitro by a number of protein phosphatases, including phosphatases 1, 2A and 2C, the intracellular phosphatases have not been identified. LIM-kinases phosphorylate cofilin and induce the stabilization of actin structures in cells (Arber et al. 1998; Yang et al. 1998). Moreover, constitutively active Rac, which had earlier been shown to stimulate the formation of lamellipodia, augmented this effect. One possible means by which these two regulatory mechanisms may be linked is the Rac-induced increase in phosphatidylinositol-4,5-bisphosphate (Rosenblatt and Mitchison 1998).

Another feature of ADF/cofilin is their preferential binding to actin-ADP. This means that "older" filaments may be more susceptible to disassembly than newly polymerized actin, which may provide an additional regulatory mechanism by selecting specific populations of filaments for turnover.

Effect of Cofilin on Actin Turnover in Cells

Cell locomotion occurs at rates up to ~30 μm min^{-1}. Assuming this process occurs solely by filament elongation at the leading edge, this corresponds to actin monomer addition ~180 subunits s^{-1}. Given the high intracellular concentration of actin, the elongation rate in vitro is sufficiently rapid to lengthen filaments at this velocity, provided filaments elsewhere disassemble fast enough to maintain the supply of monomers. However, treadmilling rates measured *in vitro* are far too slow to regenerate monomers at this rate and need to be increased at least 100-fold.

ADF accelerates pointed-end disassembly maximally by about 30 fold (Ressad et al. 1998) and by up to 125-fold in the presence of profilin (Didry et al. 1998). The overall depolymerization rate would be increased to well over 100 s^{-1} with infrequent filament severing of five or six cuts per filament (Du and Frieden 1998; Maciver et al. 1998), which increases the number of ends for disassembly (for discussion of intracellular disassembly see Zigmond 1993).

Is there evidence that cofilins might function in this way in cells? Cofilin does indeed colocalize with actin in EGF–stimulated cells (Fig. 1A,A′). Direct evidence that ADF accelerates actin-based motility is demonstrated from the rate of disappearance of actin patches in yeast in the presence of latrunculin A, a monomer-sequestering agent. Cofilin mutants known to be defective in their depolymerizing activity reduced the rate of disappearance of these patches and the magnitude of the effect correlated well with the activity of the mutants in vitro (Lappalainen and Drubin 1997; Lappalainen et al. 1997).

Cofilins also increased the turnover rate of actin comet tails in *Listeria monocytogenes*, but there was no general agreement about whether the reduction of tail length correlated precisely with motility rate (Carlier et al. 1997; Rosenblatt et al. 1997). In *Dictyostelium*, overexpression of cofilin increased membrane ruffling and cofilin accumulated with actin bundles in both phagocytic cups and nascent pseudopods of motile cells (Aizawa et al. 1996, 1997). Thus, there is considerable evidence that cofilins accelerate the actin treadmill and thereby facilitate motility (reviewed Theriot 1997; Bamburg et al. 1999).

Future Directions

Recent biochemical and structural studies have provided important insights into how ADF/cofilin regulates actin filament dynamics, but there remain a number of outstanding questions. For example, the discovery that cofilin alters actin filament twist raises the possibility of a connection between filament structure and actin dynamics. If different ADF/cofilins vary quantitatively in their effects on actin assembly, do they also differ in their effects on filament structure? Why are some, but not all, ADF/cofilins regulated by pH? Detailed studies of different members of the ADF/cofilin family will be needed to answer these questions.

An additional effect of the twist change is the possibility that it provides a novel mechanism for regulating filament interactions with other actin binding proteins. Thus there may exist in cells a pool of cofilin-decorated filaments that are inhibited from interacting with other actin-binding proteins or that show enhanced interaction by virtue of the twist change.

The observed effects of ADF/cofilins on depolymerization of *Listeria* tails, where the distribution of filaments is exponential with both time and distance from the surface of the bacteria, cannot readily be explained on the basis of current knowledge on the effects of these proteins on actin *in vitro* (Mitchison, pers. comm.). More information therefore is needed of the interplay of ADF/cofilin with other actin–binding proteins. The recent demonstration of synergistic behaviour between ADF/cofilin and profilin shows how important it is not to restrict studies to single actin-binding proteins.

The nuclear localization signals (NLS) sequences in vertebrate ADF/cofilins are too "classical" to be fortuitous accidents of nature. Their effects are well catalogued, but have been seen only in response to harsh disruptive treatment of cells. How cells do this or why they should need a NLS could well be of keen future interest.

Thus the ADF/cofilins are a remarkable family of small actin-binding proteins. They interact with both filamentous and monomeric actins, remodel filaments, and undergo nuclear translocation. Their actin-binding properties are regulated by a variety of mechanisms which perhaps makes it possible to finely tune locomotive responses in cells.

Acknowledgments. The authors thank Dr. Sabine Gonsior (MRC Laboratory of Molecular Biology, Cambridge, UK) for the images used in Fig. 1, and Drs. Jim Bamburg (Colorado State University, Fort Collins, CO) and Tim Mitchison (Harvard Medical School, Boston, MA) for helpful discussions. A.M. thanks the National Institutes of Health (GM59677) for generous support.

References

Abe H, Endo T, Yamamoto K, Obinata T (1990) Sequence of cDNAs encoding actin depolymerizing factor and cofilin of embryonic chicken skeletal muscle: two functionally distinct actin-regulatory proteins exhibit high structural homology. Biochemistry 29:7420–7425

Abe H, Obinata T, Minamide L, Bamburg J (1996) *Xenopus laevis* actin-depolymerizing factor/cofilin: a phosphorylation-regulated protein essential for development. J Cell Biol 132:871–885

Adams ME, Minamide LS, Duester G, Bamburg JR (1990) Nucleotide sequence and expression of a cDNA encoding chick brain actin depolymerizing factor. Biochemistry 29:7414–7420

Aizawa H, Fukui Y, Yahara I (1997) Live dynamics of *Dictyostelium* cofilin suggests a role in remodeling actin latticework into bundles. J Cell Sci 110:2333–2344

Aizawa H, Sutoh K, Tsubuki S, Kawashima S, Ishii A, Yahara I (1995) Identification, characterization, and intracellular distribution of cofilin in *Dictyostelium discoideum.* J Biol Chem 270:10923–10932

Aizawa H, Sutoh K, Yahara I (1996) Overexpression of cofilin stimulates bundling of actin filaments, membrane ruffling and cell movement in *Dictyostelium.* J Cell Biol 132:335–344

Arber S, Barbayannis FA, Hanser H, Schneider C, Stanyon CA, Bernard O, Caroni P (1998) Regulation of actin dynamics through phosphorylation of cofilin by LIM-kinase. Nature 393:805–809

Bamburg JR, Bray D (1987) Distribution and cellular localization of actin depolymerizing factor. J Cell Biol 105:2817–2825

Bamburg JR, Harris HE, Weeds AG (1980) Partial purification and characterization of an actin depolymerizing factor from brain. FEBS Lett 121:178–182

Bamburg JR, McGough A, Ono S (1999) Putting a new twist on actin: ADF/cofilins modulate actin dynamics. Trends Cell Biol 9:364–370

Blanchoin L, Pollard T (1998) Interaction of actin monomers with *Acanthamoeba* actophorin (ADF/cofilin) and profilin. J Biol Chem 273:25106–25111

Burtnick LD, Koepf EK, Grimes J, Jones E, Stuart D, McLaughlin P, Robinson R (1997) The crystal structure of plasma gelsolin: Implications for actin severing, capping and nucleation. Cell 90:661–670

Carlier M-F, Pantaloni D (1997) Control of actin dynamics in cell motility. J Mol Biol 269:459–467

Carlier MF, Laurent V, Santolini J, Melki R, Didry D, Xia GX, Hong Y, Chua NH, Pantaloni D (1997) Actin depolymerizing factor (ADF/cofilin) enhances the rate of filament turnover: Implication in actin-based motility. J Cell Biol 136:1307–1322

Carson M, Bugg CE (1986) Algorithm for ribbon models of proteins. J Mol Graphics 4:121–122

Chik JK, Lindberg U, Schutt C (1996) The structure of an open state of β-actin at 265 Å resolution. J Mol Biol 263:607–623

Chiu W, McGough A, Sherman M, Schmid M (1999) High-resolution electron cryomicroscopy of macromolecular assemblies. J Mol Biol 291:513–519

Didry D, Carlier M-F, Pantaloni D (1998) Synergy between actin depolymerizing factor/cofilin and profilin in increasing actin filament turnover. J Biol Chem 273:25602–25611

Drenckhahn D, Pollard T (1986) Elongation of actin filaments is a diffusion-limited reaction at the barbed end and is accelerated by inert macromolecules. J Biol Chem 261:12754–12758

Du J, Frieden C (1998) Kinetic studies on the effect of yeast cofilin on yeast actin polymerization. Biochemistry 37:13276–13284

Fedorov AA, Lappalainen P, Fedorov EV, Drubin DG, Almo SC (1997) Structure determination of yeast cofilin. Nature Struct Biol 4:366–369

Gillett GT, Fox MF, Rowe PSN, Casimir CM, Povey S (1996) Mapping of human nonmuscle type cofilin (Cfl1) to chromosome-11q13 and muscle-type cofilin (Cfl2) to chromosome-14. Ann Hum Genet 60:201–211

Goode BL, Drubin DG, Lappalainen P (1998) Regulation of the cortical actin cytoskeleton in budding yeast by twinfilin, a ubiquitous actin monomer-sequestering protein. J Cell Biol 142:723–733

Harris H, Bamburg J, Weeds A (1980) Actin filament disassembly in blood plasma. FEBS Lett 121:175–177

Hatanaka H, Ogura K, Moriyama K, Ichikawa S, Yahara I, Inagaki F (1996) Tertiary structure of destrin and structural similarity between two actin-regulating protein families. Cell 85:1047–1055

Hawkins M, Pope B, Maciver S, Weeds AG (1993) Human actin depolymerizing factor mediates a pH-sensitive destruction of actin filaments. Biochemistry 32:9985–9993

Hayden SM, Miller PS, Brauweiler A, Bamburg JR (1993) Analysis of the interactions of actin depolymerizing factor (ADF) with G- and F-actin. Biochemistry 32:9994–10004

Iida K, Matsumoto S, Yahara I (1992) The KKRKK sequence is involved in heat shock-induced nuclear translocation of the 18-kDa actin-binding protein, cofilin. Cell Struct Funct 17:39–46

Iida K, Moriyam K, Matsumoto S, Kawasaki H, Nishida E, Yahara I (1993) Isolation of a yeast essential gene, COF1, that encodes a homologue of mammalian cofilin, a low-Mr actin-binding and depolymerizing protein. Gene 124:115–120

Jans J, Moll T, Nasmyth K, Jans P (1995) Cyclin-dependent kinase site-regulated signal-dependent nuclear-localization of the SW15 yeast transcription factor in mammalian cells. J Biol Chem 270:17064–17067

Jiang CJ, Weeds AG, Hussey PJ (1997) The maize actin depolymerizing factor, ZmADF3, redistributes to the growing tip of elongating root hairs and can be induced to translocate into the nucleus with actin. Plant J 12:1035–1043

Kabsch W, Sander C (1983) Dictionary of protein secondary structure–pattern recognition of hydrogen-bonded and geometrical features. Biopolymers 22:2577–2637

Kabsch W, Mannherz HG, Suck D, Pai EF, Holmes KC (1990) Atomic structure of the actin-DNase I complex. Nature 347:37–44

Karderon D, Richardson WD, Markham AF, Smith AE (1988) Sequence requirements for nuclear localisation of simian virus 40 large-T antigen. Nature 311:499–509

Lappalainen P, Drubin D (1997) Cofilin promotes rapid actin filament turnover in vivo. Nature 388:78–82

Lappalainen P, Fedorov EV, Fedorov AA, Almo SC, Drubin DG (1997) Essential functions and actin-binding surfaces of yeast cofilin revealed by systematic mutagenesis. EMBO J 16:5520–5530

Lappalainen P, Kessels M, Cope J, Drubin D (1998) The ADF homology (ADF-H) domain: a highly exploited actin-binding module. Mol Biol Cell 9:1951–1959

Leonard SA, Gittis AG, Petrella EC, Pollard TD, Lattman EE (1997) Crystal structure of the actin-binding protein actophorin from *Acanthamoeba*. Nat Struct Biol 4:369–373

Lila T, Drubin D (1998) Evidence for physical and functional interactions among two *Saccharomyces cerevisiae* SH3 domain proteins, an adenylyl cyclase-associated protein and the actin cytoskeleton. Mol Biol Cell 8:367–385

Lopez I, Anthony R, Maciver S, Jiang C-J, Khan S, Weeds AG, Hussey P (1996) Pollen specific expression of maize genes encoding actin depolymerizing factor-like proteins. Proc Natl Acad Sci 93:7415–7420

Lorenz M, Popp D, Holmes KC (1993) Refinement of the F-actin model against X-ray fiber diffraction data by the use of a directed mutation algorithm. J Mol Biol 234:826–836

Maciver S, Pope B, Whytock S, Weeds A (1998) The effect of two actin depolymerizing factors (ADF/cofilins) on actin filament turnover: pH sensitivity of F-actin binding by human ADF but not of *Acanthamoeba* actophorin. Eur J Biochem 256:388–397

Maciver SK, Zot HG, Pollard TD (1991) Characterization of actin filament severing by actophorin from *Acanthamoeba castellanii*. J Cell Biol 115:1611–1620

Maekawa S, Nishida E, Ohta Y, Sakai H (1984) Isolation of low molecular weight actin-binding proteins from porcine brain. J Biochem 95:377–385

Markus MA, Nakayama T, Matsudaira P, Wagner G (1994) Solution structure of villin 14T, a domain conserved among actin-severing proteins. Protein Sci 3:70–81

Matsuzaki F, Matsumoto S, Yahara I, Yonezawa N, Nishida E, Sakai H (1988) Cloning and characterization of porcine brain cofilin cDNA. J Biol Chem 263:11564–11568

McGough A (1998) F-actin-binding proteins. Curr Opin Struct Biol 8:166–176

McGough A, Chiu W (1999) ADF/cofilin weakens lateral contacts in the actin filament. Mol Biol 291:513–519

McGough A, Pope B, Chiu W, Weeds A (1997) Cofilin changes the twist of F-actin: implications of actin filament dynamics and cellular function. J Cell Biol 138:771–781

McGough A, Chiu W, Way M (1998) Determination of the gelsolin binding site on F-actin: implications for severing and capping. Biophys J 126:764–772

McGough AM, Way M, DeRosier D (1994) Determination of the α-actinin-binding site on actin filaments by cryoelectron microscopy and image analysis. J Cell Biol 126:433–443

McLaughlin PJ, Gooch JT, Mannherz HG, Weeds AG (1993) Structure of gelsolin segment-1-actin complex and the mechanism of filament severing. Nature 364:685–692

Meberg PJ, Ono S, Minamide LS, Takahashi M, Bamburg JR (1998) Actin depolymerizing factor and cofilin phosphorylation dynamics: response to signals that regulate neurite extension. Cell Motil Cytoskel 2:172–190

Minamide LS, Painter WB, Schevzov G, Gunning P, Bamburg JR (1997) Differential regulation of actin depolymerizing factor and cofilin in response to alterations in the actin monomer pool. J Biol Chem 272:8303–8309

Moon A, Drubin D (1995) The ADF/cofilin proteins: Stimulus-responsive modulators of actin dynamics. Mol Biol Cell 6:1423–1431

Moon AL, Janmey PA, Louie KA, Drubin DG (1993) Cofilin is an essential component of the yeast cortical cytoskeleton. J Cell Biol 120:421–435

Morgan TE, Lockerbie RO, Minamide LS, Browning MD, Bamburg JR (1993) Isolation and characterization of a regulated form of actin depolymerizing factor. J Cell Biol 122:623–633

Moriyama K, Matsumoto S, Nishida E, Sakai H, Yahara I (1990) Nucleotide sequence of mouse cofilin cDNA. Nucl Acid Res 18:3053

Moriyama K, Yonezawa N, Sakai H, Yahara I, Nishida E (1992) Mutational analysis of an actin-binding site of cofilin and characterization of chimeric proteins between cofilin and destrin. J Biol Chem 267:7240–7244

Nishida E, Maekawa S, Sakai H (1984) Cofilin, a protein in porcine brain that binds to actin filaments and inhibits their interactions with myosin and tropomyosin. Biochemistry 23:5307–5313

Nishida E, Iida K, Yonezawa N, Koyasu S, Yahara I, Sakai H (1987) Cofilin is a component of intranuclear and cytoplasmic actin rods induced in cultured cells. Proc Natl Acad Sci 84:5262–5266

Ono S, Benian GM (1998) Two *Caenorhabditis elegans* actin depolymerizing factor/cofilin proteins, encoded by the unc-60 gene, differentially regulate actin filament dynamics. J Biol Chem 273:3778–3783

Pollard TD (1986) Rate constants for the reactions of ATP- and ADP-actin with the ends of actin filaments. J Cell Biol 103:2747–2754

Rayment I, Holden HM, Whittaker M, Yohn M, Lorenz M, Holmes KC (1993) Structure of the actin-myosin complex and its implications for muscle contraction. Science 261:58–65

Ressad F, Didry D, Xia G-X, Hong Y, Chua N-H, Pantalon D, Carlier M-F (1998) Kinetic analysis of the interaction of actin-depolymerizing factor (ADF)/cofilin with G- and F-actins. J Biol Chem 273:20894–20902

Rosenblatt J, Mitchison TJ (1998) Signal transduction–Actin, cofilin and cognition. Nature 393:739–740

Rosenblatt J, Agnew BJ, Abe H, Bamburg JR, Mitchison TJ (1997) *Xenopus* actin depolymerizing factor cofilin (XAC) is responsible for the turnover of actin filaments in *Listeria* monocytogenes tails. J Cell Biol 136:1323–1332

Schnuchel A, Wiltscheck RL, Eichinger Schleicher M, Holak TA (1995) Structure of severin domain 2 in solution. J Mol Biol 247:21–27

Schroder RR, Manstein DJ, Jahn W, Holden H, Rayment I, Holmes KC, Spudich JA (1993) Three-dimensional atomic model of F-actin decorated with *Dictyostelium* myosin S1. Nature 364:171–174

Schutt CE, Myslik JC, Rozycki MD, Goonesekere N, Lindberg U (1993) The structure of crystalline profilin β-actin. Nature 365:810–816

Steinmetz MO, Stoffler D, Muller SA, Jahn W, Wolpensinger B, Goldie KN, Engel A, Faulstich H, Aebi U (1998) Evaluating atomic models of F-actin with an undecagold-tagged phalloidin derivative. J Mol Biol 276:1–6

Sun HQ, Wooten DC, Janmey PA, Yin HL (1994) The actin side-binding domain of gelsolin also caps actin-filaments-implications for actin filament severing. J Biol Chem 269:9473–9479

Teubner A, Wegner A (1998) Kinetic evidence for a readily exchangeable nucleotide at the terminal subunit of the barbed ends of actin filaments. Biochemistry 37:7532–7538

Theriot J (1997) Accelerating on a treadmill: ADF/cofilin promotes rapid actin filament turnover in the dynamic cytoskeleton. J Cell Biol 136:1165–1168

Thompson JD, Higgins DG, Gibson TJ (1994) Clustal-W-improving the sensitivity of progressive multiple sequence alignment through sequence weighting, position-specific gap penalties and weight matrix choice. Nucleic Acid Res 22:4673–4680

VanTroys M, Dewitte D, Goethals M, Vandekerckhove J, Ampe C (1996) Evidence for an acti-binding helix in gelsolin segment 2: Have homologous sequences in segments 1 and 2 of gelsolin evolved to divergent actin-binding functions? FEBS Lett 397:191–196

VanTroys M, Dewitte D, Verschelde JL, Goethals M, Vandekerckhove J, Ampe C (1997) Analogous F-actin binding by cofilin and gelsolin segment 2 substantiates their structural relationship. J Biol Chem 272:32750–32758

Way M, Gooch J, Pope B, Weeds AG (1989) Expression of human plasma gelsolin in *E coli* and dissection of actin-binding sites by segmental deletion mutagenesis. J Cell Biol 109:593–605

Wriggers W, Tang JX, Azuma T, Marks PW, Janmey PA (1998) Cofilin and gelsolin segment 1: Molecular dynamics simulation and biochemical analysis predict a similar actin binding mode. J Mol Biol 282:921–932

Yang N, Higuchi O, Ohashi K, Nagata K, Wada A, Kangawa K, Nishida E, Mizuno K (1998) Cofilin phosphorylation by LIM-kinase 1 and its role in Rac-mediated actin reorganization. Nature 393:809–812

Yonezawa N, Nishida E, Sakai H (1985) pH control of actin polymerization by cofilin. J Biol Chem 260:14410–14412

Yonezawa N, Nishida E, Iida K, Yahara I, Sakai H (1990) Inhibition of the interactions of cofilin, destrin, and deoxyribonuclease-I with actin by phosphoinositides. J Biol Chem 265:8382–8386

Yonezawa N, Homma Y, Yahara I, Sakai H, Nishida E (1991) A short sequence responsible for both phosphoinositide binding and actin binding activities of cofilin. J Biol Chem 266:17218–17221

Zigmond SH (1993) Recent quantitative studies of actin filament turnover during cell locomotion. Cell Motil Cytoskel 25:309–316

Predicting Interaction Sites between Glycolytic Enzymes and Cytoskeletal Proteins Employing the Concepts of the Molecular Recognition Theory

R. J. Sheedy[1] and F. M. Clarke[2]

Introduction

In 1984 Blalock and Smith focused attention on certain complementary hydropathic relationships between amino acids based on the genetic code. Amino acids specified on one strand of DNA were found to be hydropathically complementary to those encoded by the opposite strand of the DNA in the same reading frame. This is so because the hydropathic character of an amino acid is determined by the second base of the triplet codon (A for hydrophilic and U for hydrophobic) with the identity of the amino acid determined by the first two bases. Consequently, peptide sequences derived from the noncoding strand of DNA, or RNA that is complementary to mRNA, will have an inverted pattern of hydropathy relative to the pattern of amino acids derived from the coding nucleic acid strand. These complementary peptides have been found more often than not to specifically bind to the partner peptide specified by the coding strand. This is the basis of the *m*olecular *r*ecognition *t*heory (MRT), as proposed by Blalock (see Blalock 1995) which hypothesizes "that complementary nucleotide sequences specify peptides or proteins that interact through complementary shapes or structures resulting from their inverted periodicity of hydrophobic and hydrophilic amino acids".

A substantial and growing literature attests to the interaction of peptides and proteins encoded by complementary nucleic acid sequences as illustrated for peptide hormones and their cognate receptors and in a wide variety of other systems (see Blalock 1995; Baranyi et al. 1995 for examples). While the physical and chemical basis of the interactions of complementary pairs of peptides is yet to be satisfactorily explained, consideration of the increasing number of observations of the phenomenon warrants that it be taken into account when investigating the molecular basis of protein–protein interactions, as it may provide the basis for predicting interacting sites between proteins and provide an understanding of the basis of their evolution.

It could be envisaged that nucleic acid and peptide complementarity were responsible for the first tenuous interactions of proteins early in the history of

[1] School of Life Sciences, Queensland University of Technology, GPO Box 2434, Brisbane 4001, Australia
[2] School of Biomolecular & Biomedical Science, Griffith University, Nathan, Queensland 4111, Australia

Results and Problems in Cell Differentiation, Vol. 32
C. dos Remedios (Ed.): Molecular Interactions of Actin

protein evolution. As biological functions were consolidated, these interaction sites eventually mutated to more substantial binding sites involving conventional interactions such as ionic and hydrophobic interactions. The mutations at the genetic level might partly obliterate the complementarity of the nucleotide sequences; however, identification of vestigial complementarity might still identify the position of the binding sites in present–day proteins. A similar scenario was specifically envisaged for the evolution of the peptide hormones and their receptors, a restricted case that has essentially been validated (Ruiz-Opazo et al. 1995).

Macromolecular organization within cells is a fundamental requirement for cell function. The evolution of viable cells required not only the evolution of functional proteins, but also, or even more so, the evolution of functional systems of interacting proteins. It is proposed that the *m*olecular *r*ecognition *t*heory may explain the essential non-randomness of the evolution of organised protein systems and assist in elucidating the molecular basis of interacting proteins systems by predicting protein-protein interaction sites. Glycolytic enzymes are abundant and of almost ubiquitous distribution and are among the most ancient of proteins. A considerable body of evidence (Clarke et al. 1985) indicates that the function of these proteins within the cell is dependent on interactions between themselves and other cellular components such as the cytoskeletal actin of eukaryotic cells.

Actin itself is currently described as a protein of intermediate antiquity but there is evidence to suggest that it may be of even greater antiquity than is currently envisaged. According to current ideas on evolution, it is thought that an RNA world preceded the familiar DNA world of today, as RNA has the ability to act as a catalyst as well as an information carrier. If this notion is correct, it is likely that the glycolytic enzymes evolved during this epoch. Any consideration that complementary nucleic acid sequences were involved in the evolution of protein recognition sites within this group of proteins should encompass the possibility that RNA was then the predominant encoding nucleic acid.

In this chapter we establish the validity of applying the *m*olecular *r*ecognition *t*heory to identify an established actin/actin binding protein interaction site. We then apply the theory to predict an interaction site between the glycolytic enzyme glyceraldehyde-3-phosphate dehydrogenase (GAPDH) and actin and analyze the involvement of the predicted site in the interaction between the two proteins.

Molecular Recognition Theory and the Interaction of Actin and Gelsolin Segment 1

There are no examples at present where the crystallographic structure of an actin-glycolytic enzyme complex has been determined. However, the structure of the actin-gelsolin segment 1 complex has been elucidated (McLaughlin et al. 1993). Gelsolin segment 1 binds to actin in a cleft at the interface of subdomains 1 and 3. McLaughlin et al. (1993) described this as an apolar patch rimmed by polar hydrogen-bonded atoms. They employed the analogy of a circular sticking plaster to describe this area, with the gauze representing the hydrophobic region and the annular adhesive part the polar rim. In the gelsolin segment 1 a corresponding apolar patch exists comprising apolar amino acids discontinuous in the sequence and centred on 11e103. This branched amino acid is buried in the hydrophobic side chains of both domains 1 and 3 of actin. lle 103 is packed by Tyr 169′, lle 345′, Leu 346′ and Leu 349′ on actin. This defined molecular interaction site was used as a test case for the applicability of the *m*olecular *r*ecognition *t*heory in identifying potential interaction sites between actin and actin-binding proteins.

The RNA sequences encoding gelsolin segment 1 and human skeletal muscle actin were analyzed for complementarity by the software analysis programs of Sheedy (1996). This software was developed after a consideration of potential difficulties that might arise in this analysis. It was anticipated that any base-pairing present between genes would be vestigial in character. Whatever the source of the original nucleotide complementarity, modification of either sequence by mutation during subsequent evolution would result in mismatches where original base pairs were present. Frame shifting caused by the deletion or addition of nucleotides in either sequence would likewise result in gaps. It is even possible that additional base pairing, related to some novel functionality, could be introduced during evolution.

Reference was made in the introduction to the possibility that the original nucleotide complementarity might have its origin in RNA structures. Extant ribonucleic acids, while possessing the required nucleotide complementarity, are characterized by other regions where the nucleotides are not involved in base pairing, such as loops and "bulge" regions, including regions of nucleotide mismatch within base-paired duplexes. If the original complementarity were derived from similar structures, mismatches and gaps might again be expected. Since the extent of evolutionary modification, and the origin of the sequences is not known, it was deemed inappropriate to employ any technique which determines secondary structure by energy minimization, as this might lead to ambiguity, since any structures represented are most likely modified. Furthermore, since subsequent evolution may have introduced additional complementarity, it may also be inappropriate to employ base pair maximization. Rather, the present software simply detects base pairing when it is present without reference to energy considerations or "gap penalties" and the like.

The encoding sequence of the gene of one protein is compared to the encoding sequence of the second protein. Any unbroken regions of standard base pairing between them (consisting of four or more nucleotides) are detected and their positions recorded. These segments can be regarded as the "building blocks" of potential complementary structures. Potential structures are then built by combining side-by-side building blocks only if their distance apart in the two sequences is less than a predetermined gap of unpaired nucleotides or "bulges". The gap used in these studies was one to ten nucleotides (maximum) in either sequence. This allows for unequal lengths of mismatching to be present between the building blocks. Also important in this analysis is the role played by the process of human cognition, and the subsequent scientific evaluation of potential structures. Since all possible structures are presented for evaluation, we hope to gain a feeling at this early stage for the quality and extent of complementarity that might be appropriate to this set of binding proteins.

The sequences gelsolin segment 1/actin and GAPDH/actin were separately analyzed in this fashion. A "sieve" program was employed to select potential complementary structures which corresponded to the main sequences at the crystallographic sites of gelsolin segment 1 and actin. A significant complementary structure was found, as shown in Table 1. The amino acid sequences encoded by these complementary regions of RNA (see Table 1) encompass several of the key residues of the actin/gelsolin segment 1 interaction site as previously identified in the crystallographic studies – a result which attests to the potential applicability of the MRT in the identification of actin/actin-binding protein interaction sites. The sieve program was then used to select potential structures in the GAPDH/actin data, where the actin sequence was common, as explained below.

Identification of a Potential Actin/GAPDH Interaction Site

Given the likelihood of molecular mimicry in the actin-binding sites amongst actin-binding proteins (see Tellam et al. 1989; Lappalainen et al. 1998), we scanned the GAPDH encoding RNA for complementarity to the actin RNA sequence encoding the gelsolin segment 1 binding site. As shown in Table 1 a significant complementary RNA structure was identified which encodes amino acids 252–268 in GAPDH. The GAPDH complementary actin nucleotide sequence overlaps the alignment of actin nucleotides already established for actin/gelsolin although it is considerably shorter (Table 1). This suggests that the nucleotide sequences of the gelsolin and GAPDH genes in these regions were in the past both highly complementary to the actin sequence and homologous to each other. Suppose, for example, that the binding site of actin/gelsolin (segment 1) was originally encoded by completely complementary sequences. The encoded protein sequences were weakly and specifically attracted as the experimental results suggest for complementary or antisense peptides. Over the period of biological time the protein sequences developed by mutation into specific binding sites of greater affinity, and employing more conventional amino acid interactions. An example in the present case is the intercalation of hydrophobic Ile 103 of gelsolin, by the hydrophobic side chains Ile 345′, Leu 346′ and Leu 349′ of actin in the observed binding site in the X-ray analysis. The effect on the once complementary nucleic acid sequences would be twofold: (1) some base pairs would disappear leading to a reduction

Table 1. A. Actin and gelsolin mRNA complementarity and encoded amino acid sequences

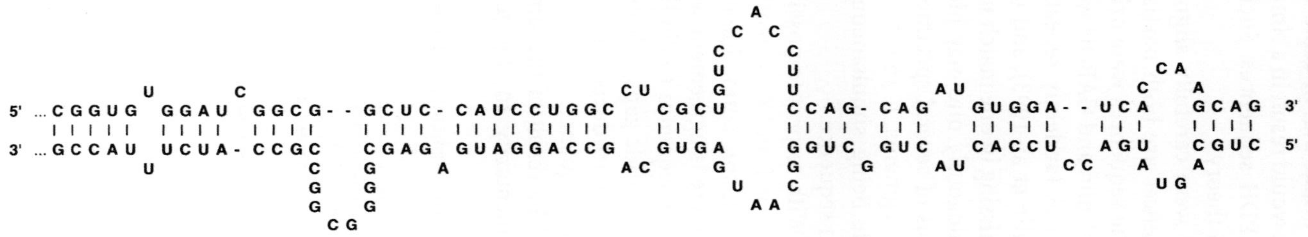

```
                                                         A
                                                       C   C
                                                      C     C
                                                      U     U                          C A          Actin
1019        U      C                       C U        G     U           A U           C   A          1086
   5' ...C G G U G  G G A U  G G C G - - G C U C - C A U C C U G G C  C G C U     C C A G - C A G  G U G G A - - U C A   G C A G  3'
         | | | | |  | | | |  | | | |    | | | |  | | | | | | | | |  | | |       | | | |  | | |  | | | | |     | | |   | | | |
   3' ...G C C A U  U C U A - C C G C  C G A G  G U A G G A C C G  G U G A     G G U C  G U C  C A C C U    A G U   C G U C  5'
316         U                C    G     A               A C        G     G        G      A U       C C   A   A            241
                             G    G                                 U   C                             U G           Gelsolin
                             G    G                                  A A
                              C G
```

								a	a			a												
Actin	S	V	W	I	G	G	S	I	L	A	S	L	S	T	F	Q	Q	M	W	I	T	K	Q		
																				b	b	b	b	b	b
Gelsolin	L	Q	Y	D	L	H	Y	W	L	G	N	E	C	S	Q	D	E	S	G	A	A	A	I	F	T

B. Actin and GAPDH mRNA complementarity and encoded amino acid sequences

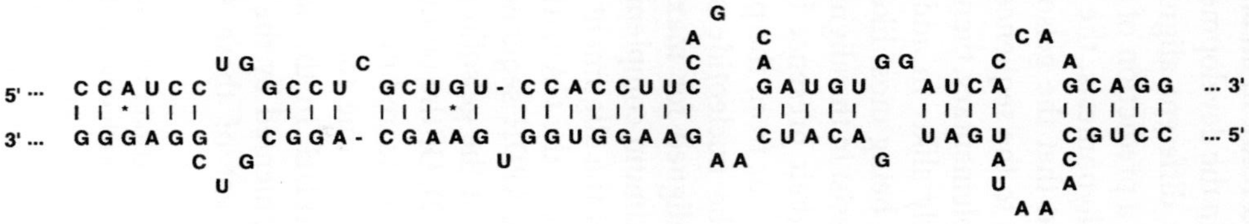

```
                                                        G
                                                      A    C                       C A         Actin
                  U G        C                        C    A           G G       C    A        1087
5' ...  C C A U C C     G C C U   G C U G U - C C A C C U U C   G A U G U    A U C A    G C A G G ... 3'
        | | * | | |     | | | |   | | | * |   | | | | | | | |   | | | | |    | | | |    | | | | |
3' ...  G G G A G G     C G G A - C G A A G   G G U G G A A G   C U A C A    U A G U    C G U C C ... 5'
                 C  G                    U                A A            G    A    C          754
                  U                                                           U    A          GAPDH
                                                                                A A
```

		a	a			a											
Actin	S	I	L	A	S	L	S	T	F	Q	Q	M	W	I	T	K	Q
GAPDH	P	A	K	Y	D	D	I	K	K	V	V	K	Q	A	S	E	G

[a] Indicates actin residues that pack ile 103 of gelsolin at actin/gelsolin interaction site.

[b] Indicates residues of the actin-binding helix of gelsolin.

of the overall complementary and (2) the appearance of gaps by insertion and deletion of nucleotides would further obscure the original complementarity. The effect would result in an alignment structure not unlike that of Table 1. A similar history in the development of the GAPDH gene would result in a similar but most likely different alignment of the actin/GAPDH sequences. Such a development is a prediction of molecular recognition theory.

To test this hypothesis, the nucleotide sequences were carefully aligned, bearing in mind that the gelsolin and GAPDH sequences are both partially complementary to the sequence of actin and that if the sequences were originally fully complementary, then the sequences from gelsolin and GAPDH were originally exactly alike. In addition, the gelsolin gene is tandemly repeated, with segment 4 being most like segment 1 (McLaughlin et al. 1993), and this fact is used to assist in the alignment process. Pieces missing in one match may be inferred by their presence in another match. Analogously one may piece together a dinosaur from the partial fossilized remains of several specimens. Consequently the nucleotide sequences of gelsolin segments 1 and 4 and GAPDH were aligned for maximum homology while being simultaneously aligned for maximum complementarity with the actin sequence.

The results of the alignment are shown in Table 2. With respect to possible homology between the GAPDH and the gelsolin segment 1 and 4 nucleotide sequences, the GAPDH sequence is scored "^" if there is nucleotide identity with either one of the gelsolin sequences or as "*" if there is agreement with both (see Table 2). Of the 60 nucleotides of the GAPDH sequence (including gaps) 33 (55%) are scored in this manner. Four one-nucleotide gaps, one two-nucleotide gap and one four-nucleotide gap have been introduced into the sequence of GAPDH. In the same region the gelsolin sequences have been slightly less fragmented by the introduction of gaps. The result of the simultaneous alignment of these sequences for complementarity to the actin

Table 2. Alignment of gelsolin segments 1 and 4 and GAPDH actin complementary regions for maximum homology to each other and complementarity to actin

Gelsolin Seg 1	3'	GCCAUUUCUA- CCGCCGGCGGGGCGAGAGUAGG- ACCGACGUGA- GUAACG- - - - GGUCGGUCAUCACCUCCAGUAUGA- - CGUC
Gelsolin Seg 4	3'	GUCAGUCCUA- CCGUCUACGUGGCUGGAGUAGG- ACCCAUCUGA- CCCGUGGGACGGUCAAUAUCUAAUUAGACGGGGA- - CC
		• • ••• •^ •• ^^ ^^•^ ^ • ^ ^^ ^• • •^^^^ • •^^^
GAPDH	3'	G- GGAGGCUGCGGA- CGAAGUGGUGGAA- G- - - - AACUACAG- U- - AGUAUAAACCGUCC
ACTIN	5'	CGGUGUGGAUCGGCG- - - - - - - - GCUC- CAUCC- UGGCCUCGCU- GUCCACCUU- CCAG- CAGAUGUGGA- - UCACCAA- - GCAGG
Complementarity with		
Gelsolin Seg 1		|||| ||| |||| |||| ||||| ||||	||	|||| ||| ||||| ||| ||||
Gelsolin Seg 4			|| |||| |||		||||| ||| ||	||||	||||					||	
GAPDH				||| |||| ||| |||||||	|||||	||| |||||
Summation of Complementarity			||||| ||||||||||	||||||| |||| |||||||||| ||||| |||||

• marks nucleotide identy between GAPDH and both gelsolin segments.

^ marks identity with either.

| marks complementarity between the gelsolin segments 1 and 4 and GAPDH sequences and the actin sequence.

nucleotide sequence is shown in the bottom part of Table 2. As expected, some base pairing is evident in the actin/segment 4 alignment which is not present in the actin/segment 1 alignment. The bottom line of Table 2 represents a logical summation of the complementary base pairing with actin in the region where all three sequences (gelsolin segments 1 and 4 and GAPDH) overlap. Of the 60 nucleotides (including gaps) 48 are scored as base pairs in the summation (80%). The result of this summation process, although not strictly an alignment, is consistent with the hypothesis presented above to explain the present appearance of the alignment of the sequences, and suggests that the nucleotide sequences of the gelsolin and GAPDH genes in these regions were in the past both more highly complementary to the actin sequence than is presently observed. It is quite possible that the degree of complementarity was high, possibly approaching that of truly complementary sequences. The alignment here is made possible by the rather fortunate circumstance of having more than one set of sequences available, with which to make the alignment. It is apparent also that the nucleotide sequences encoding GAPDH and gelsolin in these regions share a degree of homology. It is a point of the argument that they were probably once identical. It is thought likely that the demonstrated complementarity of these regions to actin at actin/gelsolin segment 1 binding site marks a possible binding site for actin in GAPDH.

Analysis of Potential Actin-Binding Site on GAPDH

The actin complementary RNA structure in GAPDH encodes amino acids 252–268. In order to test if segment 252–268 of the GAPDH was involved in the interaction of this enzyme with F-Actin, a synthetic peptide PAKYDDIKKVVKQASE corresponding to GAPDH 252–267, was prepared and examined for its ability (1) to bind to F-*a*ctin and (2) to compete with GAPDH for binding to F-*a*ctin. Peptide $GAPDH_{252-267}$ bound to F-*a*ctin in a saturable fashion with a stoichiometry of approximately one peptide per actin monomer. As shown in Fig. 1, peptide $GAPDH_{252-267}$ was also a competitive inhibitor of GAPDH binding to F-*a*ctin but did not affect the binding of another glycolytic enzyme, pyruvate kinase (PK).

This actin-binding $GAPDH_{252-267}$ peptide contains a sequence motif centred around DDIKKV which is highly homologous (see Table 3) to similar sequence motifs found in a variety of other actin-binding proteins including cofilin, destrin, ADF, actophorin, tropomyosin, glycogenin and glucokinase. In all these cases there is good experimental evidence that the motif is at or near to an actin interaction site on these proteins (Baque et al. 1997; Van Troys et al. 1998). This sequence homology with the cofilin-actin interaction site motif is interesting.

Recently, Van Troys et al. (1998) have shown that the actin-binding helix (residues 112–128) of cofilin which contains the DAIKKKL motif has structural

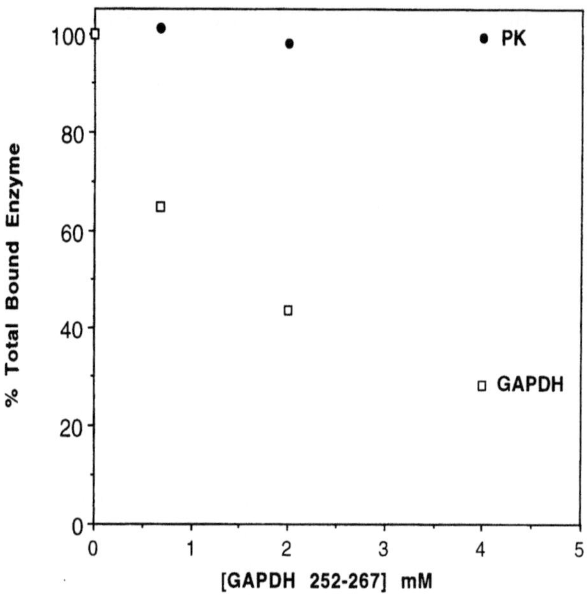

Fig. 1. Inhibition of enzyme binding to F-actin by GAPDH peptide 252–267. GAPDH and pyruvate kinase (PK) at 2 uM were separately incubated with 25 μM F-actin in 10 mM imidazole, 40 mM KCl, 1 mM $MgCl_2$, 0.5 mM DTT, pH 6.8 in the presence of increasing concentrations of GAPDH peptide 252–267. Following centrifugation to sediment the actin, the amount of each enzyme bound was determined by enzyme assay of both pellet and supernatant fractions

Table 3. Homology of potential actin-binding region of GAPDH with sequence motifs at or near the actin-binding sites of other actin-binding proteins

GAPDH human	SVVDLTCRLEKPAKYDDIKKVVKQASE	267
GAPDH rat	SVVDLTCRLEKPAKYDDIKKVVKQAAE	267
GAPDH chicken	SVVDLTCRLEKPAKYDDIKRVVKAAAD	267
GAPDH *Drosophila*	SVVDLTVRLGKGATYDEIKAKVESASK	267
GAPDH nematode	SVVDLTYRLEKPASMDDIKKVVKAAAD	267
Cofilin human	ESAPLKSKMIYASSKDAIKKKLTGIKH	133
Destrin human	ELAPLKSKMIYASSKDAIKKKFQGIKH	133
Actophorin	DSAPIKSKMMYTSTKDSIKKKLVGIQV	113
Glucokinase human	VVGLLRDAIKRRGDFEMD	198
Glycogenin human	MGADSFDNIKKKLDTYLQ	332
TM	MDAIKKKMQMLKL	8

and functional homology to the putative actin-binding helix of gelsolin segment 2, and both have structural homology to the actin-binding helix of gelsolin segment 1. While the actin interaction face of the latter helix is characterized by acidic and neutral hydrophobic residues, the actin interaction faces of the gelsolin segment 2 and cofilin helices are characterized by conserved basic residues, suggesting that these basic residues are important for interaction with the actin target site. Based on the structural and functional homologies, Hatanaka et al. (1996) suggested that the cofilin and gelsolin families may have evolved from a common ancestral actin-binding protein. The segment 252–268 of GAPDH corresponds to a surface exposed helix on the catalytic domain of the GAPDH subunit, a helix with structural homology to the known actin-binding helices of cofilin and gelsolin. This and the other observations recorded above suggest that this GAPDH segment and the actin binding regions of the gelsolin and the cofilin families may have shared a common evolutionary past.

Conclusions

These studies attest to the usefulness of the *m*olecular *r*ecognition *t*heory in predicting, and so assisting in, the identification of interaction sites between glycolytic enzymes and cytoskeletal proteins. The use of the *m*olecular *r*ecognition *t*heory is a radical new approach to determine the binding sites of protein pairs by examination of their encoding genetic sequences. Nonetheless, the theory has recently been validated for the binding of certain peptides to their receptors (Ruiz-Opazo et al. 1995). The theory proposes that the binding sites of proteins have evolved from peptides originally specified by complementary nucleotide sequences. The location of vestigial complementarity between the genes of cognate binding proteins may indicate the location of the interaction sites on the proteins. If the binding sites identified in this inve stigation are further substantiated by subsequent experimentation this will provide additional support for the *m*olecular *r*ecognition *t*heory. Further interactions between glycolytic enzymes and actin may be delineated, as well as possible interactions between the enzymes themselves. Elucidation of the binding sites will lead to the application of molecular techniques to further probe the significance of these interactions both in vitro and in vivo.

References

Baque S, Guinovart JJ, Ferrer JC (1997) Glycogenin, the primer of glycogen synthesis, binds to actin. FEBS Lett 417:355–359

Baranyi L, Campbell W, Ohshima K, Fujimoto S, Boros M, Okada H (1995) The anti-sense homology box : a new motif within proteins that encodes biologically active peptides. Nat Med 1:894–901

Blalock JE (1995) Genetic origins of protein shape and interaction rules. Nat Med 1:876–878

Blalock JE, Smith EM (1984) Hydropathic anti-complementarity of amino acids based on the genetic code. Biochem Biophys Res Commun 121:203–207

Clarke F, Stephan P, Morton D, Weidemann J (1985) Glycolytic enzyme organisation via the cytoskeleton and its role in metabolic regulation In: Beitner R (ed) Regulation of carbohydrate metabolism, vol II, CRC Press, Boca Raton, pp 1–31

Hatanaka H, Ogura K, Moriyama K, Ichikawa S, Yahara T, Inagaki F (1996) Tertiary structure of destrin and structural similarity between two actin-regulating protein families. Cell 85:1047–1055

Lappalainen P, Kessels MM, Cope MJ, Drubin DG (1998) The ADF homology (ADF-H) domain: a highly exploited actin-binding module. Mol Biol Cell 9:1951–1959

McLaughlin PJ, Gooch JT, Mannherz HG, Woods AG (1993) Structure of gelsolin segment 1-actin complex and the mechanism of filament severing. Nature 364:685–692

Ruiz-Opazo N, Akimoto K, Herrera (1995) Identification of a novel dual angiotensin II/vasopressin receptor on the basis of molecular recognition theory. Nat Med 1:1074–1081

Sheedy R (1996) Functional duality of glycolytic enzymes. PhD Thesis, Griffith University, Nathan, Australia

Tellam R, Morton DJ, Clarke FM (1989) A common theme in the amino acid sequences of actin and many actin binding proteins. Trends Biochem 14:130–133

Van Troys M, Dewitte D, Verschelde J-L, Goethals M, Vanderkerckhove J, Ampe C (1998) Analogous F-actin binding by cofilin and gelsolin segment 2 substantiates their structural relationship. J Biol Chem 272:32750–32758

Regulation of the Cytoskeleton Assembly: a Role for a Ternary Complex of Actin with Two Actin-Binding Proteins

Murat Kekic[1], Neil J. Nosworthy[1], Irina Dedova[1], Charles A. Collyer[2], and Cristobal G. dos Remedios[1]

Introduction

The actin cytoskeleton is an essential component of all eukaryotic cells. Besides its role in cell motility, it has a number of other functions, including cytokinesis, signal transduction, and the maintenance of cell shape. An essential property of the cytoskeleton is its ability to rapidly assemble and disassemble monomers of actin into F-actin filaments and this process is now known to be regulated by a number of actin-binding proteins (ABPs) of which cofilin appears to be the most widely distributed in nature. In this chapter we pose the question: does cofilin act alone in controlling actin filament assembly or is the binding of cofilin to actin modulated by other ABPs?

The Role and Distribution of Actin-Binding Proteins

Regulation of actin microfilament assembly and disassembly is usually considered in the context a 1 : 1 complex of actin and a single actin-binding protein (ABP). In eukaryotic cells, the pool of actin monomers is believed to be maintained by sequestration by one of a number of possible ABPs, of which there are many (see McGough et al. and De la Cruz, both this Vol.).

A recent analysis suggests that these ABPs fall into a small number (six or seven) of categories. These include: (1) motor proteins (like muscle and nonmuscle myosins); (2) actin filament-stabilizing proteins; (3) filament-destabilizing proteins (like cofilin); (4) filament cross-linking proteins; (5) proteins that bind to the ends of actin filaments; (6) proteins that anchor, cap or sever actin filaments; (7) regulatory proteins; and finally (8) proteins that maintain the pool of actin monomers (like DNase I). It is probable that other categories exist but have yet to be characterized. This concept is discussed in more detail in another chapter in this volume by Janmey et al. Each prokaryotic cell seems to contain at least one protein from each of the above categories and so, in principle at least, it is likely that cooperative interactions can occur between ABPs.

[1] Muscle Research Unit, Institute for Biomedical Research F13, Department of Anatomy and Histology, The University of Sydney, Sydney 2006, Australia

[2] Department of Biochemistry G08, The University of Sydney, Sydney 2006, Australia

Results and Problems in Cell Differentiation, Vol. 32
C. dos Remedios (Ed.): Molecular Interactions of Actin

Atomic Structure of Actin

The propensity of actin to self assemble into filaments rather than into a defined oligomeric form probably explains why actin has never been crystallized without an ABP. In 1977, a short report (Oriol et al. 1977) was published describing the formation of crystals of actin in the absence of any other proteins. However, the crystals were small and apparently not capable of growth to the minimum size suitable for X-ray diffraction at that time. Since then, no one has succeeded in growing crystals of actin in the absence of an ABP. It is not a matter of idle curiosity that we should seek the structure of actin in an uncomplexed state, since we and others have reported that actin undergoes a conformational change when it binds actin-binding proteins.

Atomic Structure of Actin and Actin-Binding Proteins

The precise binding site on actin is known for only three ABPs. The first actin-ABP structure was solved by Kabsch et al. in 1990 using cocrystals of actin with bovine pancreatic DNase I (Kabsch et al. 1990). Three years later, an actin structure was solved using gelsolin segment 1 as the ligand (McLaughlin et al. 1993), and in the same year a third structure was solved with bovine spleen profilin (Schutt et al. 1993). A second structure of the actin-profilin complex was subsequently solved and the two are known as the open and closed forms.

The conclusion most commentators draw from the four known crystallographic structures is that the secondary structure of actin varies very little between complexes with various ABPs. Some have concluded that there is some room for conformational changes, particularly for shear movement between the major domains (Page et al. 1998), but most seem to ignore one important point, namely that the structure of actin alone (uncomplexed to an ABP) remains unknown. This fact leaves open the possibility that there *may* be small but significant structural changes between monomeric actin and actin complexed to an ABP. It is for this principal reason that we began looking for such evidence.

Cofilin is our ABP of choice because, not only is it apparently universally expressed by virtually all eukaryotes, but it also appears to play a central role in regulating the actin in cytoskeleton. Cofilin appears to bind to the barbed end of monomers (i.e., to actin subdomains 1 and 3) but this location is subtly different to the location of either gelsolin segment-1 (McLaughlin; also see Burtnick, this Vol.) or profilin.

Where Does Cofilin Bind to Actin?

Cofilin belongs to a family of proteins that includes actin-depolymerizing factor (ADF), destrin, actophorin and depactin. These proteins are small (about

20 kDa) and their binary complex with actin has been reasonably well characterized. The ADF family is capable of either assembling or disassembling actin depending on the pH (Moon and Drubin 1995). There are subtle differences between the members of this family and these are discussed in more detail by McGough et al., this Vol. For example, cofilin binds to the sides of F-actin whereas destrin is a pH-independent actin-depolymerizing protein.

In mammals, cofilin is expressed as two closely related muscle and non-muscle isoforms whose amino acids sequences are strongly conserved. It is probable that cofilin is involved in the regulation of assembly of actin in both muscle and nonmuscle cells (Ono et al. 1994). The carboxyl terminus of cofilin exhibits a sequence homology with gelsolin and profilin and the latter competes with actophorin for binding to actin monomers.

Cofilin plays a vital role in depolymerizing actin filaments (Lappalainen and Drubin 1997). Recent evidence suggests that the activity of cofilin is directly regulated via phosphorylation at Ser 3. (Moriyama et al. 1996). Phosphorylation inactivates cofilin and local concentrations may be mediated by cofilin kinases and phosphatases. Arber et al. (1998) demonstrated that phosphorylation of cofilin by LIM-kinase 1 inhibited actin binding and resulted in a buildup of actin filaments. This interaction has been described by the molecular modeling of actin in complex with yeast cofilin (Wriggers et al. 1998).

The predicted structure of cofilin shows a strong interaction between the N-terminus of actin and cofilin. A salt bridge can form between Asp 1 on actin and Arg 3 on yeast cofilin that is blocked by phosphorylation at Ser 4 resulting in the inhibition of actin depolymerizing activity.

While there are no reported crystal structures of the cofilin-actin complex, chemical cross-linking studies have shown that cofilin binds to both the amino terminus (residues 1–12) and the carboxyl terminus (residues 357–375) of actin. Depactin is closely related to cofilin and can be cross-linked to actin residues 361, 363 and 364, all of which are located in subdomain 1.

Where Does DNase I Bind to Actin?

The location of DNase I on actin is known to a resolution of 2.8 Å from the cocrystal structure (Kabsch et al. 1990), although the structure of DNase I itself had already been solved at higher resolution. DNase I binds to the so-called top of the actin monomer (subdomains 2 and 4) and the contact sites include residues 39, 41, 43, 45, 61 and 63 of subdomain 2 and residues 203, and 207 in subdomain 4. A mutagenesis study of actin revealed that substitutions near its C-terminus (Asp363His and Glu364Lys) decreased the affinity of actin for DNase I, although neither substitution was near the DNase I-binding site (Drummond et al. 1992). Furthermore, these authors suggested that long-range (presumably allosteric) effects might form a significant part of the normal function of actin.

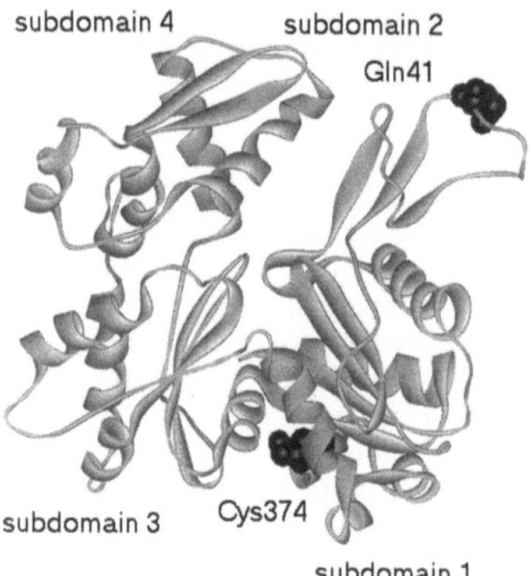

Fig. 1. The structure of the actin monomer-DNase I complex where the DNase I has been subtracted. Note that the loop at the top of subdomain 2 (containing glutamine-41) is extended to a position where it made contact with DNase I (not shown here). Cysteine 374 (near the C terminus) is located on the back surface of the lower portion of subdomain 1

DNase I was initially thought to be a key cytoskeletal protein (Hitchock et al. 1976), and there were hopes that it would prove to be widely distributed in nature. The DNA-cleaving activity of DNase I was known to be strongly inhibited by a DNase I inhibitor that subsequently turned out to be monomeric actin (Lazarides and Lindberg 1974). DNase I belongs to a family of DNases, not all of which are inhibited by monomeric actin. The relationship between the inhibition of DNase activity and its inhibition by actin (and the actin-related proteins) suggested that it may play an important biological function. However, today the extent of its cellular distribution is still not known (U. Lindberg, pers. comm.). These two proteins form a tight complex (K_d is approximately 10^{-9} M, Mannherz et al. 1980; Carter et al. 1997) and although the affinity of DNase I for F-actin is much lower (ca. 10^{-4} M) it can depolymerize F-actin in vitro (unpub. data).

Ternary Complexes of DNase I and Actin

Actin is capable of forming a ternary complex with DNase I and a small (5-kDa) ABP, thymosin β4 (Discher et al. 1995). This protein/peptide is a major actin-sequestering factor in eukaryotic cells. The K_d of thymosin β4 for the actin-DNase I complex was not significantly different from actin alone, and the binding of thymosin β4 to actin does not appear to alter its ability to inhibit DNase I activity (Discher et al. 1995). These findings suggest that the binding of DNase I and thymosin β4 to actin are essentially independent. However, the recent reported competition between thymosin β4 and gelsolin (including its N-terminal segment 1) suggests that these two proteins bind to similar, perhaps even identical, sites on actin (Ballweber et al. 1997). DNase I dissociates both the actin-thymosin β4 complex and the actin-gelsolin complex, indicating both negatively cooperative binding and an allosteric effect. This result is interesting because it explained how actin monomers sequestered by thymosin β4 could be released into the pool of assembly-competent actin by gelsolin.

Our initial suspicions that there may be a ternary complex of DNase I-actin-cofilin arose from a two-dimensional PAGE study of actin bound to an affinity matrix (Mini-Leak by Kem-en-Tek) (Coumans and dos Remedios 1998). In this study we compared 2D PAGE gels of the proteins captured by an actin affinity Mini-Leak column with those isolated by chromatography using a DNase I-actin affinity column. We observed an increase in the density of a protein spot tentatively identified as cofilin. We surmised that a probable explanation was that the affinity of cofilin for actin was increased by the presence of DNase I.

Other Pointed-End-Binding Proteins

The treadmilling of actin filaments involves addition of monomers to the "barbed" ends and their removal from the pointed ends. However, few proteins are known to bind to the pointed end of actin filaments. Tropomodulin binds with equimolar stoichiometry and blocks monomer addition and loss at the pointed end. Weber et al. (1994) showed that tropomodulin acts in unison with tropomyosin to cap the ends of actin filaments. However, tropomodulin binds only weakly to actin monomers.

Spectrin also binds to the so-called pointed end of F-actin and, like tropomodulin, it may play a role in controlling filament length. The widespread distribution of spectrin in nature has increased the interest in the role of this pointed end-binding protein (see Schafer and Cooper 1995).

What Do We Know About Ternary Complexes of Actin, Cofilin and Other ABPs?

One of the first suggestions that ABPs can bind to the complex of actin with DNase was made in the original paper by Lazarides and Lindberg (1974). These authors identified actin as the naturally ocurring but previously unidentified inhibitor of DNase I. Passing a cell extract over a DNase I affinity column, they found there were minor components contaminating the actin eluted from the DNase-agarose column. They speculated that they may have some relationship to the DNase I-actin system. Carlsson et al. (1977) hypothesized that since profilin could not depolymerize F-actin but could bind to G-actin, a depolymerizing factor must exist in cells. They suggested it could be DNase I.

Bretscher and Weber (1980) purified villin using a DNase I affinity column, thus taking advantage of its ability to bind tightly to actin. They passed a cell extract over a DNase I affinity column and found that the villin-actin complex remained bound to the DNase I column. This complex was resistant to high ionic strength, but villin was released from the column when free Ca^{2+} was removed by EGTA. Thus, while Ca^{2+} is needed for the villin-actin complex, actin was not removed from the column unless 3 M GuHCl was added (Lazarides and Lindberg 1974). This observation suggested that the ternary complex of villin, actin and DNase I was very stable.

Since then, several other ABPs have been purified using a DNase I affinity column. Nishida et al. (1981) isolated an 88-kDa protein that could be eluted from the DNAse I column by an EGTA wash. Unlike villin, this protein eluted with actin bound. A ternary gelsolin-actin-DNase I complex was formed on a DNase I affinity matrix (Markey et al. 1982) and could be eluted by EGTA as a gelsolin-actin complex.

Adducin is a filament-bundling protein that binds to spectrin-actin forming a ternary complex. This association is thought to occur through an initial binary complex of spectrin with actin followed by the association of adducin. The binding of adducin is thought to recruit a second spectrin molecule (Gardner and Bennett 1987).

Maekawa et al. (1984) purified profilin using a DNase I affinity column as well as several ABPs with MWs of 26, 21 and 19 kDa. These proteins were eluted with high salt while low salt and 2 M urea was needed to elute profilin. The 21-kDa protein was later identified as cofilin (Nishida et al. 1984). These findings suggested that cofilin bound to DNase I-agarose through actin (Muneyuki et al. 1985). Daoud et al. (1988) showed that actin cross-linked to chicken ADF could still bind to a DNase I column. This demonstrated that the ADF-binding site did not appear to overlap with the DNase I site.

DNase I affinity chromtography is now routinely employed in the preparation of a number of cytoskeletal proteins including: actolinkin, adseverin, destrin (a member of the the ADF family), radixin and villin (Isenberg 1995). Many, if not all, of these preparations probably involve ternary DNase I-actin-ABP complexes.

Preparation of Actin and Cofilin

Actin was prepared using the method of Spudich and Watt (1971) with minor modifications (Barden and dos Remedios 1984). The concentration of actin monomers was determined spectrophotometrically using an extinction coefficient (0.1%, 290 nm) of $0.63\,cm^{-1}$ (Lehrer and Kerwar 1972). Cofilin was prepared as a recombinant protein using a cDNA sequence derived from chick embryo and kindly supplied by Takashi Obinata. The protein was overexpressed in *E. coli* using a pGEX vector and the GST-fusion protein was applied to a glutathione-Sepharose affinity column. Following cleavage of cofilin from GST using thrombin, the product was further purified on a Mono-S ion cation-exchange column in 10 mM PIPES, pH 6.8 and eluted with a (0–0.5 M) NaCl gradient. The resulting cofilin was obtained with a purity of >98% and its concentration was determined using an extinction coefficient (0.1%, 280 nm) of 0.93 cm^{-1}. We determined the pI of the cofilin to be 8 using Pharmacia isoelectric focusing gels and determined a molecular mass value of 18,808 Da for cofilin using electrospray mass spectrometry (data not shown). DNase I was purchased from Worthington Biochemicals as the DPRF grade and its concentration was determined using an extinction coefficient (0.1%, 280 nm) of 1.10 cm^{-1}.

Native Gel Electrophoresis

Native polyacrylamide gels were prepared according to the method of Laemmli (1970). A resolving gel of 10% acrylamide was combined with a discontinuous 5% stacking gel. Native gel electrophoresis is well established as a method for observing the interaction between monomeric actin and DNase I (Edgar 1989). As a consequence, proteins do not migrate according to their relative molecular mass but move in the field according to their charge density. Since most proteins are negatively charged, they migrate towards the cathode of the apparatus, but proteins with a net positive charge or having an isoelectric point near the pH of the running buffer may not enter the resolving gel. In isocratic gels, the mobility of proteins (and their complexes) is determined by the number of negative charges they carry as well as by their relative molecular mass (i.e., charge density). In native gels, cofilin remains in the stacking gel because it is positively charged under the pH conditions of the stacking gel.

Native gel electrophoresis is not generally used as an analytical tool because the effective resolution of protein bands in native gels is usually lower than that of SDS-PAGE gels. Native gel separations are usually run at higher than room temperature. We found that running a gel cooled in an ice jacket significantly improved the resolution of actin with cofilin and DNase I. Accordingly, we performed binding assays of ABPs with actin at or near 0 °C. The method used in the present study is based on one kindly supplied by Dr Danuta Szczesna. Details of these procedures are published elsewhere (Kekic and dos Remedios 1999).

Titration of Cofilin into the Binary Actin-DNase I Complex

To our knowledge, the first study of native PAG electrophoresis of actin and DNase I was reported by Edgar (1989). Figure 2 illustrates a native PAGE gel (for details of gel and protein loading see figure legend) run at 0 °C. Note the

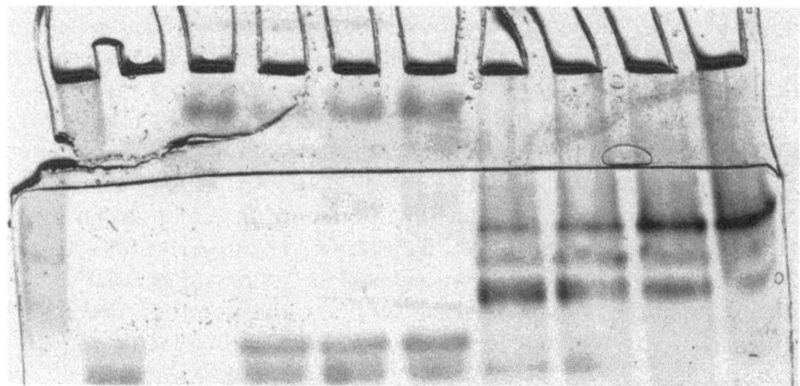

Fig. 2. A native PAGE gel containing a 4.8% stacking gel and 10% running gel. The gel was run at approximately 4°C and stained with Coomassie Brilliant G. The lanes contain: *1* G-actin; *2* DNase I; *3* cofilin; *4–6* constant DNase I with increasing cofilin; *7–10* constant actin-DNase I with increasing cofilin. The *horizontal line* marks the separation between the stacking and running gels

actin monomer band and the trailing dimers, trimers (identified by us and other as the "nucleus" for actin polymerization, Barden et al. 1982), and higher-order oligomers. We conclude that the buffer conditions in the running gel contain insufficient ionic strength (Tris is a zwitterionic buffer) to prevent the formation of actin filaments.

Lane 2 contains Worthington DPRF grade DNase I that runs essentially as a single, somewhat diffuse band. Lane 3 contains purified cofilin. Note that this protein remains in the stacking gel rather than entering the running gel, confirming that the net charge of this protein is positive. Lanes 4–6 contain DNase I mixed with cofilin in a molar ratio up to 1 : 1 (lane 6). Note that these two proteins do not appear to form a complex in the absence of actin. In lanes 7–10, 1 : 1 actin and DNase I are present and increasing cofilin was added in the ratios 0.2, 0.4, 1.0 and 2.0 cofilin/actin. The slowest migrating band is the ternary complex, that increases in proportion to the amount of added cofilin. Below this is a band whose density does not change. The next-fastest band corresponds to the actin : DNase I complex and is progressively reduced as the molar ratio of cofilin increases.

Titration of Increasing DNase I into the 1 : 1 Actin-Cofilin Complex

An equivalent result was obtained using a preformed actin-cofilin complex to which DNase I was progressively added. Figure 3 illustrates a native gel containing: actin only (lane 1); cofilin only (lane 2); DNase I only (lane 3);

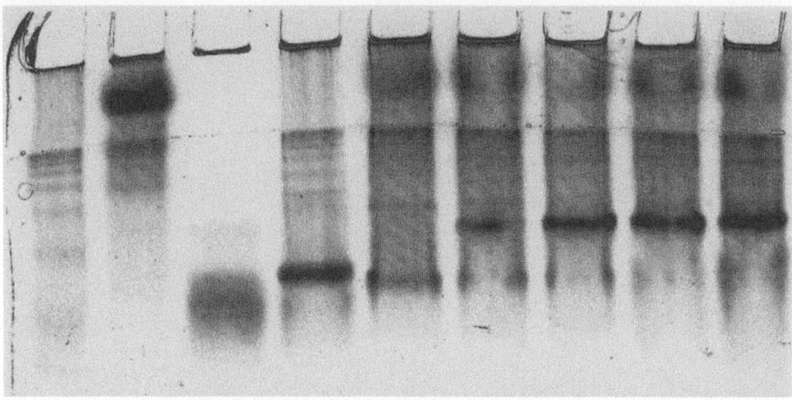

Fig. 3. A native PAGE gel containing a 4.8% acrylamide stacking gel and 10% running gel. The gel was run at approximately 4°C and stained with Coomassie Brilliant G. The lanes contain: *1* G-actin (10 μM); *2* cofilin (10 μM); *3* DNase I (10 μM); *4* actin-DNase I (10 μM); *5–9* constant actin-cofilin with increasing (2, 4, 10 and 20 μM) DNase I

preformed 1 : 1 actin : DNase I complex (lane 4); 1 : 1 actin and cofilin are present in lane 5 and then increasing molar ratios of DNase I are shown in lanes 6–9. In these latter lanes the addition of DNase I increases the intensity of the ternary complex band while at the same time the intensity of the actin : cofilin band progressively decreases. These results suggest that the stability of the cofilin : actin : DNase I ternary complex is greater than either binary complex.

Identification of the Proteins in the Ternary Complex

The approximate M_r of proteins can be determined by migration of the proteins in gradient native gels until migration essentially stops. Such gradient gels are commercially available from Pharmacia marketed as Phast native 8–25% gradient gels. The proteins were subjected to prolonged electrophoresis (400 volt h) until the protein migration was effectively arrested. Under these conditions, proteins in the range 50–200 kDa display a linear mobility relative to the log of their MWs.

Figure 4 shows a typical gel in which four standard proteins (phycoerythrin 150 kDa, phosphorylase B 94 kDa, bovine serum albumin 68 kDa, and ovalbumin 43 kDa) demonstrate there is a linear relationship ($r^2 = 0.997$) between log Mr and migration distance. Note that the actin-DNase I complex migrates almost precisely (actin 43 kDa + DNase I 31 kDa) at its predicted position of 74 kDa. More importantly, the titration of cofilin (20 kDa) into the actin-DNase I complex produces increasing amounts of a complex with an apparent MW of 94 Da. We therefore conclude that this band, that corresponds exactly with those in the linear native gels, is indeed the ternary complex of actin, DNase I and cofilin.

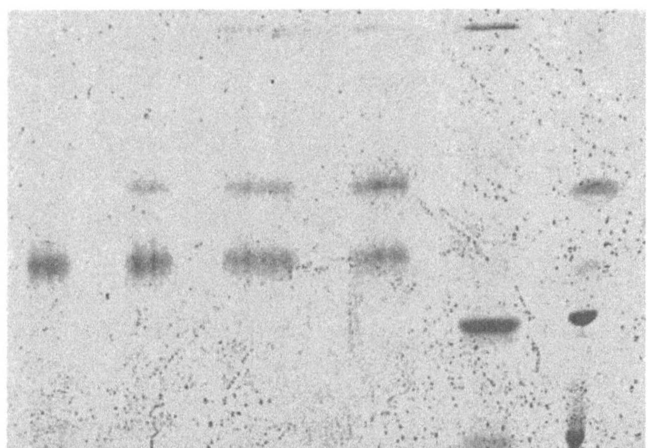

Fig. 4. A native Pharmacia PAGE Phast gel containing a gradient from 8–25% acrylamide. The gel was run for 400 volt h at 4°C. The lanes contain: *1* actin (10 μM) + DNase I (10 μM); *2–4* 10 μM actin-DNase I with increasing amounts (2, 4 and 6 μM) cofilin; *5* Gradipore native MW protein markers; *6* low MW standard proteins (Pharmacia LMW marker kit)

Positive Cooperativity in Ternary Complexes

The actin-spectrin-protein 4.1 interaction is known to form a very tight ternary complex ($K_a \sim 1 \times 10^{12} M^{-1}$). In contrast, the binary complex of spectrin dimers with F-actin forms a very weak association ($K_a \sim 5 \times 10^{3} M^{-1}$) and the interaction of protein 4.1 with actin is also weak. The association of protein 4.1 and spectrin is very stable ($K_a \sim 0.5\text{–}8.6 \times 10^{6} M^{-1}$) (Cohen and Foley 1984). More recent cross-linking evidence has suggested a cooperative binding of protein 4.1 to actin (Becker et al. 1990).

Adducin is a filament-bundling protein that binds to spectrin-actin to form a ternary complex. The association is thought to follow the following pathway: spectrin and actin bind, adducin then associates with the spectrin-actin complex; the binding of adducin may then recruit a second spectrin molecule to the ternary complex (Gardner and Bennett 1987).

Negative Cooperativity Between Thymosin β4 and Actin

Thymosins are small (5-kDa) intracellular actin sequestering proteins that stabilise the monomeric actin pool (see review by Safer and Nachmias 1994). A recent study has analyzed the binding of thymosin β-4 to actin complexed with DNase I, gelsolin or gelsolin segment 1 using either native gel electrophoresis or cross-linking (Ballweber et al. 1997). The results showed competition between (no ternary complex formation visualized with native gels or cross-linking) thymosin β-4 and gelsolin/gelsolin segment 1, indicating similar binding sites. A ternary complex was formed with DNase I/actin only in the presence of a chemical cross-linker. The dissociation of the actin-thymosin β-4 complex by DNase I suggests there is a negative cooperativity between these two ABPs, despite the fact that they bind to distinctly different sites on actin.

The binding of thymosin β-4 to actin does not appear to alter its ability to inhibit DNase I activity (Discher et al. 1995). These findings suggest that the binding of DNase I and thymosin β-4 to actin is essentially independent. This is understandable considering that DNase I binds to subdomains 2 and 4, whereas gelsolin and profilin bind to similar sites on subdomains 1 and 3.

Evidence for a Conformational Change in G-Actin

Two loci on actin, Gln-41 and Cys-374, can be specifically and uniquely labeled with fluorescent probes. These residues are located at (or very close to) the DNase I and cofilin-binding sites, respectively. We placed a fluorescent donor probe at Gln-41 (dansyl-cadaverine) and observed its spectral characteristics when cofilin was added in a 1.5 molar excess over actin. Two spectral changes suggested an allosteric conformational change occurs in actin. Firstly, the peak of the dansyl fluorescence emission underwent a blue shift of about 4–8 nm (N = 3). Secondly, the intensity of the emission peak increased (20–29%) significantly. These findings are illustrated in Fig. 5.

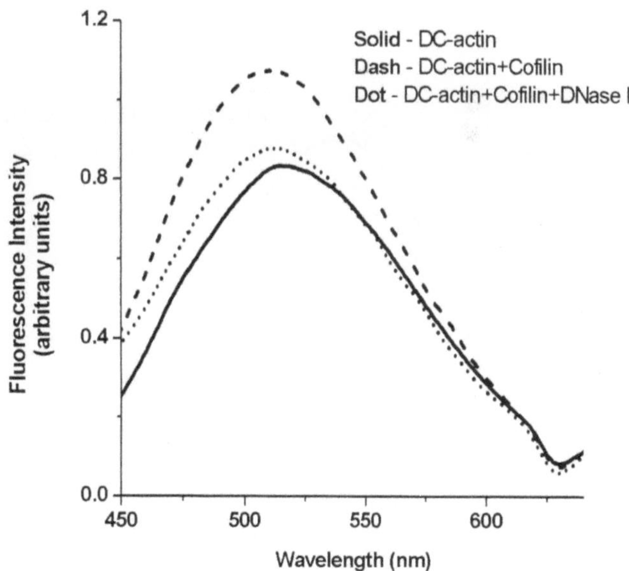

Fig. 5. Fluorescence emission spectra of DC-G-actin in the absence (*solid line*) and presence (*dashed line*) of cofilin. The *dotted line* represents the changes in DC emission due to DNase I binding to the actin-cofilin complex. Conditions are: G buffer (2 mM Tris pH 8, 0.2 mM $CaCl_2$, 0.2 mM ATP, 0.2 mM dithiothreitol). Actin concentration was 3 μM, cofilin and DNase were 4.5 μM. λ_{ex} was 330 nm

The significant increase in the emission intensity of the dansyl probe is also consistent with increased hydrophobicity of the environment of Gln-41. These changes are unlikely to be induced by a direct steric mechanism. Cofilin probably binds to subdomains 1 and/or 3, i.e., remote from the Gln-41 region of subdomain 2. It is known that cofilin competes with profilin and gelsolin segment 1, both of which have been cocrystallized with actin, as discussed earlier.

The reverse sequence of experiments was also done. We began with the spectrum of dansyl-labeled G-actin and added excess (1 : 1.5 molar ratio) of DNase I and then added 1 : 1.5 cofilin. These data, shown in Fig. 6, are consistent with the data obtained in Fig. 5. The emission characteristics of the dansyl probe at Gln-41 are hardly altered by the addition of DNase I, but its presence almost entirely negates the large spectral changes induced by cofilin binding. Addition of DNase I to the actin-cofilin complex resulted in a decrease in fluorescence intensity of the dansyl probe. This suggests that the DNase I shifts the probe back to a more hydrophilic environment.

These experiments strongly suggest that cofilin appears to induce a conformational change in the structure of the actin monomer, whereas DNase I binding alone does not induce an equivalent change. However, is there direct experimental evidence for cofilin inducing a conformational change in G-actin? FRET spectroscopy can provide an answer to this question. Using the dansyl label as a donor at Gln-41, we also labeled the actin monomers with the

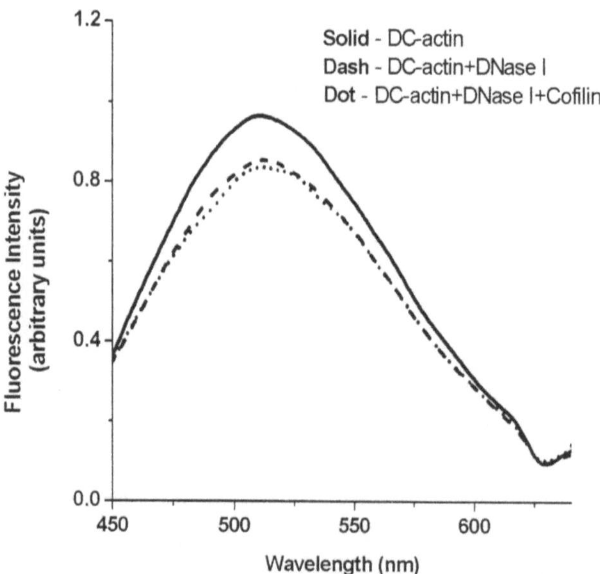

Fig. 6. Changes in fluorescence emission of DC-G-actin (*solid line*), actin-DNase I (*dashed line*) and actin-DNase I-cofilin (added in this sequence) (*dotted line*). Sample conditions are the same as in Fig. 5. λ_{ex} was 330 nm

non-fluorescent acceptor, N-(4-dimethylamino-3,5-dinitrophenyl-maleimide (DDPM). We confirmed that the FRET efficiency of this pair in the absence of an ABP gave a distance of 34.4 Å Moraczewska et al. (1996) found only a small distance decrease (~1 Å) when DNase I bound to G-actin. A similar distance (30 Å) was obtained using dansyl cadaverine and a different acceptor (5-[iodoacetamido] fluorescein), but these authors observed a larger (3 Å) decrease in the distance when DNase I bound. Addition of cofilin (in the absence of acceptor) caused a >20% increase in fluorescence intensity. However, in the presence of the DDPM acceptor we observed either no increase or a slight decrease in intensity. This observation is consistent with an increase in FRET efficiency, i.e., cofilin induces a decrease in the donor-acceptor distance.

Addition of DNase I to the actin-cofilin complex has the opposite effect. In other words, DNase I binding increases the distance between the acceptor located in subdomain 1 and the donor in subdomain 2. This is consistent with our earlier report on the effect of DNase I on G-actin (dos Remedios et al. 1994).

Conclusions

Cofilin is widely distributed in eukaryotic cells and plays a crucial role in regulating the assembly and disaasembly of the actin cytoskeleton. This chapter presents evidence that cofilin may not act alone in this regulatory behavior. Using a model system consisting of actin, DNase I and cofilin, we show that, although these two actin-binding proteins independently bind to parts of the actin monomer, they nevertheless act allosterically, increasing the binding of cofilin to actin and vice versa. The evidence is based on increased stability of the DNase-actin-cofilin complex observed in native polyacrylamide gel electrophoresis. Conformational changes in actin are detected based on changes in the spectral properties of fluorescent probes attached to glutamine-41 and cysteine-374 of actin as well as on fluorescence resonance energy transfer spectroscopy which senses distances between these two loci. We suggest that this in vitro cooperative behavior may be important in regulating the state of assembly of the actin cytoskeleton.

References

Arber SFA, Barbayannis FA, Hanser H, Schneider C, Stanyon CA, Bernard O, Caroni P (1998) Regulation of actin dynamics through phosphorylation of cofilin by lim-kinase. Nature 393:805–809

Ballweber E, Hannappel E, Huff T, Mannherz HG (1997) Mapping the binding site of thymosin β4 on actin by competition with G-actin binding proteins indicates negative cooperativity between binding sites located on opposite subdomains of actin. Biochem J 327: 787–793

Barden JA, dos Remedios CG (1984) The environment of the high affinity cation binding site on actin and the separation between cation and ATPase sites as revealed by proton NMR and fluorescence spectroscopy. J Biochem 96:913–921

Barden JA, Grant NJ, dos Remedios CG (1982) Identification of the nucleus of actin polymerisation. Biochem Int 5:685–692

Becker PS, Schwartz MA, Morrow JS, Lux SE (1990) Radiolable-transfer cross-linking demonstrates that protein 4.1 binds to the N-terminal region of beta-spectrin and to actin in binary interactions. Eur J Biochem 193:827–836

Bretscher A, Weber K (1980) Vilin is a major protein of the microvillus cytoskeleton which binds both G and F actin in a calcium-dependent manner. Cell 20:839–847

Carlsson L, Nyström L-E, Sundkvist I, Markey F, Lindberg U (1977) Actin polymerizability is influenced by profilin, a low molecular weight protein in non-muscle cells. J Mol Biol 115:465–483

Carter LK, Christopherson RI, dos Remedios CG (1997) Analysis of the binding of deoxyribonuclease I to G-actin by capillary electrophoresis. Electrophoresis 18:1054–1058

Cohen CM, Foley SF (1984) Biochemical characterization of complex formation by human erythrocyte spectrin, protein 4.1, and actin. Biochemistry 23:6091–6098

Coumans J, dos Remedios CG (1998) Actin-binding proteins in mouse C2 myoblasts and myotubes – a combination of affinity chromatography and two-dimensional gel electrophoresis. Electrophoresis 19:826–833

Discher DE, Winardi R, Schischmanoff PO, Parra M, Conboy JG, Mohandas N (1995) Mechanochemistry of protein 4.1's spectrin-actin-binding domain: ternary complex interactions, membrane binding, network integration, structural strengthening. J Cell Biol 130:897–907

dos Remedios CG, Kiessling PC, Hambly BD (1994) DNase I binding induces a conformational changes in the actin monomer. In: Synchrotron radiation in the biosciences. Oxford Science Publications, Oxford, 418–425 pp

Daoud EW, Hayden SM, Bamberg JR (1988) Inhibition of deoxyribonuclease I activity by actin covalently cross-linked to chick brain actin depolymerizing factor through exposed sulfhydryls. Biochem Biophys Res Commun 155:890–894

Drummond DR, Hennessey ES, Sparrow JC (1992) The binding of mutant actins to profilin, ATP and DNase I. Eur J Biochem 209:171–179

Edgar AJ (1989) Gel electrophoresis of native actin and the actin-deoxyribonuclease I complex. Electrophoresis 10:722–725

Gardner K, Bennett V (1987) Modulation of spectrin-actin assembly by erythrocyte adducin. Nature 328:359–362

Hitchcock SE, Carlsson L, Lindberg U (1976) Depolymerization of F-actin by deoxyribonuclease I. Cell 7:531–542

Isenberg G (1995) Cytoskeletal proteins. A purification manual. Springer, Berlin, Heidelberg, New York

Kabsch W, Mannherz HG, Suck D, Pai EF, Holmes KC (1990) Atomic structure of the actin-DNase I complex. Nature 347:37–44

Kekic M, dos Remedios CG (1999) Electrophoretic monitoring of pollutants: effect of cations and organic compounds on proteins interaction monitored by native gel electrophoresis. Electrophoresis 20:2053–2058

Laemmli UK (1970) Cleavage of structural proteins during the assembly of the head of bacteriophage T4. Nature 227:680–685

Lappalainen P, Drubin DG (1997) Cofilin promotes rapid actin filament turnover in vivo. Nature 388:78–82

Lazarides E, Lindberg U (1974) Actin is the naturally occurring inhibitor of deoxyribonuclease I. Proc Natl Acad Sci USA 71:4724–4746

Lehrer SS, Kerwar G (1972) Intrinsic fluorescence of actin. Biochemistry 11:1211–1217

MacLean-Fletcher S, Pollard TD (1980) Identification of a factor in conventional muscle actin preparations which inhibits actin filament self-association. Biochem Biophys Res Commun 96:18–27

Maekawa S, Nishida E, Ohta, Sakai H (1984) Isolation of low molecular weight actin-binding proteins from porcine brain. J Biochem 95:377–385

Mannherz HG, Goody RS, Konrad M, Nowak E (1980) The interaction of bovine pancreatic deoxyribonuclease I and skeletal muscle actin. Eur J Biochem 104:367–379

Markey F, Persson T, Lindberg U (1982) A 90,000-dalton actin-binding protein from platelets. Comparison with vilin and plasma brevin. Biochim Biophys Acta 709:122–133

McLaughlin PJ, Gooch JT, Mannherz HG, Weeds AG (1993) Structure of gelsolin segment-1-actin complex and the mechanism of filament severing. Nature 364:685–692

Moon A, Drubin DG (1995) The ADF/cofilin proteins: stimulus-responsive modulators of actin dynamics. Mol Biol Cell 6:1423–1431

Moraczewska J, Strzelecka-Golaszewska H, Moens P, dos Remedios CG (1996) Structural changes in subdomain 2 of G-actin observed by fluorescence spectroscopy. Biochem J 317:605–611

Moriyama K, Nishida E, Yonezawa N, Sakai H, Matsumoto S, Iida K, Yahara I (1990) Destrin, a mammalian actin-depolymerizing protein, is closely related to cofilin. Cloning and expression of porcine brain destrin cDNA. J Biol Chem 265:5768–5773

Moriyama K, Iida K, Yahara I (1996) Phosphorylation of Ser-3 of cofilin regulates its essential function on actin. Gene Cell 1:73–86

Muneyuki E, Nishida E, Sutoh K, Sakai H (1985) Purification of cofilin, a 21,000 molecular weight actin-binding protein, from porcine kidney and identification of the cofilin-binding site in the actin sequence. J Biochem 97:563–568

Nishida E, Kuwaki T, Maekawa S, Sakai H (1981) A new regulatory protein that affects the state if actin polymerization. J Biochem 89:1655–1658

Nishida E, Maekawa S, Sakai A (1984) Cofilin, a protein in porcine brain that binds to actin filaments and inhibits their interactions with myosin and tropomyosin. Biochemistry 23:5307–5313

Ono S, Minami N, Abe H, Obinata A (1994) Characterization of a novel cofilin isoform that is predominantly expressed in mammalian skeletal muscle. J Biol Chem 269:15280–15286

Oriol C, Dubord C, Landon F (1977) Crystallization of native striated-muscle actin. FEBS Lett 73:89–91

Page R, Lindberg U, Schutt CE (1998) Domain motions in actin. J Mol Biol 280:463–474

Safer D, Nachmias VT (1994) Beta thymosins as actin binding peptides. BioEssays 16:473–479

Schafer DA, Cooper JA (1995) Control of actin assembly at filament ends. Annu Rev Cell Devel Biol 11:497–518

Schutt CE, Myslik JC, Rozycki MD, Goonesekere N, Lindberg U (1993) The structure of crystalline profilin-beta-actin. Nature 365:810–816

Spudich JA, Watt S (1971) The regulation of rabbit skeletal muscle contraction. I Biochemical studies of the interaction of tropomyosin-troponin complex with actin and the proteolytic fragments of myosin. J Biol Chem 246:4866–4871

Weber A, Pennise CR, Babcock CR, Fowler V (1994) Tropomodulin caps the pointed ends of actin filaments. J Cell Biol 127:1627–1635

Wriggers WJ, Tang X, Azuma T, Janmey PA (1998) Cofilin and gelsolin segment-1 – molecular dynamics simulation and biochemical analysis predict a similar actin binding mode. J Mol Biol 282:921–932

Actin Filament Networks

Paul A. Janmey[2], Jagesh V. Shah[3], Jay X. Tang[4], and Thomas P. Stossel[1]

Introduction

Since its discovery as an essential component of the contractile machinery of striated muscle, the ability of purified actin to increase the viscosity of its aqueous solutions has been the object of many studies. Biochemical and biophysical studies soon established that the chemical stimuli, usually addition of monovalent or divalent salts, that increased viscosity did so coincident with promoting the polymerization of globular (G-) actin into filamentous polymers (F-actin). Defining how the properties of the individual filaments and their interactions with each other account for the macroscopic mechanical properties of F-actin solutions is an active area of research, with several fundamental questions remaining unanswered.

A quantitative characterization of F-actin viscoelasticity will help explain the influence of actin on cell mechanics. Genetic and cell biologic evidence show that altering the function of actin-binding proteins that cross link F-actin or that alter actin filament length can have large effects on cell stiffness and motility (Cunningham et al. 1992; Schindl et al. 1995; Rivero et al. 1996). In a different context, defining the viscoelasticity of purified F-actin solutions has utility for testing and developing theories of polymer physics that seek to define the properties of materials made from the relatively rigid, or semiflexible, polymers that are common in biology and that have properties fundamentally different from those of the well-characterized rubber like networks common in synthetic materials (MacKintosh et al. 1995; Isambert and Maggs 1996; Kroy and Frey 1996; Satcher and Dewey 1996; Maggs 1997; Morse 1998).

Before reviewing the studies of actin networks and the viscoelasticity of polymerized actin, we begin this chapter by defining two key terms, network

[1] Hematology Division, Brigham and Women's Hospital, Harvard Medical School, 221 Longwood Ave., LMRC 301, Boston, Massachussetts 02115, USA and Harvard-MIT Division of Health Sciences and Technology, Cambridge, Massachussetts 02138, USA

[2] Institute for Medicine and Engineering (IME), University of Pennsylvania, 1010 Vagelos Laboratories, 3340 Smith Walk, Philadelphia, PA 19104, USA

[3] Ludwig Institute for Cancer Research, University of California, San Diego, 9500 Gilman Drive, CMM-East 3080 – MAIL CODE 0660, La Jolla, CA 92093, USA

[4] Physics Department, Indiana University, Swain West 165, 727 East Third St, Bloomington, IN 47405, USA

Results and Problems in Cell Differentiation, Vol. 32
C. dos Remedios (Ed.): Molecular Interactions of Actin

and gel. These entities have frequently been used interchangeably, as mutually exclusive concepts, or often with vague definition. Evans (1993) provides a definition of networks as "condensed phases where dynamics are governed by equations of motion for collective (mechanical) behavior.... Network segments are kinetically restricted by cross-links and entanglements". This definition is primarily structural. Correspondingly, we propose as a simple definition that a network exists if individual linear polymers cannot diffuse without interfering with the thermal motions of others. The nature of this interdependence, and the resulting collective motions, depends on geometric factors like the length, stiffness, and concentration of the filaments, as well as on specific bonds holding filaments together. A minimal condition required for such network to exist, based on percolation, has been described for F-actin by Forgacs (1995).

The term gel, defined by the viscoelastic properties of the material, arose long before any information about its molecular structure (see (Stossel 1990) for review). Qualitatively a gel is a solution with a coherent solute structure that confers elasticity on the viscosity of the dominant liquid component. Almdal et al. (1993) have surveyed previous definitions of gel in the parlance of rheological science, to provide a rheological characteristic that any solvent/solute system must have to be a gel. They posit that the shear storage modulus (G′) should be greater than the loss modulus (G″) when measured over a range of frequencies or times accessible to natural observation, namely on the order of 0.1 to 10 s. The shear modulus, G, during constant shear deformation is the ratio of the deforming force per unit area (or stress) to the degree of deformation (or strain). When measurements are made by applying oscillatory deformations, there are two components to this ratio: the ratio of stress in phase with the strain (G′) and that out of phase with strain (G″). Since G′ is a measure of the energy stored elastically in the material when it deformed, and G″ a measure of the energy dissipated during deformation, intuitively this definition means that a gel is more elastic than viscous, although it will almost certainly have both properties.

This definition is less strict than one introduced by Ferry (1980) who proposed that a gel, as distinct from a viscoelastic fluid, will have a finite elastic modulus at infinite time, because of the practical inability to distinguish very slow relaxation from no relaxation. Often however, the stricter definition of gel is retained, implying that a true gel has a critical number of permanent bonds holding the network strands together into a stable network that can only partially relax. According to this more strict definition, it is redundant to say "cross linked" gel, but a network may be cross linked or not. However, the number of cross links and their geometric arrangement would still have large effects on the viscoelasticity of the gel. As the technology for defining molecular characteristics of viscoelastic solutions and gels has advanced, the stricter definition becomes increasingly appropriate.

Viscoelastic Characterization of Actin Networks

Viscosity of F-Actin

Most early studies of actin rheology measured F-actin solution viscosity, rather than elasticity, which requires more delicate instruments for detection. Straub et al. (Straub 1942; Straub and Feuer 1950) found that, in contrast to simple viscous fluids, F-actin viscosity was highly non-Newtonian, i.e., the measured viscosity depended on how fast the solution was made to flow. By extrapolation, the viscosity appeared to diverge to infinity in the limit of zero shear rate. This characteristic alone is evidence for network formation, and the loss of viscosity at higher shear rates may be due to breakage of filaments, alignment of filaments in the flow field, or breakage of contacts (cross links) between filaments. An unusual property of actin noted in such studies was that when left undisturbed after flow, the apparent viscosity rose with time. The apparent self-repairing of a viscoelastic material was called thixotropy, and the molecular basis for its appearance in actin was the subject of several studies (Maruyama et al. 1974; Buxbaum et al. 1987; Kerst et al. 1990).

Maruyama et al. (1974) showed that, when measured over a range of shear rates from $0.0005\,s^{-1}$ to $3\,s^{-1}$, the viscosity of F-actin dropped from 45 Pa.s to 0.008 Pa.s. This large and regular decrease over the entire range of shear rates corresponds to a power law relation of $\eta \propto (d\gamma/dt)^{-0.98}$. Since η is defined by $\sigma/(d\gamma/dt)$, the stress required to deform the material at a given shear rate is $\sigma = \eta \times (d\gamma/dt)$, and if $\eta \propto (d\gamma/dt)^{-1}$, then the stress required to deform the material becomes independent of the shear rate. In other words, it takes no more force for the actin solution to flow fast than to flow slowly. This unusual material property was studied in greater detail for both F-actin and microtubule solutions by Buxbaum et al. (1987), who concluded that when actin was subjected to shear flow at any practically attainable rate, stress was primarily dissipated in the breaking of solid-like domains that after breakage flowed past each other with little resistance. Figure 1 shows that this relationship between apparent viscosity and shear rate holds for F-actin over at least eight decades of shear rates, consistent with the idea that, under these conditions, F-actin behaves like a solid that flows only after it has been ruptured.

While this rheologic feature has been well documented and is highly reproducible, its molecular origin remains obscure. One hint as to its molecular basis comes from the finding that when the length of the actin filaments is limited to approximately 1.6 µm the shear rate dependence is lower over the entire range of measured rates, and a range of very low flow rates appears at which the apparent viscosity becomes constant. The cross over from shear rate dependence to independence occurs at a flow rate equivalent to the rotational diffusion constant of 1.6-µm–long filaments in semidilute solution (Janmey et al.

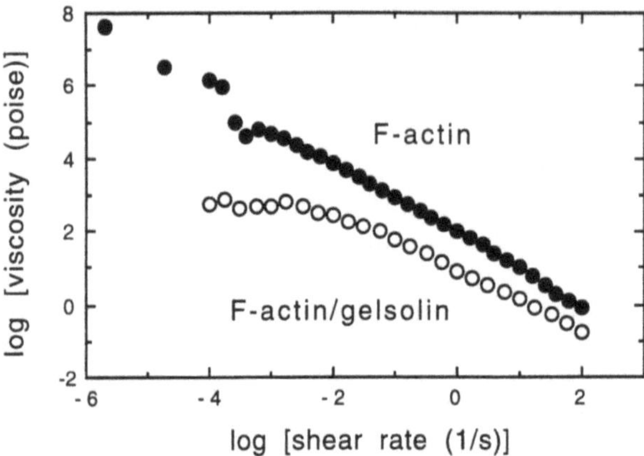

Fig. 1. Apparent viscosity of 2 mg ml^{-1} F-actin (*filled circles*) or 2 mg ml^{-1} actin : gelsolin (500 : 1 molar ratio) measured over a range of shear rates

1988), suggesting that one interpretation of the anomalous shear rate dependence of actin is that F-actin filaments are so long that their rotational relaxation is so slow that no relaxation without severe distortion of the filaments is possible at any practically achievable shear rate.

Elasticity of F-Actin

Accompanying the viscous dissipation of mechanical stress by actin is an elastic storage of mechanical energy that depends on the frequency and magnitude of the deformation or strain. In the absence of actin-binding proteins that regulate filament length or cross linking, the measured elastic moduli reported for purified actin have varied over a wide range, but some consistent trends have emerged. Early studies by Kasai et al. (1960) and Maruyama et al. (1974; 1975), established that actin filament preparations, later shown to contain filaments with a length of 1–3 microns (Kawamura and Maruyama 1972), had storage shear moduli G′ measured by oscillatory deformation of a few dyne cm^{-2} for concentrations of 1–4 mg ml^{-1}. Later studies by Jen et al. (1982) reported higher moduli of 10 dyne cm^{-2} for conventionally purified actin of 1.7 mg ml^{-1} and showed that further purification by gel filtration increased the moduli to near 50 dyne cm^{-2} when measured at frequencies between 1 and 10 Hz. Most later studies of purified actin report similar values between 1 and 10 dyne cm^{-2} for F-actin between 1 and 2 mg ml^{-1} (summarized in Xu et al. 1998a). Some studies report much higher values of 1000 dyne cm^{-2} for F-actin (Janmey et al. 1988), and these values appear to result from formation of crosslinks between the filaments due to oxidation or other modifications of the actin (Xu et al. 1998a, b; Tang et al. 1999).

The most consistent quantitative description for the viscoelasticity of 1 mg ml^{-1} uncross linked actin networks appears to include a rather narrow plateau in a plot of G′ vs frequency at which the modulus is relatively stable at a value on the order of 1–10 dyne cm^{-2} and where the mechanical loss (ratio of loss to storage moduli G″/G′) is significantly less than 1, followed by a frequency-dependent increase in both G′ and G″ at higher frequencies. For very dilute F-actin, a terminal relaxation zone, corresponding to the zero shear rate viscosity measurements, can be detected at shear rates below 10^{-3} rad s^{-1} (Müller et al. 1991). A factor that complicates the process of making reproducible actin preparations and comparing experimental results with theories is the inability, in the absence of actin binding proteins, to regulate the rate-determining nucleation step of polymerization, leading to uncertainty in the average filament length as well as the wide distribution of lengths at steady state. A related experimental problem is that any contamination by as little as 0.1% of filament nucleating or cross linking proteins can strongly alter actin rheology. As a result, the most consistent rheological measurements have resulted from studies where actin is polymerized with defined concentrations of actin-binding proteins.

Actin-Binding Proteins that Alter Network Formation

Cross Linking Proteins

Gelation and superprecipitation of actomyosin, the mixture of F-actin and myosin filaments enabled the identification and characterization of both polymers responsible for muscle contraction. Abe and Maruyama (1974) showed that the elastic modulus of 4 mg ml^{-1} F-actin was increased by a factor of 1000 to between 4000 and 5000 dyne cm^{-2} when heavy meromyosin was added in the presence of ATP. The molecular basis for actomyosin gelation is the crosslinking of actin filaments by the divalent HMM, and now many proteins in addition to myosin have been found to crosslink actin networks in a variety of geometries. Another muscle protein, α–actinin, was also found in early studies to possess the ability to increase actin elasticity (Ebashi et al. 1964; Abe and Maruyama 1973).

The finding of actin in nonmuscle cells (Hatano and Oosawa 1966; Miki and Oosawa 1969; Pollard et al. 1970) and the formation of actin-dependent cytoplasmic extract gels (Kane 1976), led to identification of many proteins that affect the network formation of actin and its resulting rheological properties. The first protein found to promote actin gelation in nonmuscle cells was ABP, a $M_r = 280\,000$ actin-binding protein purified from macrophages (Hartwig and Stossel 1975; Hartwig et al. 1980), and a related protein, filamin, was purified from skeletal and smooth muscle (Wang 1977; Wang and Singer 1977). Gelation of actin by ABP was very efficient, and could be accounted for by theories for cross linked polymer network formation developed by Flory (1953; Hartwig and Stossel 1979, 1981). The relation between actin filament concentration, ABP concentration and the onset of the sol-gel transition is given by the relation

$$[X]_c = [\text{actin}]/\text{DOP}_{(\text{F-actin})}, \tag{1}$$

where [X]c is the minimal concentration of cross linker to form a network from a sample of actin with concentration of [actin] and an average degree of polymerization, or number of subunits per filament of $\text{DOP}_{(\text{F-actin})}$.

Application of this fundamental relation to actin means that gelation, and the resulting increase in elasticity, depend on formation of a continuous multifilament complex where each filament is linked at two sites by ABP to other filaments. The structure of ABP as a highly extended dimer with actin-binding sites at the ends of the extensions and a rigid L-shaped central domain would bias crosslinks to large angles, and so allow network formation with a minimum of filaments and allow a maximum separation between filaments. High-angle branching of F-actin by ABP has been documented by electron microscopy (Hartwig et al. 1980; Niederman et al. 1983). The efficiency of other proteins to gel a solution of actin filaments would depend on both the affinity of these cross linkers for the actin and the geometry with which the filaments are linked. Proteins such as fimbrin, which, like ABP, are dimeric complexes with homologous actin-binding domains, fail to gel F-actin efficiently (Matsudaira 1994) because the two actin-binding sites are spatially arranged to promote lateral association of filaments into bundles instead of an open meshwork. Other filament cross linkers have variable ability to form gels depending on the angle at which filament-filament cross links are made, and the intermediate abilities of α-actinin, muscle filamin (Ruddies et al. 1993), and gelation factor (ABP-120) (Janssen et al. 1996) are consistent with this model.

The rheology of ABP cross linked actin gels resembles that of covalently linked viscoelastic solids, characterized by a very high shear modulus on the order of 1000 dyne cm^{-2} for 1 mg ml^{-1}, and a nearly infinite stress relaxation time. Whatever relaxation is observed could be due to the breakage of filaments held under stress, rather than slippage of cross links or diffusion of filaments. These rheological properties are similar to those of actin gels in which biotinylated actin subunits are cross linked by avidin tetramers where the rate of dissociation of the cross linkers is practically zero (Janmey et al. 1990; Wachsstock et al. 1993).

An additional or alternative mechanism for the rheologic effects of different cross linkers derives from studies of α-actinin. Nonmuscle, as well as muscle, isoforms of α-actinin share sequence homology in their actin-binding domains with ABP, but their ability to form gels of F-actin differs from ABP as well as from one isoform of α-actinin to another (Sato et al. 1987; Meyer and Aebi 1990; Wachsstock et al. 1993, 1994; Tempel et al. 1996). Some isoforms of α-actinin form actin bundles under conditions where others form isotropic gels, and the balance between gelation and bundling depends on the ratio of actin to α-actinin. The elasticity of isotropic actin networks cross linked by some α-actinin isoforms also depends on the rate of deformation. At high frequencies, α-actinin forms actin gels with elastic moduli >400 dyne cm^{-2}, near the levels of gels cross linked by ABP. However, while the elastic modulus of ABP cross linked gels is relatively constant over a large range of frequencies, the modulus of α-actinin cross linked actin falls steeply at frequencies <100 rad/s and the gels show much more stress relaxation than those cross linked by ABP or biotin/avidin (Sato et al. 1987; Wachsstock et al. 1993, 1994; Tempel et al. 1996; Xu et al. 1998b). The interpretation of these effects is that the dominant viscoelastic relaxation depends on dissociation of the α-actinin crosslink from the filament. Direct measurements of dissociation rates of α-actinin from F-actin (Kuhlman et al. 1994) are consistent with this interpretation and with the effects of temperature on α-actinin's gelation efficiency (Xu et al. 1998), but the similar rate constants measured for the actin–binding domains of α-actinin and filamin (Goldmann et al. 1994) suggest contrast with their different rheologic effects, and also that geometric factors as well as binding kinetics can contribute to the rheologic effects of these cross linkers.

Actin Severing and Capping Proteins

Observations that the gelation of cell extracts was inhibited or reversed by micromolar Ca^{2+} concentrations led to the discovery of proteins that severed F-actin and capped their fast-growing barbed end, thereby regulating filament length (Mimura and Asano 1978; Yin and Stossel 1979). The first such protein, purified from macrophages, was called gelsolin to reflect its ability to dissolve actin gels (Mimura and Asano 1978; Yin and Stossel 1979; Yin et al. 1980). Gelsolin is a member of a large group of related proteins that share the ability to lower filament length, reduce viscosity, and affect the actin cytoskeleton in vivo. The regulation of these proteins, usually by Ca^{2+} or lowered pH, and their inactivation by inositol lipids suggests that they are involved in cytoskeletal remodeling initiated by several types of intracellular signals (see Stossel 1994; Janmey et al. 1998 for reviews).

The solation activity of filament severing proteins is evident from the relation shown in Eq (1), because a decrease in length at constant polymer concentrations implies an increase in filament number, and therefore an increase in the crosslinker concentration required to form the gel. In uncrosslinked networks, filament length is also a key parameter, and its effect is evident in the relations specifying the transition from dilute to semi dilute, to concentrated polymer solutions. Three regimes of length (L) and concentration C define three different physical states of rod-like or semi–flexible polymers like F-actin.

$$C << 1/L^3 \text{ dilute solution} \quad (2)$$

$$1/L^3 << c << 1/dL^2 \text{ semi-dilute solution} \quad (3)$$

$$C >> 1/dL^2 \text{ highly entangled network or liquid crystalline phase} \quad (4)$$

At concentrations and lengths sufficiently low to allow each polymer to execute rotation without steric contact with its neighbors Eq (2), the solution viscosity is only slightly higher than that of the solvent, and there is no significant elastic contribution.

In the semidilute solution, defined by Eq (3), rotational diffusion is vastly retarded, as is translational diffusion of the filament perpendicular to its long axis, because of collisions with neighboring filaments, and therefore longrange diffusion requires collective motion of many polymers within the solution. As a result, solution viscosity increases much more strongly when either the length or concentration is increased to produce a semidilute solution (Janmey et al. 1986). These features satisfy some of the requirements cited by Evans (1993) for network formation, and some studies of actin rheology suggested that the steric constraints present in such solution may be equivalent to the entanglements seen in flexible polymers and therefore may form elastically effective cross links that produce gels with large shear moduli. However, under these conditions, motion of filaments along their long axis is largely unimpeded, and the filaments move as though trapped within a tube by a motion called reptation (deGennes 1976; Doi and Edwards 1986). Polymer reptation was first observed in a single filament by studies of fluorescently labeled DNA or F-actin (Perkins et al. 1994; Käs et al. 1994) and this ability to observe single polymers within a meshwork has made such biopolymers powerful test systems for analyzing general properties of polymers in solution. For F-actin of average length 1 micron, semidilute solution conditions are present at a concentration well below 1 $mg\,ml^{-1}$ (Janmey et al. 1985).

At higher concentrations or longer lengths, defined by Eq (4), the longitudinal diffusion of actin filaments also becomes restricted because the finite volume of the filament occupies sufficient space to require collective motions for even the end of a filament to leave the space of the virtual tube in which it is contained. Under such conditions, which are generally met for actin filaments of the length (0.1 to 1 micron) and concentration (10 $mg\,ml^{-1}$) of the cytoskeleton, the filaments are thermodynamically driven to form a nematic liquid crystalline phase in which filaments are aligned, and such a transition of F-actin from isotropic networks to liquid crystals is observed for actin to depend on length and concentration, approximately as predicted from a relation first developed by Onsager (1949; Coppin and Leavis 1992; Furukawa et al. 1993; Suzuki et al. 1996).

In uncross linked actin networks, both viscosity and elasticity are strongly decreased by lowering the filament length (Janmey et al. 1986; 1994). In contrast, with sufficient cross linker concentration, the elasticity of the network is independent of the average filament length until the length becomes too short to satisfy the criterion for semidilute solutions or to ensure two cross links per filament (Brotschi et al. 1978; Janmey et al. 1990).

Comparison with Theories for Semiflexible Polymers

Early studies of actin rheology interpreted the viscoelasticity in terms of theories derived for rubber-like polymers in which the elastic response arises from the loss of entropy as the randomly coiled polymer strands between cross links or points of entanglement are extended from their equilibrium configuration. Networks of F-actin and other stiff biopolymers are fundamentally different because the network strands between junction points are approximately straight, and therefore the physical origin of the elastic response is different. Because actin filaments are much stiffer than most conventional polymers, the classic theories for flexible polymer chains are no longer applicable. Several theories of semiflexible polymer networks have been recently developed in part to meet the experimental challenge of explaining how semiflexible filaments such as F-actin can form gels at much lower volume fractions than required for flexible polymers, and to estimate the magnitude of the elastic modulus of such gels.

A theoretical treatment by Morse (1998) extends the reptation in a tube model (deGennes 1976; Doi and Edwards 1986) to treat the viscoelastic behavior of a semiflexible polymer network following a detailed calculation of the stress tensor, which describes local force and torque at every position. The response to an oscillatory deformation of a semiflexible filament network may be separated into several terms. First, a rapid distortion to the network will cause stretch or compression to a slightly curved filament, and such stretch or compression gives rise to a large tensional stress (tension in Fig. 2). The tensional stress term for semiflexible polymers was treated by MacKintosh et al. (1995). Second, the deformation of the network causes bending of the filaments, thus causing a curvature stress curve in Fig. 2. Third, the deformation causes changes in filament orientation, thus giving rise to an orientational stress (orient in Fig. 2). Both the curvature stress and orientational stress have been analyzed by Maggs (Isambert and Maggs 1996; Maggs 1997), who predicted these terms based on dimensional analysis. Figure 2 depicts the three stress terms acting on a single filament confined in a virtual tube formed by the surrounding filaments. Also indicated on the cartoon are the tube diameter D and the entanglement length, L_e, which is the average distance between the collision points of the test filaments to its neighbours, or equivalently, its surrounding tube.

These three stresses have disparate values and decay times. The tension modulus has a very large initial value for an F-actin network, but for an uncross linked network the decay time is predicted to be less than 1ms, and therefore is significant only in response to either a rapid deformation or a high frequency oscillatory strain (>1000 rad sec^{-1}). The curvature stress has an initial value on

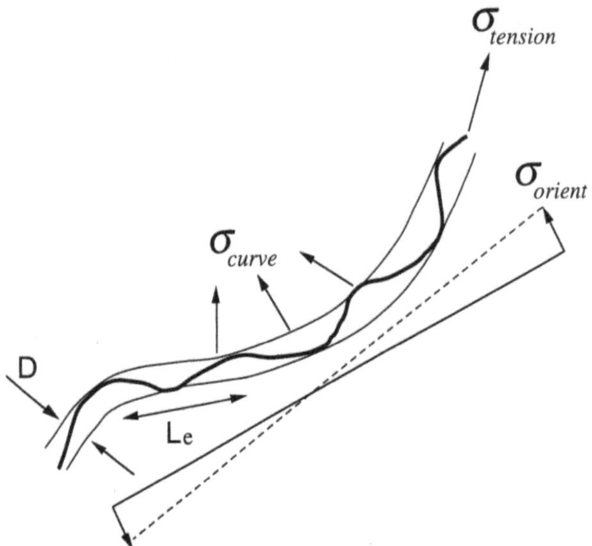

Fig. 2. Schematic representation of an actin filament within a virtual tube of diameter *D* showing the three types of stresses applied to the filament when the network is strained. Other symbols are defined above in section Comparison with Theories for Semiflexible Polymers

the order of 1–10 dyne cm^{-2} for 1 mg ml^{-1} F-actin, and a characteristic decay time due to reptation that depends on filament length. Since the reptation time t_{rep} of long F-actin in a 1 mg ml^{-1} network is on the order of minutes or hours, the curvature stress remains practically constant in a frequency range of 0.01–10 rad sec^{-1}. The orientational stress is smaller than the curvature stress by a factor of L_e/L, with a decay time L_p^2/L_{trep}^2, where L_p is persistence length of F-actin. Although in principle such a stress term may cause a second plateau modulus, all reported rheologic data show only one plateau. This is perhaps not surprising due to the fact that the average filament length is comparable to the persistence length, hence the two plateau would actually overlap. In addition, the extreme polydispersity of F-actin would also smear out the second plateau to a broader and effectively negligible level in comparison with the curvature stress.

An estimate of curvature stress that dominates within the frequency range of a typical mechanical rheometer gives a value on the order of 1 dyne/cm^2 for 1 mg ml^{-1} actin. This value is at the low end of the collective experimental data from many groups (Jen et al. 1982; Janmey et al. 1994; Xu et al. 1998a). Although the much larger tension stress is not expected due to its rapid decay, slight cross linking between even a small portion of the filaments may drastically slow down the decay of tensional stress, hence increasing the measured modulus significantly. In addition to cross linking proteins that might copurify with actin, recent experimental work suggests strongly that some conformations of actin can themselves allow branched structures in F-actin (Steinmetz et al. 1997) and that oxidation of a small percentage of actin causes a dramatic increase in G′ (Tang et al. 1999). Although the mechanism of cross linking by oxidized actin is at present unclear, the detectable amounts of actin dimers and oligomers due to oxidation may cause the cross links of actin networks.

Recent theoretical treatments (Maggs 1997; Gittes and MacKintosh 1998; Hinner et al. 1998; Morse 1998) also predict that the plateau modulus of a semi-flexible polymer network varies with concentration with a power of 7/5. Such a prediction agrees with the experimental data by Xu et al. (1998a) and Hinner et al. (1998), although the absolute values at all concentrations differ by an order of magnitude between these two groups. Some earlier data (Maruyama et al. 1974; Janmey et al. 1988) show an increase with concentration by powers between 2.1 and 2.5. The steeper concentration dependence is consistent with the theory of MacKintosh et al. (1995; Mackintosh and Janmey 1997) which predicts a power of close to 2.25, assuming that the tensional stress is the dominant response. A slightly cross linked network would most likely validate such an assumption in light of the most recent work of Morse (1998).

Satcher and Dewey (1996) proposed a simpler mechanical model to estimate the shear modulus of a cross linked actin network. Their model assumes a simple limit of orthogonally cross linked network with cubic unit cells. Each side of the cube is a segment of a single actin filament and the cage size is determined by the protein concentration. Using the bending modulus for F-actin derived from measurements of persistence length, the Young's modulus of such an idealized actin network at 10 mg ml^{-1} is estimated to be on the order of 10^5 dyne cm^{-2}. The value of the shear modulus is estimated to be on the same order of magnitude. Such a simple mechanical model predicts a dependence on protein concentration to the power of two, and hence predicts a shear modulus on the order of 1000 dyne cm^{-2} for 1 mg ml^{-1} F-actin. This estimate is rather close to the high limit of experimentally measured actin network at 1 mg ml^{-1}, particularly when the network is strongly cross linked or immobilized (Janmey et al. 1988, 1994; Wachsstock et al. 1994; Xu et al. 1998).

Interaction with Other Polymers

Actin networks play an important role in determining the mechanical properties of the cell but the cell also relies on other intracellular polymer networks which, together with F-actin, comprise the cytoskeleton. In addition to F-actin, microtubules (MTs) and intermediate filaments (IFs) are the main constituents of the cytoskeleton. MTs and IFs each have a structure and rigidity distinct from F-actin, resulting in networks with distinct mechanical properties. MTs are much stiffer than actin filaments, whereas IFs are more flexible. Within the cell, these three polymer systems interpenetrate and can be cross linked to each other, forming a composite network. The resulting mechanical properties of such a composite network are expected to be unlike any of the networks

Table 1. Linkage of F-actin to other network-forming biopolymers

Microtubules	Intermediate filaments	Other biopolymers
Microtubule-associated proteins, MAP2c, MIP-90	Direct, plectin (vimentin) filamin, calponin (desmin) BPAG1n (NFs)	Fibrin DNA

of single composition. Table 1 summarizes the cross linking proteins that have been implicated thus far in linking F-actin to other polymers of the cytoskeleton.

Actin and Microtubules

The spatial organization of F-actin and microtubules in motile cells, neuronal growth cones and dividing cells indicates a link between these polymers. For example, the motion of MTs is modulated by the F-actin network at the lamellae of motile cells (Waterman-Storer and Salmon 1997), and the interaction between MTs and F-actin in the neuronal growth cone is implicated in the guidance of growth cone to their targets (Lin and Forscher 1993). MAP2c, a neuron-specific microtubule-binding protein can partially substitute for lack of filamin expression in mutant melanoma tumor cell lines, and may help to coordinate MTs and F-actin when expressed in these cells (Cunningham et al. 1997). In vitro mixtures of F-actin and MT could increase in viscosity upon addition of microtubule-associated proteins (Griffith and Pollard 1982). In most cases the molecules responsible for F-actin-MT interactions have not been identified, but candidate proteins such as MIP-90 (Gonzalez et al. 1998) have recently been reported.

Actin and Intermediate Filaments

F-actin and IFs are colocalized in many cell types (Tint et al. 1991; Goldman et al. 1996; Yang et al. 1996), and a number of proteins have been shown to link F-actin to IF networks. In the case of vimentin intermediate filaments found in cells of mesenchymal origin, binding to F-actin can be either direct, or modulated by cross linking proteins. The C-terminal domain of vimentin has been shown to interact with F-actin structures directly (Cary et al. 1994), and plectin may also bind vimentin filaments to F-actin (Foisner and Wiche 1991). In muscle cells desmin IFs have been reported to interact with F-actin through calponin (Mabuchi et al. 1997) and filamin (Brown and Binder 1992). In neuronal cells, F-actin may be linked to neurofilaments by the protein BPAG1n (Yang et al. 1996), whose absence results in dystonia and degeneration of motor neurons in mice.

Actin and Noncytoplasmic Biopolymers

In cases of cell death, e.g., at sites of infection and cell damage, intracellular F-actin is exposed to the extracellular environment, where it can interact with molecules not present in the cytoplasm. In blood clots, F-actin can bind the coagulation polymer fibrin, resulting in modified clot mechanics and resistance to clot degradation (Janmey et al. 1992). The severing of F-actin by gelsolin, which is also present in blood plasma, reduced the effect, thereby implicating the entanglement of actin filaments into the fibrin network as a mechanism of modulating the mechanical properties of the clot. F-actin is also reported to interact with DNA, possibly mediated by binding to histones (Lestourgeon et al. 1975; Magri et al. 1978). In the course of cell death DNA is exposed to the extracellular environment, and in cases of chronic infection, such as seen in the lungs of cystic fibrosis patients, the resulting actin-DNA complexes (Sheils et al. 1996) may contribute to the abnormally high viscosity of affected sputum. The use of gelsolin to shorten actin filaments reduces the viscosity of the actin-DNA network and has been proposed as a treatment for cystic fibrosis (Vasconcellos et al. 1994).

Roles for Network Formation Independent of Mechanical Strength

While the primary cellular role of the actin cytoskeleton may be to provide mechanical strength, the fact that actin filaments together with microtubules and intermediate filaments, traverse the entire cell volume and connect distant structures within the cytoplasm suggests a variety of possible functions in signal transduction and spatial organization.

One such effect of the actin network may be to provide a lattice onto which enzymes and other proteins involved in transmitting signals may dock. The surface area of the cytoskeleton is much greater than the surface area of lipid bilayers exposed to the cytoplasm. A typical cell with diameter 20 μm contains on the order of 700 μm^2 plasma membrane and perhaps ten times this of internal area, but the surface area of 10 $mg\,ml^{-1}$ F-actin within such a cell is 47 000 μm^2. This large surface, along with the high negative electrostatic charge density of all cytoskeletal filaments (Tang and Janmey 1996; Tang et al. 1997) presents a strong potential for localization and immobilization of cytoplasmic components (see Janmey 1998 for a review). Two examples of such localization include the requirement of an intact actin network for transport of PKCa to the nucleus of NIH 3T3 fibroblasts treated with phorbol ester (Schmalz et al. 1996) and the localization of enzymes involved in inositol lipid signaling to the actin cytoskeleton following stimulation of platelets by thrombin (Nahas et al. 1989; Grondin et al. 1991; Banno et al. 1996; Hinchliffe et al. 1996).

In addition to altering the balance between soluble and cytoskeleton-bound pools of signaling proteins or enzymes and substrates, the radial distribution of the actin network from the perinuclear space to the cell membrane suggests models of biased transport of materials or signals. A role for the formation of percolating actin networks, and the connections to other cytoskeletal networks in signal transduction, have recently been presented in a series of works by Forgacs and colleagues (Forgacs 1995).

Actin Networks in Vivo

The direct relevance of the mechanics of actin networks in live cells has been shown by the effects of mutations in actin-binding proteins in cells of higher and lower eukaryotes. Cells from a human melanoma tumor line devoid of ABP-280 expression differ from ABP–positive counterparts by lacking the flat organelle-excluding periphery typically associated with the cortical actin network. When stimulated to move, these cells extend spherical blebs rather than the flat lammellipodia typical of this cell type, and as a result they can neither polarize nor move unidirectionally. They also have a whole cell shear modulus less than half that of ABP–positive cells (Cunningham et al. 1992) and fail to reinforce focal adhesions at sites where integrins are mechanically stressed (Glogauer et al. 1997, 1998). ABP-280 has biochemical functions in addition to its actin gelation effect, and the hypothesis that the phenotype of the ABP–cells is related to insufficient actin cross linking is strengthened by the finding that microinjection of MAP2c, a potent gelation factor for F-actin in vitro, but sharing no other feature with ABP, also reverses the mutant phenotype of ABP cells (Cunningham et al. 1997).

In *Dictyostelium discoidium*, a number of actin gelation proteins have been eliminated by chemical mutagenesis or homologous recombination. Cells lacking expression of α-actinin and ABP120 are significantly less resistant to shear deformation under a wide range of conditions (Eichinger et al. 1996). While these cells are able to locomote nearly normally on highly adhesive surfaces, on less favorable surfaces they exhibit gross defects in motility related to weaker elasticity at the cytoskeletal/membrane interface (Schindl et al. 1995). They are also defective in several responses to mechanical stresses including changes in temperature and osmotic pressure (Fisher et al. 1997; Rivero et al. 1996).

Conclusions

The unusual viscoelastic properties of polymerized actin, first recognized more than 50 years ago, have motivated studies of actin filament structure that suggest how the networks formed by these filaments produce the resulting macroscopic rheologic features. From a biological perspective, the relation between single filament structure and network properties is important to predict how activation of specific actin–binding proteins might alter the stiffness of cellular actin networks, and so modulate motility and shape change. From the perspective of material science, F-actin represents a class of semi-flexible polymers that form networks with viscoelastic properties distinct from those of most common synthetic, flexible polymers. The tools developed to visualize actin filaments (see Ishiwata et al. this Vol.) have enabled studies of single polymer conformation and motion within actin networks that are not yet possible with synthetic polymers.

References

Abe S, Maruyama K (1973) Effect of α-actinin on F-actin: a dynamic viscoelastic study. J Biochem 73:1205–1210

Abe SI, Maruyama K (1974) Dynamic viscoelastic study of acto-heavy meromyosin in solution. Biochim Biophys Acta 160:160–174

Almdal K, Dyre J, Hvidt S, Kramer O (1993) Towards a phenomenological definition of the term gel. Poly Gels Networks 1:5–17

Banno Y, Nakashima S, Ohzawa M, Nozawa Y (1996) Differential translocation of phospholipase C isozymes to integrin-mediated cytoskeletal complexes in thrombin-stimulated human platelets. J Biol Chem 271:14989–14994

Brotschi E, Hartwig J, Stossel T (1978) The gelation of actin by actin-binding protein. J Biol Chem 253:8988–8993

Brown KD, Binder LI (1992) Identification of the intermediate filament-associated protein gyronemin as filamin Implications for a novel mechanism of cytoskeletal interaction. J Cell Sci 102:19–30

Buxbaum RE, Dennerll T, Weiss S, Heidemann SR (1987) F-actin and microtubule suspensions as indeterminate fluids. Science 235:1511–1514

Cary RB, Klymkowsky MW, Evans RM, Domingo A, Dent JA, Backhus LE (1994) Vimentin's tail interacts with actin-containing structures in vivo. J Cell Sci 107:1609–1622

Coppin C, Leavis P (1992) Quantitation of liquid-crystaline ordering in F-actin solutions. Biophys J 63:794–807

Cunningham CC, Gorlin JB, Kwiatkowski DJ, Hartwig JH, Janmey PA, Byers HR, Stossel TP (1992) Actin-binding protein requirement for cortical stability and efficient locomotion. Science 255:325–327

Cunningham CC, Leclerc N, Flanagan LA, Lu M, Janmey PA, Kosik KS (1997) Microtubule-associated protein 2c reorganizes both microtubules and microfilaments into distinct cytological structures in an actin-binding protein-280-deficient melanoma cell line. J Cell Biol 136:845–857

deGennes PG (1976) Dynamics of entangled polymer solutions I The Rouse model. Macromolecules. 9:587–593

Doi M, Edwards SF (1986) The theory of polymer dynamics. Clarendon, Oxford

Ebashi S, Ebashi F, Maruyama K (1964) A new protein factor promoting contraction of actomyosin. Nature 203:645–646

Eichinger L, Koppel B, Noegel AA, Schleicher M, Schliwa M, Weijer K, Witke W, Janmey PA (1996) Mechanical perturbation elicits a phenotypic difference between *Dictyostelium* wild-type cells and cytoskeletal mutants. Biophys J 70:1054–1060

Evans E (1993) New physical concepts for cell amoeboid motion. Biophys J 64:1306–1322

Ferry J (1980) Viscoelastic properties of polymers. John Wiley, New York

Fisher PR, Noegel AA, Fechheimer M, Rivero F, Prassler J, Gerisch G (1997) Photosensory and thermosensory responses in *Dictyostelium* slugs are specifically impaired by absence of the F-actin cross-linking gelation factor (abp-120). Curr Biol 7:889–892

Flory P (1953) Principles of polymer chemistry Cornell University Press, Ithaca

Foisner R, Wiche T (1991) Intermediate filament-associated proteins. Curr Opin Cell Biol 3:75–81

Forgacs G (1995) On the possible role of cytoskeletal filamentous networks in intracellular signaling: an approach based on percolation. J Cell Sci 108:2131–2143

Furukawa R, Kundra R, Fechheimer M (1993) Formation of liquid crystals from actin filaments. Biochemistry 32:12346–12352

Gittes F, MacKintosh F (1998) Dynamic shear modulus of a semiflexible polymer network. Phys Rev E 58:R1241–R1244

Glogauer M, Arora P, Yao G, Sokholov I, Ferrier J, Mcculloch C (1997) Calcium ions and tyrosine phosphorylation interact coordinately with actin to regulate cytoprotective responses to stretching. J Cell Sci 110:11–21

Glogauer M, Arora P, Chou D, Janmey P, Downey G, McCulloch CAG (1998) The role of ABP-280 in integrin-mediated mechanoprotection. J Biol Chem 273:1689–1698

Goldman RD, Khuon S, Chou YH, Opal P, Steinert PM (1996) The function of intermediate filaments in cell shape and cytoskeletal integrity. J Cell Biol 134:971–983

Goldmann WH, Senger R, Isenberg G (1994) Analysis of filamin-actin binding and cross-linking/bundling by kinetic method. Biochem Biophys Res Commun 203:338–343

Gonzalez M, Cambiazo V, Maccioni RB (1998) The interaction of Mip-90 with microtubules and actin filaments in human fibroblasts. Exp Cell Res 239:243–253

Griffith LM, Pollard TD (1982) The interaction of actin filaments with microtubules and microtubule-associated proteins. J Biol Chem 257:9143–9151

Grondin P, Plantavid M, Sultan C, Breton M, Mauco G, Chap H (1991) Interaction of pp60c-src phospholipase C, inositol-lipid, and diacyglycerol kinases with the cytoskeletons of thrombin-stimulated platelets. J Biol Chem 266:15705–15709

Hartwig J, Stossel T (1975) Isolation and properties of actin, myosin, and a new actin-binding protein in rabbit alveolar macrophages. J Biol Chem 250:5696–5705

Hartwig J, Stossel T (1981) The structure of actin-binding protein molecules in solution and interacting with actin filaments. J Mol Biol 145:563–581

Hartwig J, Tyler J, Stossel T (1980) Actin-binding protein promotes the bipolar and perpendicular branching of actin filaments. J Cell Biol 87:841–848

Hartwig JH, Stossel TP (1979) Cytochalasin B and the structure of actin gels. J Mol Biol 134:539–554

Hatano S, Oosawa F (1966) Isolation and characterization of plasmodium actin. Biochim Biophys Acta 127:488–498

Hinchliffe KA, Irvine RF, Divecha N (1996) Aggregation-dependent, integrin-mediated increases in cytoskeletally associated PtdInsP2 (4,5) levels in human platelets are controlled by translocation of PtdIns 4-P 5-kinase C to the cytoskeleton. EMBO J 15:6516–6524

Hinner B, Tempel M, Sackmann E, Kroy K, Frey E (1998) Entanglement, elasticity and viscous relaxation of actin solutions. Phys Rev Lett 81:2614–2617

Isambert H, Maggs A (1996) Dynamics and rheology of actin solutions. Macromolecules 29:1036–1040

Janmey PA (1998) The cytoskeleton and cell signaling–component localization and mechanical coupling. Physiol Rev 78:763–781

Janmey PA, Lind SE, Yin HL, Stossel TP (1985) Effects of semi-dilute actin solutions on the mobility of fibrin protofibrils during clot formation. Biochim Biophys Acta 841:151–158

Janmey PA, Peetermans J, Zaner KS, Stossel TP, Tanaka T (1986) Structure and mobility of actin filaments as measured by quasielastic light scattering, viscometry, and electron microscopy. J Biol Chem 261:8357–8362

Janmey PA, Hvidt S, Peetermans J, Lamb J, Ferry JD, Stossel TP (1988) Viscoelasticity of F-actin and F-actin/gelsolin complexes. Biochemistry 27:8218–8227

Janmey PA, Hvidt S, Lamb J, Stossel TP (1990) Resemblance of actin-binding protein/actin gels to covalently crosslinked networks. Nature 345:89–92

Janmey PA, Lamb JA, Ezzell RM, Hvidt S, Lind SE (1992) Effects of actin filaments on fibrin clot structure and lysis. Blood 80:928–936

Janmey PA, Hvidt S, Kas J, Lerche D, Maggs A, Sackmann E, Schliwa M, Stossel TP (1994) The mechanical properties of actin gels Elastic modulus and filament motions. J Biol Chem 269:32503–32513

Janmey PA, Stossel TP, Allen PG (1998) Deconstructing gelsolin–identifying sites that mimic or alter binding to actin and phosphoinositides. Chem Biol 5:81–85

Janssen KP, Eichinger L, Janmey PA, Noegel AA, Schliwa M, Witke W, Schleicher M (1996) Viscoelastic properties of F-actin solutions in the presence of normal and mutated actin-binding proteins. Arch Biochem Biophys 325:183–189

Jen C, McIntire L, Bryan J (1982) The viscoelastic properties of actin solutions. Arch Biochem Biophys 216:126–132

Kane RE (1976) Actin polymerization and interaction with other proteins in temperature induced gelation of sea urchin egg extracts. J Cell Biol 71:704–714

Käs J, Strey H, Sackmann E (1994) Direct imaging of reptation for semiflexible actin filaments. Nature 368:226–229

Kasai M, Kawashima H, Oosawa F (1960) Structure of F-actin solutions. J Polymer Sci XLIV:51–69

Kawamura M, Maruyama K (1972) Length distribution of F-actin transformed from Mg-polymer. Biochim Biophys Acta 267:422–434

Kerst A, Chmielewski C, Livesay C, Buxbaum RE, Heidemann SR (1990) Liquid crystal domains and thixotropy of filamentous actin suspensions. Proc Natl Acad Sci USA 87:4241–4245

Kroy K, Frey E (1996) Force-extension relation and plateau modulus for wormlike chains. Phys Rev Lett 77:306–309

Kuhlman PA, Ellis J, Critchley DR, Bagshaw CR (1994) The kinetics of the interaction between the actin-binding domain of alpha-actinin and F-actin. FEBS Lett 339:297–301

Lestourgeon WM, Forer A, Yang YZ, Bertram JS, Pusch HP (1975) Contractile proteins Major components of nuclear and chromosome non-histone proteins. Biochim Biophys Acta 379:529–552

Lin CH, Forscher P (1993) Cytoskeletal remodeling during growth cone-target interactions. J Cell Biol 121:1369–1383

Mabuchi K, Li B, Ip W, Tao T (1997) Association of calponin with desmin intermediate filaments. J Biol Chem 272:22662–2266

MacKintosh F, Käs J, Janmey P (1995) Elasticity of semiflexible biopolymer networks. Phys Rev Lett 75:4425–4428

Mackintosh FC, Janmey PA (1997) Actin gels. Curr Opin Solid State Mater Sci 2:350–357

Maggs AC (1997) Two plateau moduli for actin gels. Phys Rev A 55:7396–7400

Magri E, Zaccarini M, Grazi E (1978) The interaction of histone and protamine with actin. Biochem Biophys Res Commun 82:1207–1210

Maruyama K, Kaibara M, Fukada E (1974) Rheology of actin I Network of F-actin in solution. Biochim Biophys Acta 371:20–29

Maruyama K, Abe S, Ishii T (1975) Dynamic viscoelastic study of the effect of beta-actinin on theinteraction between F-actin and heavy meromyosin. J Biochem 77:131–136

Matsudaira P (1994) Actin crosslinking proteins at the leading edge. Semin Cell Biol 5:165–174

Meyer R, Aebi U (1990) Bundling of actin filaments by α-actinin depends on its molecular length. J Cell Biol 110:2013–2024

Miki NT, Oosawa F (1969) An actin-like protein of the sea urchin eggs I Its interaction with myosin from rabbit striated muscle. Exp Cell Res 56:224–232

Mimura N, Asano A (1978) Actin-related gelation of Ehrlich tumour cell extracts is reversibly inhibited by low concentrations of Ca^{2+}. Nature 272:273–276

Morse D (1998) Viscoelasticity of tightly entangled solutions of semiflexible polymers. Phys Rev E 58:R1237–R1240

Muller O, Gaub H, Baermann M, Sackmann E (1991) Viscoelastic moduli of sterically and chemically cross-linked actin networks in the dilute to semidilute regime – measurements by an oscillating disk rheometer. Macromolecules 24:3111–3120

Nahas N, Plantavid M, Mauco G, Chap H (1989) Association of phosphatidylinositol kinase and phosphatidylinositol 4-phosphate kinase activities with the cytoskeleton in human platelets. FEBS Lett 264:30–34

Niederman R, Amrein PC, Hartwig JH (1983) Three-dimensional structure of actin filaments and of an actin gel made with actin-binding protein. J Cell Biol 96:1400–1403

Onsager L (1949) The effects of shape on the interaction of colloidal particles. Ann NY Acad Sci 51:627–659

Perkins TT, Smith DE, Chu S (1994) Direct observation of tube-like motion of a single chain. Science 264:819–822

Pollard TD, Shelton E, Weihing RR, Korn ED (1970) Ultrastructural characterization of F-actin isolated from *Acanthamoeba castellanii* and identification of cytoplasmic filaments as F-actin by reaction with rabbit heavy meromyosin. J Mol Biol 50:91–97

Rivero F, Koppel B, Peracino B, Bozzaro S, Siegert F, Weijer CJ, Schleicher M, Albrecht R, Noegel AA (1996) The role of the cortical cytoskeleton – f-actin crosslinking proteins protect against osmotic stress, ensure cell size, cell shape and motility, and contribute to phagocytosis and development. J Cell Sci 109:2679–2691

Ruddies R, Goldmann WH, Isenberg G, Sackmann E (1993) The viscoelastic moduli of actin/filamin solutions: a micro-rheologic study. Biochem Soc Trans 21:37S

Satcher RL Jr, Dewey CF Jr (1996) Theoretical estimates of mechanical properties of the endothelial cell cytoskeleton. Biophys J 71:109–118

Sato M, Schwarz WH, Pollard TD (1987) Dependence of the mechanical properties of actin/alpha-actinin gels on deformation rate. Nature 325:828–830

Schindl M, Wallraff E, Deubzer B, Witke W, Gerisch G, Sackmann E (1995) Cell-substrate interactions and locomotion of *Dictyostelium* wild-type and mutants defective in three cytoskeletal proteins: a study using quantitative reflection interference contrast microscopy. Biophys J 68:1177–1190

Schmalz D, Kalkbrenner F, Hucho F, Buchner K (1996) Transport of protein kinase C a into the nucleus requires intact cytoskeleton while the transport of a protein containing a canonical nuclear localization signal does not. J Cell Sci 109:2401–2406

Sheils CA, Käs J, Travassos W, Allen PG, Janmey PA, Wohl ME, Stossel TP (1996) Actin filaments mediate DNA fiber formation in chronic inflammatory airway disease. Am J Pathol 148: 919–927

Steinmetz M, Goldie K, Aebi U (1997) A correlative analysis of actin filament assembly, structure and dynamics. J Cell Biol 138:559–574

Stossel TP (1990) How cells crawl. Am Sci 78:408–423

Stossel TP (1994) The machinery of cell crawling. Sci Am 271:54–63

Stossel TP, Hartwig JH (1976) Interaction of actin, myosin, and a new actin-binding protein of rabbit pulmonary macrophages II Role in cytoplasmic movement and phagocytosis. J Cell Biol 68:602–614

Straub FB (1942) Actin. Stud Szeged 2:3–15

Straub FB, Feuer G (1950) Adenosinetriphosphate the functional group of actin. Biochim Biophys Acta 4:455–470

Suzuki A, Yamazaki M, Ito T (1996) Polymorphism of F-actin assembly 1 A quantitative phase diagram of F-actin. Biochemistry 35:5238–5244

Tang JX, Janmey PA (1996) The polyelectrolyte nature of F-actin and the mechanism of actin bundle formation J Biol Chem 271:8556–8563

Tang JX, Ito T, Tao T, Traub P, Janmey PA (1997) Opposite effects of electrostatics and steric exclusion on bundle formation by F-actin and other filamentous polyelectrolytes. Biochemistry 36:12600–12607

Tang JX, Janmey PA, Stossel TP, Ito T (1999) Thiol oxidation of actin produces dimers that enhance the elasticity of the F-actin network. Biophys J 76:2208–2215

Tempel M, Isenberg G, Sackmann E (1996) Temperature-induced sol-gel transition and microgel formation in alpha-actinin cross-linked actin networks – a rheological study. Phys Rev E 54:1802–1810

Tint IS, Hollenbeck PJ, Verkhovsky AB, Surgucheva IG, Bershadsky AD (1991) Evidence that intermediate filament reorganization is induced by ATP-dependent contraction of the actomyosin cortex in permeabilized fibroblasts. J Cell Sci 98:375–384

Vasconcellos CA, Allen PG, Wohl ME, Drazen JM, Janmey PA, Stossel TP (1994) Reduction in viscosity of cystic fibrosis sputum in vitro by gelsolin. Science 263:969–971

Wachsstock DH, Schwartz WH, Pollard TD (1993) Affinity of alpha-actinin for actin determines the structure and mechanical properties of actin filament gels. Biophys J 65:205–214

Wachsstock DH, Schwarz WH, Pollard TD (1994) Cross-linker dynamics determine the mechanical properties of actin gels. Biophys J 66:801–809

Wang K (1977) Filamin, a new high-molecular-weight protein found in smooth muscle and nonmuscle cells Purification and properties of chicken gizzard filamin. Biochemistry 16:1857–1865

Wang K, Singer S (1977) Interaction of filamin with F-actin in solution. Proc Natl Acad Sci USA 74:2021–2025

Waterman-Storer CM, Salmon ED (1997) Actomyosin-based retrograde flow of microtubules in the lamella of migrating epithelial cells influences microtubule dynamic instability and turnover and is associated with microtubule breakage and treadmilling. J Cell Biol 139:417–434

Xu JY, Schwarz WH, Käs JA, Stossel TP, Janmey PA, Pollard TD (1998a) Mechanical properties of actin filament networks depend on preparation, polymerization conditions, and storage of actin monomers. Biophys J 74:2731–2740

Xu JY, Wirtz D, Pollard TD (1998b) Dynamic cross-linking by alpha-actinin determines the mechanical properties of actin filament networks. J Biol Chem 273:9570–9576

Yang Y, Dowling J, Yu QC, Kouklis P, Cleveland DW, Fuchs E (1996) An essential cytoskeletal linker protein connecting actin microfilaments to intermediate filaments. Cell 86:655–665

Yin HL, Stossel TP (1979) Control of cytoplasmic actin gel-sol transformation by gelsolin, a calcium-dependent regulatory protein. Nature 281:583–586

Yin HL, Zaner KS, Stossel TP (1980) Ca^{2+} control of actin gelation. J Biol Chem 255:9494–9500

Structure and Function of Gelsolin

Leslie D. Burtnick[1], Robert C. Robinson[2], and Senyon Choe[2]

The Biochemical Actions of Gelsolin

Gelsolin is a protein that exerts a variety of influences both within the cytoplasm and in extracellular fluids (reviewed in Yin 1987; Janmey et al. 1998). Intracellular gelsolin participates in regulation of cellular architecture and motility through its severing, capping and nucleating activities on actin filaments. Gelsolin itself is subject to control by calcium ions and polyphosphoinositide metabolites. While gene knockout experiments suggest that gelsolin is not essential for survival, it is necessary for rapid responses of such dynamic cells as fibroblasts, as during the process of wound healing, and platelets, as during clotting (Witke et al. 1995). Fibroblasts in which gelsolin has been overexpressed display increased motility (Cunningham et al. 1991). The secreted form of gelsolin, exemplified by that found in blood plasma, is identical in amino acid sequence to that found in the cytosol except that it incorporates a short peptide extension at the N-terminus of the cytoplasmic sequence (Kwiatkowski et al. 1986; Koepf et al. 1998). Alternative transcription initiation and selective RNA processing permit a single gelsolin gene in each species to produce distinct mRNA messages that code for both cytoplasmic and secreted forms of the protein.

Plasma gelsolin plays a different and less complicated role than its cytoplasmic counterpart, acting as one half of a two-protein actin scavenging system (Haddad et al. 1990; Vasconcellos and Lind 1993). As a consequence of cellular death and lysis, actin is released into extracellular space. Much of this actin ends up in blood plasma where ionic strength, pH and temperature favour the formation of elongated F-actin filaments that could interfere with blood flow through microcirculatory vessels.

Gelsolin in the plasma acts rapidly to block adverse effects by binding to, severing and capping F-actin to yield short gelsolin-capped oligomers that are of minimal consequence. In response to the low concentration of free G-actin in plasma, the gelsolin-capped oligomers lose monomers from their uncapped ends. These in turn are sequestered by plasma vitamin D-binding protein (DBP), the second member of the actin scavenging team. DBP-actin complexes

[1] Chemistry Department, University of British Columbia, Vancouver, V6T 1Z1, Canada
[2] Structural Biology Laboratory, Salk Institute, La Jolla, California 92186-5800, USA

Results and Problems in Cell Differentiation, Vol. 32
C. dos Remedios (Ed.): Molecular Interactions of Actin

are recognized by receptors in the liver and cleansed from the circulation (Herrmannsdoerfer et al. 1993).

The influence of gelsolin extends beyond direct restructuring of actin filaments, although actin most likely is involved in these activities. Gelsolin is an activator of DNase I (Davoodian et al. 1997; also see Kekic et al. this Vol.) and it is implicated in the functions of lipids and phospholipases (Steed et al. 1996; Chen et al. 1996; Baldassare et al. 1997). Gelsolin enhances permeabilized mast cell secretion (Borovikov et al. 1995), it has been correlated with cancer and cellular transformation in various instances (Tanaka et al. 1995; Fujita et al. 1995; Asch et al. 1996), and it participates in caspase-3-mediated apoptosis (Kothakota et al. 1997).

The Structure of Gelsolin

Gelsolin is one member of a family of actin-binding proteins that contain from one to six repeats of a common domain approximately 120 amino acids in length (Vandekerckhove 1990; Hartwig and Kwiatkowski 1991). Analysis of the amino acid sequence of gelsolin (Kwiatkowski et al. 1986; Way and Weeds 1988) suggests that its six similar repeats, S1 through S6, may have arisen from gene triplication of the ancestral domain, followed by gene duplication.

Despite extensive biochemical analysis of gelsolin and gelsolin fragments, and structural analysis of the single S1 domain bound to actin, the mechanisms by which whole gelsolin modulates the architecture of F-actin filaments are understood only at a rudimentary level. The crystal structure of Ca^{2+}-free, inactive, intact horse plasma gelsolin to a resolution of 2.5 Å (Burtnick et al. 1997) provides a firm foundation for construction of new models of gelsolin action and will guide future development and testing of those models.

The crystal structure of gelsolin reveals the protein to be compact and globular, with dimensions of approximately $85 \times 36 \times 55$ Å. The six similarly folded but distinct domains can be readily distinguished (Fig. 1) and are connected in the intact molecule by less-structured linker regions (Fig. 2). Each domain is folded in a manner similar to that reported for human gelsolin S1 in a complex with G-actin (McLaughlin et al. 1993). A central five- or six-stranded β-sheet is sandwiched between a 3.5–5 turn helix that runs approximately parallel to the sheet and a shorter 1–2 turn helix that lies roughly perpendicular to it. The structural relations among the domains are that S1 and S4 are most similar, followed by S3 and S6, then S2 and S5 (Table 1).

Our structure is devoid of Ca^{2+} ions. The structure of human gelsolin S1 in a complex with G-actin had two bound Ca^{2+} ions (McLaughlin et al. 1993). While horse and human gelsolins share 94% sequence identity (Koepf et al. 1998) and could be expected to have very similar structures, it is important to note that the Ca^{2+}-free and Ca^{2+}-bound forms of S1 are virtually identical. There are 108 equivalent α-carbons and an overall rms-deviation between α-carbon positions of only 0.64 Å.

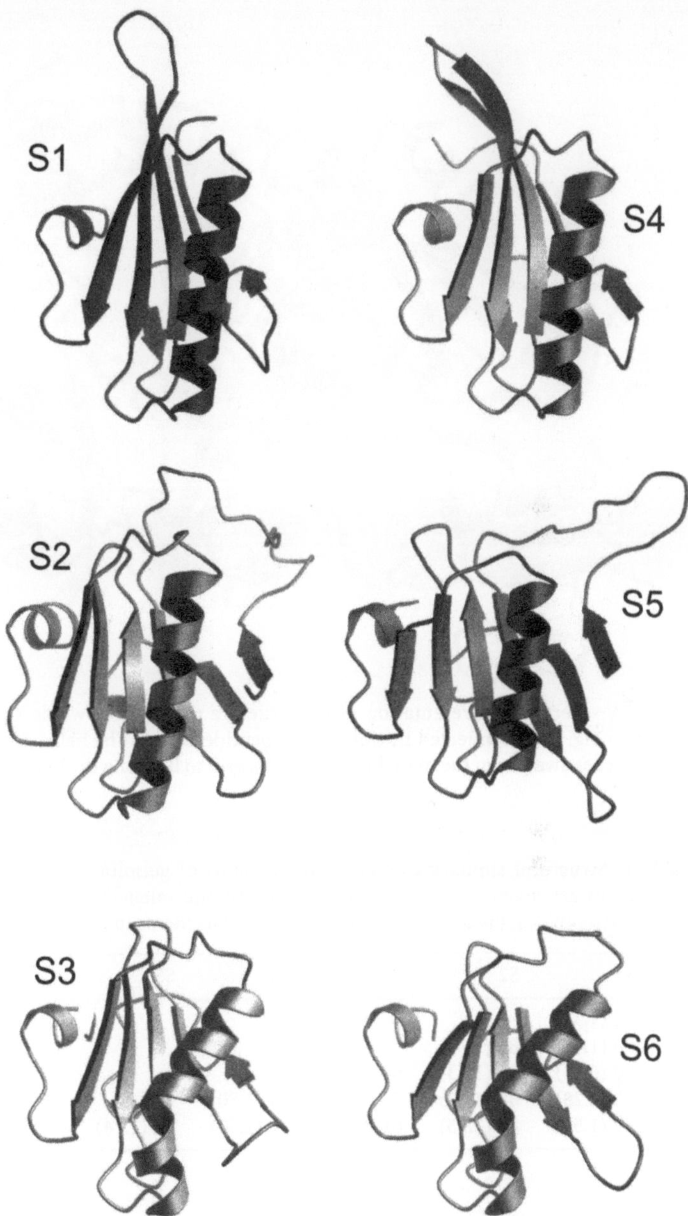

Fig. 1. A schematic representation of the six domains of gelsolin in similar orientations. The colors used to represent each segment are maintained in subsequent color figures: *S1 red*; Pro39-Tyr133; *S2 light green*, Gly137-Gly248, *S3 yellow*, Ala271-Gln364; *S4 pink* Gln419-Leu511; *S5 dark green* Pro516-Leu618; *S6 orange* Pro640-Gly731. This and subsequent ribbon diagram figures were prepared with the program MOLSCRIPT (Kraulis 1991)

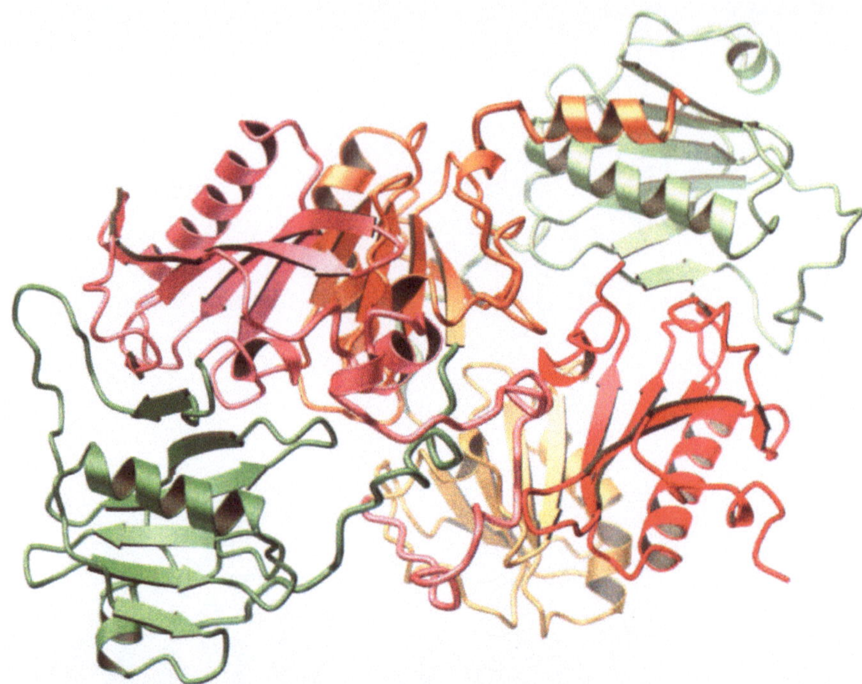

Fig. 2. A schematic representation of the structure of Ca^{2+}-free whole gelsolin. The segments shown in Fig. 1 are connected by linking polypeptide chains. The helix at the C-terminus of gelsolin reaches away from the main body of S6 (*orange*) to lie along the long helix of S2 (*light green*)

Table 1. Structural similarities among the domains of gelsolin. Similarities are given as the number of α-carbon equivalencies and, in parentheses, the average α-carbon rms displacements in Å

	S1	S2	S3	S4	S5
S2	87 (1.37)				
S3	80 (1.47)	83 (1.32)			
S4	90 (0.97)	83 (1.33)	80 (1.54)		
S5	80 (2.28)	88 (1.82)	76 (1.88)	74 (1.80)	
S6	79 (1.54)	83 (1.55)	86 (0.80)	77 (1.50)	71 (1.74)

Ca^{2+} binding does not have a dramatic effect on the structure of this individual domain. Similar suggestions emerge from NMR structures of the ± Ca^{2+} forms of isolated domains from proteins closely related to gelsolin, namely the domain of villin (Markus et al. 1994) that corresponds to S1 and that of severin (Schnuchel et al. 1995) that corresponds to S2.

A topological diagram (Fig. 3) reveals how the six domains are connected. The N- and C-terminal halves of the protein are nearly identical, except that a

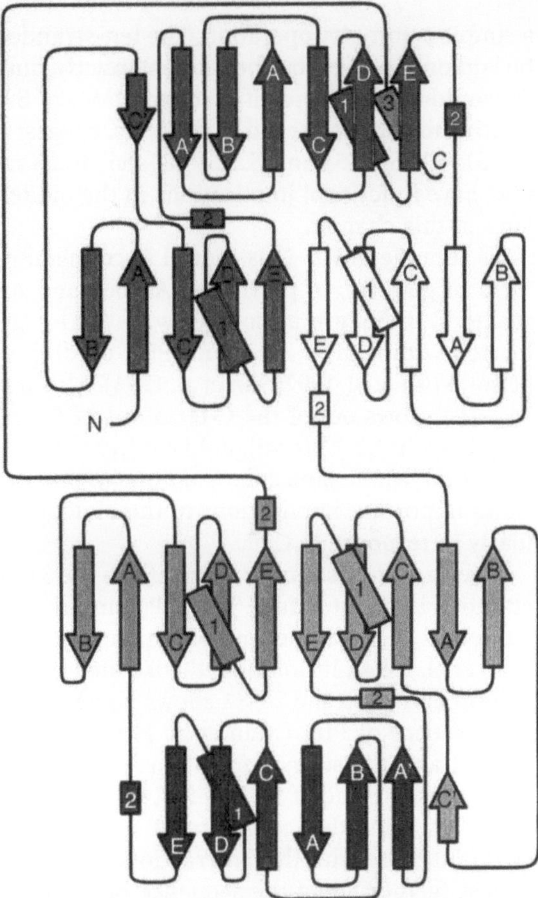

Fig. 3. A topological diagram for the structure of Ca^{2+}-free whole gelsolin. β-strands are represented as *arrows* and corresponding strands in different segments are identified by *identical letters*. α-Helices are represented as *rectangles* and corresponding helices in different segments are identified by *identical numbers*. The C-terminal helix is identified as the third helix from the start of S6

unique helical tail at the C-terminus of the protein reaches back to clasp the long helix of S2 in the first half of the molecule.

In the first half of the molecule, S1 runs into S2, which then is connected by a 20-residue linking sequence to S3. The segments are arranged so that a 10-stranded continuous β-sheet is formed as a result of contacts between S1 and S3. The C′ strand from S1 makes contact with the edge of the central β-sheet of S2. In the three-dimensional structure (Fig. 2) the kinked helix of S3 averts collision with the long helix of S1.

The two halves of gelsolin are joined by an extended 50-residue linker polypeptide between S3 and S4, and are almost independent of each other. S4 runs into S5, which is connected to S6 by another long polypeptide linker, much as in the first half of the molecule; but the structures are not exactly related by a simple symmetry operation. The ten-stranded sheets of S1–S3 and S4–S6 can be laid on top of each other almost exactly, but the resulting overlap of S2 with S5 would be poor. The different positions of S2 and S5 in the crystal structure, relative to the ten-stranded β-sheets, suggest that alternate configurations of the S1–S2 (S4–S5) and S2–S3 (S5–S6) links might be possible. This, in turn, may have functional implications in the models for gelsolin activity discussed later in this chapter.

S6 (orange in Fig. 2) is central in connecting the two halves of the Ca^{2+}-free form of gelsolin. In particular, Asp669 and Asp670, just prior to the start of strand C in S6, are in contact with Arg169-Arg170, a pair of basic residues located within the F-actin and PIP_2-binding sites near the S1–S2 junction in gelsolin (Yu et al. 1992; Sun et al. 1994). Also, a helical tail that involves residues 745–754 grows out of the C-terminus of S6 and reaches back to lie along the long helix of S2. This tail, unlike other secondary structural features of either half of the whole molecule, is unique to the C-terminal half of gelsolin. We feel it has important implications in the biological functioning of gelsolin, particularly in response to Ca^{2+}.

The part of gelsolin that constitutes the S1 to S2 junction has received much attention in the literature. Two PIP_2-binding sites are found here (Yu et al. 1992; Sun et al. 1994), the second of which overlaps a putative F-actin-binding site (Sun et al. 1994). In solution, the isolated peptide that corresponds to the A′ to A loop and A strand of S2 shows some tendency toward helix formation in the presence of PIP_2 (Xian et al. 1995). In our Ca^{2+}-free structure, no helix is evident and the strands form part of the central sheet of S2, which, in turn, abuts the C′ strand of S1. The flat, surface-exposed part of this sheet does have appeal as a site that could bind F-actin and at which competition from PIP_2 could interfere with that interaction.

Just further along the sequence of S2 from the F-actin-binding site is the single disulfide bond (Cys188 to Cys201) found in plasma gelsolins (Wen et al. 1996), and the mutation site (Asp187) that has been identified in gelsolin from sufferers of Finnish-type familial amyloidosis (Maury et al. 1990). The amyloidogenic breakdown product of this gelsolin mutant extends from the N-terminus of the B strand of the central β-sheet, a novel proteolytic site in the mutant, through the long helix and to the C-terminus of S2. Asp187 in the structure of S2 is within contact distance of three residues, Gln164 (3.24 Å), Lys166 (2.97 Å) and Asn184 (2.96 Å). The loss of these interactions on mutation of Asp187 to Asn or Tyr may destabilize this portion of the core β-sheet and expose the otherwise masked proteolytically sensitive site.

Interactions with Actin

Fragments of gelsolin that contain from one to five of its domains can be generated by limited proteolytic digestion (Kwiatkowski et al. 1985; Chaponnier et al. 1986; Bryan and Hwo 1986; Yin et al. 1988; Pope et al. 1997) or by expression in bacterial cell lines (Way et al. 1989; Pope et al. 1991). Investigations of the activities of these products have led to the assignment of discrete functions to individual domains. S1 binds actin monomers in the absence of Ca^{2+} and the complex can cap F-actin filaments.

The C-terminal half of gelsolin, S4–S6, constitutes a second, Ca^{2+}-dependent, actin monomer-binding fragment that, specifically through S4, competes for the same binding site on actin as S1. S2 contains a Ca^{2+}-independent F-actin-binding site and is able to decorate actin filaments.

A fragment consisting of S2–S6 can nucleate pointed-end growth of actin filaments, whereas the N-terminal half of gelsolin, S1–S3, retains F-actin-severing activity. S3 may contain a polyphosphoinositide-binding site (Fujita et al. 1995) and it, as well as S6, may participate in calcium regulation by shielding the actin-binding sites on S1 and S4, respectively, in low calcium environments.

The helical tail at the C-terminus of S6 latches the second half of the protein to the first until released by Ca^{2+} (Burtnick et al. 1997). Dissociation of actin-gelsolin complexes can be induced by the binding of polyphosphoinositides to a fragment consisting of S1 and S2.

The crystal structure of isolated human S1 bound to monomeric actin (McLaughlin et al. 1993) provided details of the interaction of S1 with actin. The longer of the two α-helices of S1 binds in a cleft in the actin monomer that would be located at the fast-growing, barbed end of the Holmes model of F-actin (Holmes et al. 1990). Two Ca^{2+} ions are associated with this structure, one exclusively bound to gelsolin and another bound at the actin/gelsolin interface.

Intact, Ca^{+}-free gelsolin has the long helices of all six domains facing outward. Helices from S1, S3, S4 and S6 are presented on one side of the structure, with those from S2 and S5 on the other (Fig. 2). This enables the docking of S1 as part of whole gelsolin onto the actin protomer at the barbed end of a short length of F-actin in a manner consistent with that in which G-actin binds human S1 (McLaughlin et al. 1993). This docking procedure (Fig. 4) puts much of gelsolin in steric conflict with positions that would be occupied by other actins in an F-actin filament structure. It clearly demonstrates gelsolin to be the appropriate size and shape to cap an actin filament. However, closer inspection of this model suggests why the Ca^{2+}-free form of gelsolin does not bind actin.

Firstly, there is significant steric conflict between S3 (yellow in Fig. 4) and the same actin monomer to which S1 is docked. Secondly, S2 (green in Fig. 4) is responsible for binding gelsolin to F-actin. It is not in contact with actin at all, but points away from the filament instead. Left exposed in this model are two regions on the surface of actin that are thought to play a part in the binding of S2, the region near its N-terminus (Feinberg et al. 1995), and the binding site for α-actinin (Way et al. 1992); and thirdly, while the C-terminal half of gelsolin (not shown in Fig. 4) would lie within striking distance of a second actin at the end of the filament, S4, which contains the third actin-binding site in gelsolin (Pope et al. 1995), is not in a position to contact actin.

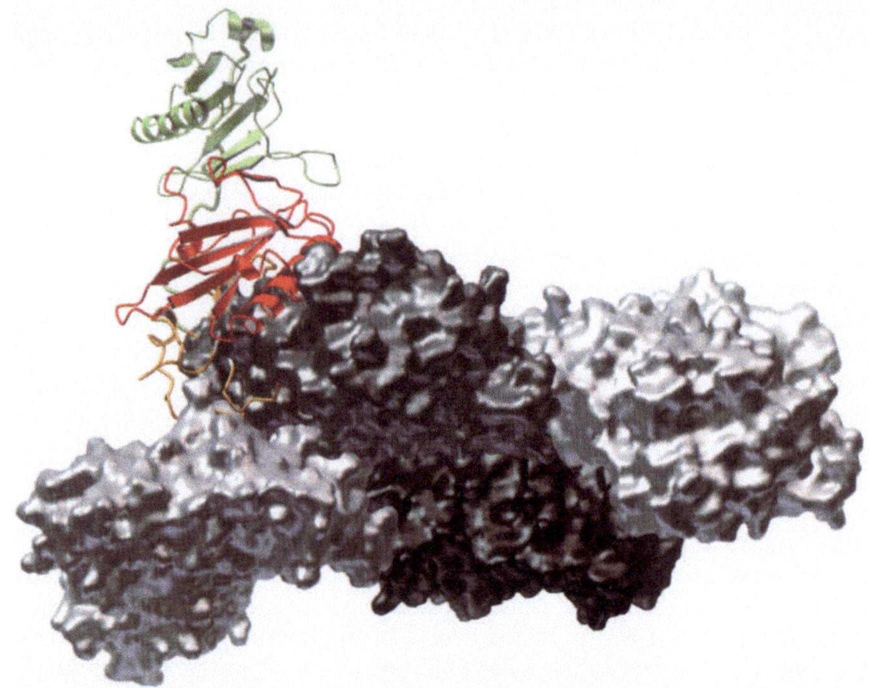

Fig. 4. A model showing S1 through S3 docked to the S1-binding site on an actin monomer at the end of a short stretch of F-actin consisting of four actin monomers positioned according to Holmes et al. (1990). The surfaces of the four actin units are portrayed in *different shades of gray* using the program GRASP (Nicholls and Honig 1991). In this model, S3 suffers severe steric clashes with the same actin unit to which S1 is bound, while S2 makes no contact with actin at all

These discrepancies, taken with the apparent lack of effect of Ca^{2+} on the secondary and tertiary structures of individual gelsolin-type domains (Markus et al. 1994; Schnuchel et al. 1995; Koepf and Burtnick 1996), strongly suggest that there must be shifts in the relative positions of the domains in order to permit the known interactions between gelsolin and actin to occur.

To accommodate these structural data, and to be consistent with the known effects of Ca^{2+} on the function of gelsolin (Pope et al. 1997), we present a model for the activation of gelsolin by Ca^{2+}. Firstly, Ca^{2+} binding to the second half of gelsolin results in release of the interactions between the C-terminal tail and S2. This allows the structure to open up, consistent with light-scattering data (Patkowski et al. 1990; Hellweg et al. 1993), to expose the previously masked F-actin-binding site on S2. This enables the initial contact between gelsolin and F-actin, which takes place prior to the actual severing event (Kinosian et al. 1996).

Further structural rearrangement, facilitated by the binding of a second Ca^{2+} ion, could free S1, the G-actin-binding domain, from inhibition by S3. Two possibilities now occur. Each domain may maintain its individual structural integrity, with motion or flexion only at the interdomain linkers. This would limit S1 to bind to its identified binding site on the same actin unit to which S2 is attached. Alternatively, and more consistent with the data of McGough and Way 1995; McGough et al. 1998), some unraveling of the S1 to S2 junction could be induced, possibly by extraction of the A′ strand from S2, leaving behind a 5-stranded core β-sheet as found in S1, S3, S4 and S6. This would lengthen the S1 to S2 linker to 17 residues, comparable to that between S2 and S3, and permit S1 to reach its binding site on the next actin unit down longitudinally from the one initially contacted by S2. Either case would require disconnection of the actin unit adjacent to the one that binds S1, i.e., severing would occur.

The second half of gelsolin, once the initial Ca^{2+}-binding event has released the interaction between the C-terminal helix and S2, is sufficiently independent to seek and bind to another actin unit across the filament from where the first half is attached. Presumably this occurs through an interaction between S4 and actin that mirrors the S1-actin-binding mode. This would require relief of the inhibitory effect of S6 on the actin-binding ability of S4, perhaps by intervention of another Ca^{2+}.

The movement implied by this model is well within the range of the strand that connects S3 to S4 (Burtnick et al. 1997). Such a series of events would firmly cap the new barbed end of the freshly severed F-actin. Given the appropriate environmental conditions, this gelsolin-capped section of filament could serve as a nucleus for extension at its free pointed end.

Conclusions

The actin-related activities of gelsolin depend upon its actin-binding sites being occluded in the absence of activating conditions. Concealment is achieved by using the tertiary structures of S3, S6, and the C-terminal helix to mask the actin-binding surfaces of S1, S4 and S2, respectively. Subsequent successive binding of at least two Ca^{2+} causes the structure to open up. The individual domains are then able to shift in position relative to one another to reveal the three actin-binding surfaces. Sequential interactions with S2, then S1 and S4, sever and then cap the newly exposed barbed end of the cut filament.

Acknowledgments. This work has been funded by grants to LDB from the Heart and Stroke Foundation of British Columbia and the Yukon.

References

Asch HL, Head K, Dong Y, Natoli F, Winston JS, Connolly JL, Asch BB (1996) Widespread loss of gelsolin in breast cancers of humans, mice and rats. Cancer Res 56:4841–4845

Baldassare J, Henderson PA, Tarver A, Fisher G (1997) Thrombin activation of human platelets dissociates a complex containing gelsolin and actin from phosphatidylinositide-specific phospholipase C. Biochem J 324:283–287

Borovikov YS, Norman JC, Price LS, Weeds A, Koffer A (1995) Secretion from permeabilised mast cells is enhanced by addition of gelsolin: contrasting effects of endogenous gelsolin. J Cell Sci 108:657–666

Bryan J, Hwo S (1986) Definition of an amino-terminal actin-binding domain and a carboxyl-terminal Ca^{2+}-regulatory domain in human brevin. J Cell Biol 102:1439–1446

Burtnick LD, Koepf EK, Grimes JM, Jones EY, Stuart DI, McLaughlin PJ, Robinson RC (1997) The crystal structure of plasma gelsolin: implications for actin severing, capping and nucleation. Cell 90:661–670

Chaponnier C, Janmey PA, Yin HL (1986) The actin filament-severing domain of gelsolin. J Cell Biol 103:1473–1481

Chen P, Murphy-Ullrich HE, Wells A (1996) A role for gelsolin in actuating epidermal growth factor receptor-mediated cell motility. J Cell Biol 134:689–698

Cunningham CC, Stossel TP, Kwiatkowski DJ (1991) Enhanced mobility in NIH 3T3 fibroblasts that overexpress gelsolin. Science 251:1233–1236

Davoodian K, Ritchings BW, Ramphal R, Bubb M (1997) Gelsolin activates DNase I in vitro and in cystic fibrosis sputum. Biochemistry 36:9637–9641

Feinberg J, Benyamin Y, Roustan C (1995) Definition of an interface implicated in gelsolin binding to the sides of actin filaments. Biochem Biophys Res Commun 209:426–432

Fujita H, Laham LE, Janmey PA, Kwiatkowski DJ, Stossel TP, Banno Y, Nozawa Y, Müllauer L, Ishizaki A, Kuzumaki N (1995) Functions of [His321]gelsolin isolated from a flat revertant of ras-transformed cells. Eur J Biochem 229:615–620

Haddad JG, Harper KD, Guoth M, Pietra GG, Sanger SW (1990) Angiopathic consequences of saturating the plasma scavenger system for actin. Proc Natl Acad Sci USA 87:1381–1385

Hartwig JH, Kwiatkowski DJ (1991) Actin-binding proteins. Curr Opin Cell Biol 3:87–97

Hellweg T, Hinssen H, Eimer W (1993) The Ca^{2+}-induced conformational change of gelsolin is located in the carboxyl-terminal half of the molecule. Biophys J 65:799–805

Herrmannsdoerfer AJ, Heeb PJ, Feustel PJ, Estes JE, Keenan CJ, Minnear FL, Selden L, Giunta C, Flor JR, Blumenstock FA (1993) Vascular clearance and organ uptake of G- and F-actin in the rat. Am J Physiol 265:G1071–G1081

Holmes KC, Popp D, Gebhard W, Kabsch W (1990) Atomic model of the actin filament. Nature 347:44–49

Janmey P, Stossel T, Allen P (1998) Deconstructing gelsolin: identifying sites that mimic or alter binding to actin and phosphoinositides. Chem Biol 5:R81–R85

Kinosian HJ, Selden LA, Estes JE, Gershman LC (1996) Kinetics of gelsolin interaction with phalloidin-stabilized F-actin. Biochemistry 35:16550–16556

Koepf EK, Burtnick LD (1996) Multiple pathways for denaturation of horse plasma gelsolin. Biochem Cell Biol 74:101–107

Koepf EK, Hewitt J, Vo H, MacGillivray RTA, Burtnick LD (1998) *Equus caballus* gelsolin: cDNA sequence and protein structural implications. Eur J Biochem 251:613–621

Kothakota S, Azuma T, Reinhard C, Klippel A, Tang J, Chu K, McGarry TJ, Kirschner MW, Koths K, Kwiatkowski DJ, Williams LT (1997) Caspase-3-generated fragment of gelsolin: effector of morphological change in apoptosis. Science 278:294–298

Kraulis P (1991) MOLSCRIPT: a program to produce both detailed and schematic plots of protein structures. J Appl Crystallogr 24:46–950

Kwiatkowski DJ, Janmey PA, Mole JE, Yin HL (1985) Isolation and properties of two actin-binding domains in gelsolin. J Biol Chem 260:15232–15238

Kwiatkowski DJ, Stossel TP, Orkin SH, Mole JE, Colten H, Yin HL (1986) Plasma and cytoplasmic gelsolins are encoded by a single gene and contain a duplicated actin-binding domain. Nature 323:455–458

Markus MA, Nakayama T, Matsudaira P, Wagner G (1994) Solution structure of villin 14T, a domain conserved among actin-severing proteins. Protein Sci 3:70–81

Maury CPJ, Alli K, Baumann M (1990) Finnish hereditary amyloidosis. Amino acid sequence homology between the amyloid fibril protein and human plasma gelsolin. FEBS Lett 260:85–87

McGough A, Way M (1995) Molecular model of an actin filament capped by a severing protein. J Struct Biol 115:144–150

McGough A, Chiu W, Way M (1998) Determination of the gelsolin binding site on F-actin: implications for severing and capping. Biophys J 74:764–772

McLaughlin P, Gooch JT, Mannherz H-G, Weeds AG (1993) Structure of gelsolin segment 1-actin complex and the mechanism of severing. Nature 364:685–692

Nicholls A, Honig B (1991) A rapid finite difference algorithm, utilizing successive over-relaxation to solve the Poisson-Boltzmann equation. J Comp Chem 12:435–445

Patkowski A, Seils J, Hinssen H, Dorfmüller Th (1990) Size, shape parameters, and Ca^{2+}-induced conformational change of the gelsolin molecule: a dynamic light scattering study. Biopolymers 30:427–434

Pope B, Way M, Weeds AG (1991) Two of the three actin-binding domains of gelsolin bind to the same subdomain of actin. FEBS Lett 280:70–74

Pope B, Maciver S, Weeds A (1995) Localization of the calcium-sensitive actin monomer-binding site in gelsolin to segment 4 and identification of calcium binding sites. Biochemistry 34:1583–1588

Pope B, Gooch J, Weeds A (1997) Probing the effects of calcium on gelsolin. Biochemistry 36:15848–15855

Schnuchel A, Wiltscheck R, Eichinger L, Schleicher M, Holak TA (1995) Structure of severin domain 2 in solution. J Mol Biol 247:21–27

Steed PM, Nagar S, Wennogle LP (1996) Phospholipase D regulation by a physical interaction with the actin-binding protein gelsolin. Biochemistry 35:5229–5237

Sun H-Q, Wooten DC, Janmey PA, Yin HL (1994) The actin side-binding domain of gelsolin also caps actin filaments. Implications for actin filament severing. J Biol Chem 269:9473–9479

Tanaka M, Müllauer L, Ogiso Y, Fujita H, Moriya S, Furuuchi K, Harabayashi T, Shinohara N, Koyanagi T, Kuzumaki N (1995) Gelsolin: a candidate for suppressor of human bladder cancer. Cancer Res 55:3228–3232

Vandekerckhove J (1990) Actin binding proteins. Curr Opin Cell Biol 2:1–50

Vasconcellos CA, Lind SE (1993) Coordinated inhibition of actin-induced platelet aggregation by plasma gelsolin and vitamin D-binding protein. Blood 82:648–3657

Way M, Weeds AG (1988) Nucleotide sequence of pig plasma gelsolin. Comparison of protein sequence with human gelsolin and other actin-severing proteins shows strong homologies and evidence for large internal repeats. J Mol Biol 203:1127–1133

Way M, Gooch J, Pope B, Weeds AG (1989) Expression of human plasma gelsolin in *Escherichia coli* and dissection of actin binding sites by segmental deletion mutagenesis. J Cell Biol 109:593–605

Way M, Pope B, Weeds AG (1992) Evidence for functional homology in the F-actin binding domains of gelsolin and actinin: implications for the requirements of severing and capping. J Cell Biol 119:835–842

Wen D, Corina K, Chow E, Miller S, Janmey P, Pepinsky R (1996) The plasma and cytoplasmic forms of human gelsolin differ in disulfide structure. Biochemistry 35:9700–9709

Witke W, Sharpe AH, Hartwig JH, Azuma T, Stossel TP, Kwiatkowski DJ (1995) Hemostatic, inflammatory, and fibroblast responses are blunted in mice lacking gelsolin. Cell 81:41–51

Xian W, Vegners R, Janmey P, Braunlin W (1995) Spectroscopic studies of a polyphosphoinositide-binding peptide from gelsolin: behavior in solutions of mixed solvent and anionic micelles. Biophys J 69:2695–2702

Yin H (1987) Gelsolin: calcium and polyphosphoinositide-regulated actin-modulating protein. BioEssays 7:176–179

Yin H, Iida K, Janmey P (1988) Identification of a polyphosphoinositide-modulated domain in gelsolin which binds to the sides of actin filaments. J Cell Biol 106:805–812

Yu F-X, Sun H-W, Janmey PA, Yin HL (1992) Identification of a polyphosphoinositide-binding sequence in an actin monomer-binding domain of gelsolin. J Biol Chem 267:14616–14621

Arps: Actin-Related Proteins

Laura M. Machesky and Robin C. May[1]

Introduction

The Arps (actin-related proteins) constitute a recently discovered family of proteins related to actin in sequence and probably 3D structure. Although their roles in the cell are diverse, many of the Arps function as subunits of multi-protein complexes. Arp1 is found in a 20-S complex (with at least seven different proteins) which regulates cytoplasmic dynein-based microtubule-membrane interactions. Arp2 and Arp3 function in an 8.5-S protein complex that can initiate actin polymerization, as well as cross-linking and capping actin filaments. Arp7 and Arp9 are components of the 2000-kDa SWI/SNF complex that is thought to regulate chromatin structure to alleviate transcriptional repression. There are several other members of the Arp family (up to ten family members in budding yeast) whose functions are less well-characterized (Fig. 1), but it seems that Arps generally have diverse functions and in some cases may be related to actin only in sequence and structure rather than in function. The Hsc70 family and some sugar kinases also show a significant structural similarity to actin, but are not included as Arps because they do not share significant primary sequence homology (Schroer et al. 1994).

Arp1: the Dynactin Complex

Microtubules provide the cell with a means to establish polarity and maintain the organization of its contents. They provide a structural scaffolding for membrane-bound organelles, such as the Golgi and endoplasmic reticulum. Additionally, microtubules serve as tracks which motor proteins use to transport vesicles to various destinations in the cell. When a cell divides, microtubules form a spindle, along which chromosomes and intracellular membranes are divided between the daughter cells. Various motor proteins in the cell generate force between microtubules and membranes. The microtubule-based motor dynein has roles in spindle orientation and assembly, vesicle transport and nuclear migration. Dynein activity in these events requires the dynactin complex, a large complex containing at least seven different proteins including

[1] School of Biosciences, The University of Birmingham, Edgbaston, Birmingham B15 2TT, UK

Results and Problems in Cell Differentiation, Vol. 32
C. dos Remedios (Ed.): Molecular Interactions of Actin

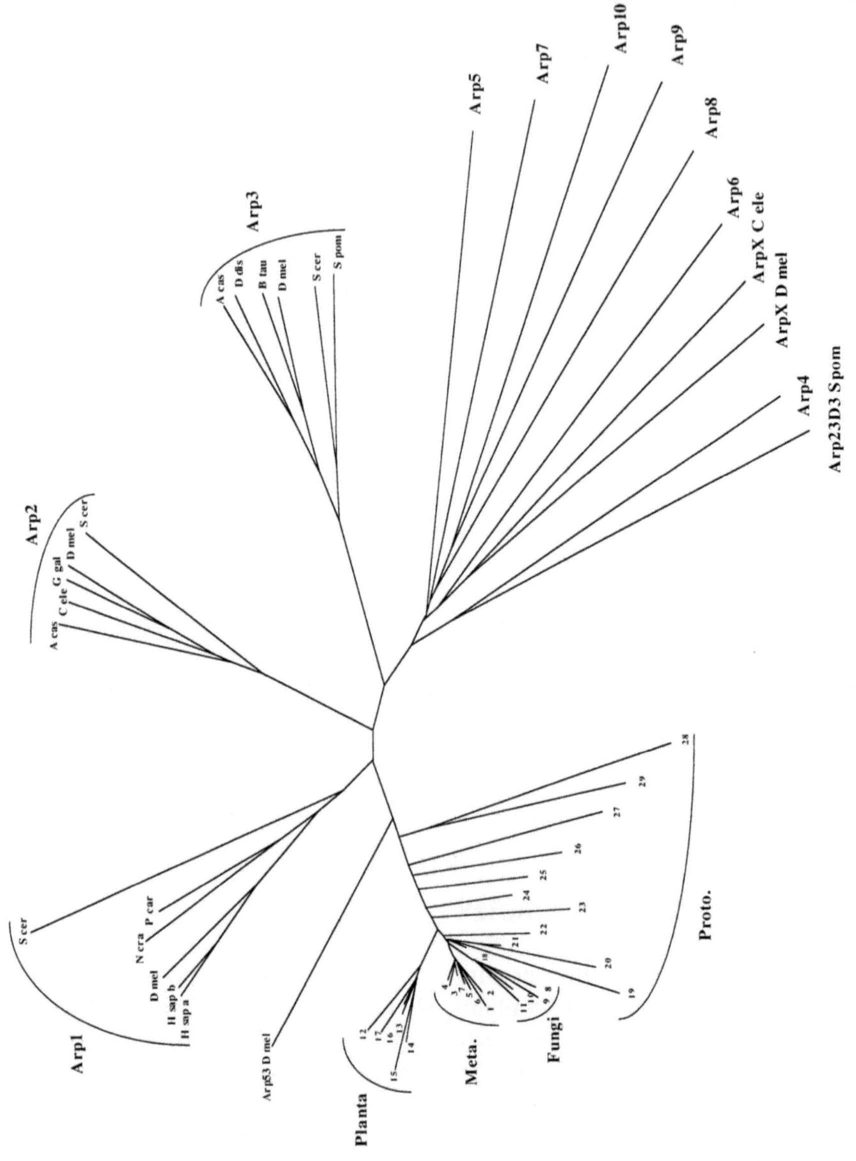
Arp1
S cer
N cra
P car
D mel
H sap b
H sap a
Arp2
A cas
C ele
G gal
D mel
S cer
Arp3
Arp5
Arp7
Arp10
Arp9
Arp8
Arp6
ArpX C ele
ArpX D mel
Arp4
Arp23D3 Spom
Proto.
Fungi
Meta.
Planta
Arp53 D mel

the actin-related protein Arp1 (Holleran et al. 1998). The dynactin complex was originally discovered as an activator of dynein-based vesicle motility (hence the name *dyn*ein-*act*ivator).

Structure of the Dynactin Complex

Electron microscopic and biochemical studies have yielded a picture of how subunits of the dynactin complex are organized and of how they might connect to the dynein motor and to cell membranes (Fig. 2A, B). Subunits of 150, 62, 50, 45, 37, 32, 27 and 22 kDa are present in a stoichiometry of 2 : 1 : 5 : 10 : 1 : 1 : 1 : 1 in the dynactin complex (Gill et al. 1991; Holleran et al. 1996; Paschal et al. 1993; Schafer et al. 1994). The 150 kDa subunit, $p150^{Glued}$, binds both to dynein and to microtubules (Karki and Holzbaur 1995; Vaughan and Vallee 1995; Waterman-Storer et al. 1995). p62 is an apparently novel protein which localizes to one end of a short Arp1 filament in the dynactin complex (Schafer et al. 1994).

p45, originally called centractin, is now termed Arp1. Arp1 filaments contain 8–13 subunits and are similar in appearance to actin filaments (Schafer et al. 1994). p50 is named dynamitin because its overexpression disrupts the dynactin complex (Burkhardt et al. 1995). p37 and p32 are the α- and β2-subunits of actin-capping protein, which apparently have a dual role in the cell, as they also bind to and cap the fast-growing ends of conventional actin filaments (Schafer et al. 1994a, b). In the dynactin complex, the capping protein localizes to one end of the short Arp1 filament (Schafer et al. 1994), but has not yet been shown to bind directly to Arp1. The 22 kDa subunit has no homology to other proteins and its function in the complex is not yet known (Holleran et al. 1998).

Fig. 1. Bootstrapped nonrooted phylogenetic tree comparing the amino acid sequences of actin-related proteins and 29 actins of diverse phyla. The tree was generated by the computer program CLUSTALW (Thompson et al. 1994). Swissprot databank mnemonics of the sequences are (1) for actin sequences: *1/* Acts_Human; *2/* Act4_Drome; *3/* Act3_Limpo; *4/* Actb_Xenbo; *5/* Actm_Aplca; *6/* Act1_Podca; *7/* Act3_Bommo; *8/* Act_Yeast; *9/* Act_Thela; *10/* Act_Schpo; *11/* Act_Cryne; *12/* Act1_Maize; *13/* Act3_Pea; *14/* Act1_Dauca; *15/* Act2_Dauca; *16/* Act2_Pea; *17/* Act3_Soybn; *18/* Act1_Acaca; 19/ Actd_Phypo; *20/* Act4_Dicdi; *21/* Act3_Dicdi; *22/* Act_Enthi; *23/* Act_Phyme; *24/* Act1_Naefo; *25/* Act1_Plafa; *26/* Act2_Plafa; *27/* Act_Leima; 28/ Act_Eupcr; *29/* Act1_Oxyn, Arp53D D mel/Acty_Drome; (2) for ARP1 sequences: *S cer/* Act5_Yeast; *H sap a/* Actz_Human; *H sap b/* Acty_Human; *D mel/* Actz_Drome; *P car/* Actz_Pneca; *N cra/* Actz_Neucr; (3) for ARP2 sequences: *S cer/* Act2_Yeast; *A cas/* Actw_Acaca; *C ele/* Actw_Caeel; *G gal/* Actw_Chick; *D mel/* Actw_Drome; for ARP3 sequences: *S pom/* Act2_Schpo; *S cer/* Act4_Yeast; *D mel/* Actv_Drome; *B tau/* Actv_Bovin; *D dis/* Actv_Dicdi; *A cas/* Actv_Acaca; (4) for distantly related ARPs: *ARP4/* Act3_Yeast; *Arp23D3 S pom/* Yae9_Schpo; ARP5/ Ynf9_Yeast; *ArpX D mel/* Actu_Drome; *ArpX C ele/* Yp86_Caeel. The other *S cerevisiae* ARP sequences are taken from the PIR databank: *ARP6* Accession number S64917; *ARP7* S54508; *ARP8* S61695; *ARP9* S53950; *ARP10* S52672. (Permission Barbara Winsor)

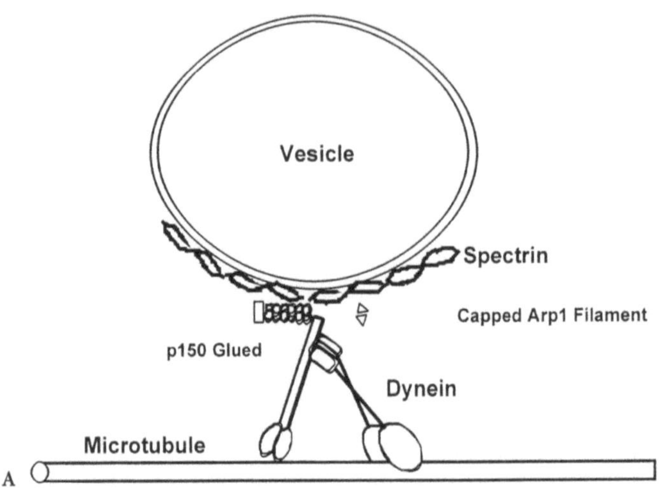

I. Nuclei move during cell division

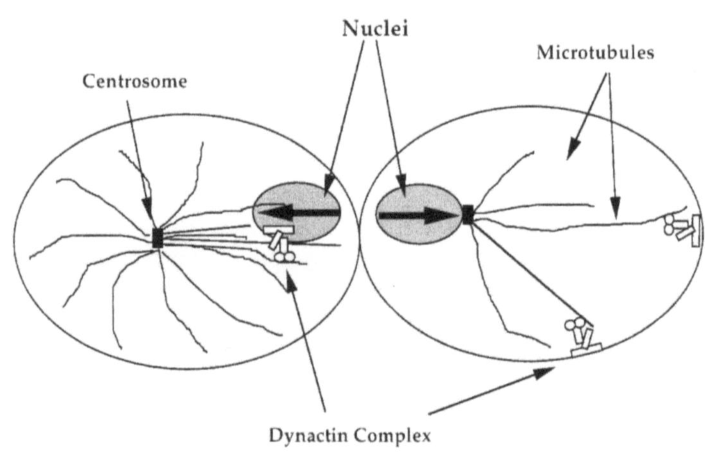

II. Nuclei move to distribute evenly in hyphae

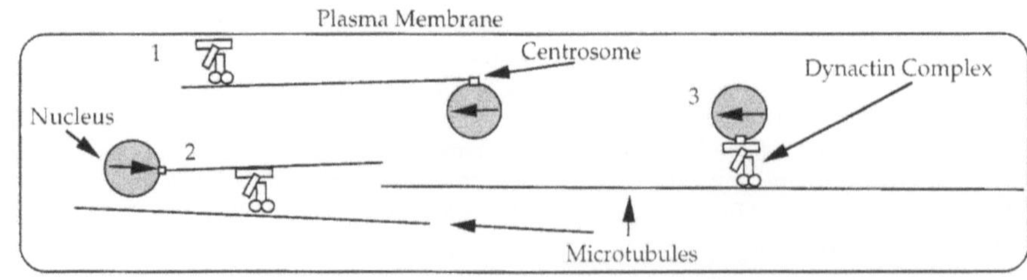

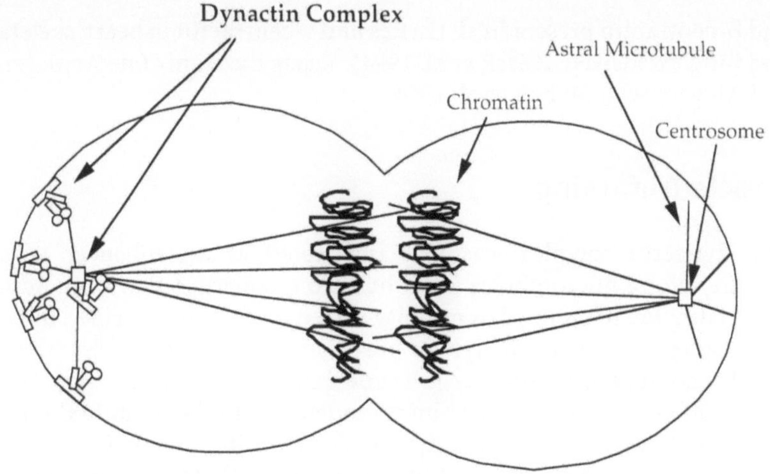

Fig. 2A–C. The dynactin complex is important for vesicle trafficking nuclear migration and mitotic spindle assembly. A Dynactin complex connects membrane-bound organelles to microtubules to create force and movement. This model suggests how the dynactin complex might tether dynein microtubules and organelles together based on evidence of protein interactions and electron microscopic studies (Holleran et al. 1998). The short Arp1 filament capped at one end by capping protein and at the other by p62 may bind to the spectrin network surrounding the organelle. p150^{Glued} forms a dimer and binds to microtubules dynein and Arp1 providing multiple connections in this complex. The *small inset* shows purified dynactin complex visualized by rotary shadowing electron microscopy. (Photo. Dorothy Schafer, John Heuser, Trina Schroer). B When cells divide the nucleus moves into a central position in newly formed cell (I). The dynactin complex may attach directly to the nucleus or the cytoskeletal network surrounding the nucleus and use microtubule-mediated dynein motor activity to pull the nucleus toward the center of the cell (shown *on the left side*). Alternatively the centrosome might capture microtubules which are tethered dynactin complexes in the cell cortex and thus pull the nucleus to the correct position (shown *on the right side*). Hyphae of filamentous fungi contain many nuclei all evenly distributed throughout the long filaments (II). Dynactin and dynein are required to establish and maintain this arrangement. Three possible models are shown for how the dynactin complex might position nuclei. *1* Cortically attached dynactin could produce force on a microtubule attached to the centrosome and thus pull the nucleus by "reeling in" the microtubule. *2* The dynactin complex may be able to generate microtubule sliding thus pulling a nucleus attached by its centrosome. *3* Dynactin complex might attach directly to the nucleus or surrounding cytoskeleton and pull the nucleus as cargo along microtubules. C The dynactin complex is involved in spindle assembly and orientation. In this model the dynactin complex connects to the cell cortex via its short Arp1 filament. Dynein and p150^{Glued} interact with astral microtubules in the spindle and can create force to position the spindle correctly in the dividing cell

Arp1 appears to be well conserved among phyla in both function and structure. Arp1 genes have been characterized in yeast, *Drosophila, Pneumocystis carinii* (Christopher et al. 1995) and vertebrates. Humans have at least three isoforms of Arp1, termed α-centractin, β-centractin and γ-centractin. Both α- and β-centractin are present in the dynactin complex, while γ-centractin is less well characterized. These show tissue-specific localizations, with α-centractin and β-centractin present in all tissues and γ-centractin in heart, skeletal muscle and lung exclusively (Clark et al. 1994). Yeasts have only one Arp1 gene (Clark and Meyer 1994; Muhua et al. 1994).

Vesicle Trafficking

The dynactin complex was first implicated as an enhancer of organelle motility along microtubules in an in vitro reconstituted system. Adding dynactin complex to an in vitro motility system containing purified microtubules and dynein restores full dynein-based movement of vesicles along microtubules compared to motility in crude extracts (Gill et al. 1991). Dynactin is an essential component, since immunodepletion (Gill et al. 1991) or addition of antibodies which block the dynein-dynactin interaction (Waterman-Storer et al. 1995) disrupts the motility of vesicles along microtubules in vitro. This suggests that dynactin is essential for at least some forms of vesicle trafficking.

The 150 kDa subunit of dynactin is homologous to the gene mutated in a well-characterized *Drosophila* mutant *glued* (Plough and Ives 1935). Glued mutants show aberrant formation of the optic lobe and eye and express a truncated Glued protein (McGrail et al. 1995). The truncated product is thought to act as a dominant negative, because it still binds to cytoplasmic dynein, but no longer associates with other components of the dynactin complex (Waterman-Storer and Holzbaur 1996).

Current structural data suggest that the dynactin complex acts as a tether between the vesicle and the microtubule, with the Arp1 filament bound to the vesicle membrane-cytoskeleton, and dynein moving along the microtubule (Fig. 2A). Holleran et al. (1996) showed that Arp1 associates directly with spectrin, providing a link to the membrane cytoskeleton. The properties of Arp1 filaments are consistent with a role in linking the complex to membrane-bound organelles. Arp1 can also form long macromolecular structures of 0.62–22 μm when over-expressed, which are likely to be bundles or aggregates. These Arp1 structures do not bind to the conventional actin-binding reagents phalloidin or cytochalasin D, nor do they contain any detectable conventional actin (Holleran et al. 1996).

The dynactin complex is important for maintaining the organization of the vesicle trafficking pathway in mammalian cells, as evidenced by the dramatic disruption caused by overexpressing components of the complex. Arp1 overexpression disrupts the Golgi complex, possibly by recruiting Golgi-derived vesicles to Arp1 filament aggregates (Holleran et al. 1996). Overexpression of p50-dynamitin causes a disruption of endoplasmic reticulum to Golgi transport, and a redistribution of endosomes and lysosomes in the cell from the perinuclear regions to the cell periphery (Burkhardt et al. 1997).

Nuclear Migration

The position of the nucleus in cells is maintained by interactions with microtubule-based motors. During cell division, for example, nuclei must reposition in each new daughter cell. A more extreme example comes from hyphal fungi, where nuclei position themselves along the hyphae so that they are approximately evenly distributed. Both in yeast and filamentous fungi, the dynactin complex has been implicated in nuclear migration (Clark and Meyer 1994; Muhua et al. 1994; Planmann et al. 1994).

Both Arp1 and cytoplasmic dynein are required for normal nuclear distribution in the filamentous fungus *Neurospora crassa*. Mutants in these two genes show identical phenotypes, with nuclei concentrating in large clumps along hyphae. Hyphal growth is also retarded in these mutants, despite the presence of normal microtubule networks (Planmann et al. 1994). Thus, the dynactin complex is not required for microtubule structure or function, but is required for nuclei to distribute along microtubules.

In budding yeast, Arp1 null mutants display an increased number of anucleate and binucleate cells, indicating a defect in nuclear migration during division (Muhua et al. 1994). Cytoplasmic dynein mutants also have a similar phenotype, as do cells overexpressing Arp1. Strong evidence for the requirement of dynein in nuclear migration along microtubules in comes from the in vitro reconstitution studies of Reinsch and Karsenti using *Xenopus* egg extracts (Reinsch and Karsenti 1997). Figure 2B shows three possible models for how cytoplasmic dynein and the dynactin complex could move nuclei along microtubules.

Spindle Assembly and Orientation

Dynactin is also required for spindle assembly and orientation during mitosis (Fig. 2C). Immunodepeletion of either p150^{Glued} or cytoplasmic dynein prevents spindle assembly in mitotic extracts (Gaglio et al. 1996). Overexpression of p50 (dynamitin) causes an increased percentage of cells to be blocked in mitosis and show an aberrant spindle morphology (Echeverri et al. 1996). Orientation of the spindle may be regulated by interactions between the dynactin complex and the cortical actin cytoskeleton. In *S. cerevisiae*, mutants in Arp1 or capping protein showed defective spindle orientation, and double mutants showed a more severe phenotype (Muhua et al. 1994).

It has been proposed that the dynactin complex could interact with yeast cortical actin patches via the Arp1 filament, which binds to capping protein in the patches. Dynein bound to the astral microtubules in the spindle could thus contact the cell cortex and orient the spindle relative to the rest of the cell (Fig. 3C; Muhua et al. 1994). This model fits well to studies in *C. elegans*, where patches containing capping protein and actin localize to places where astral microtubules contact the cortex to effect spindle rotation during specific cell

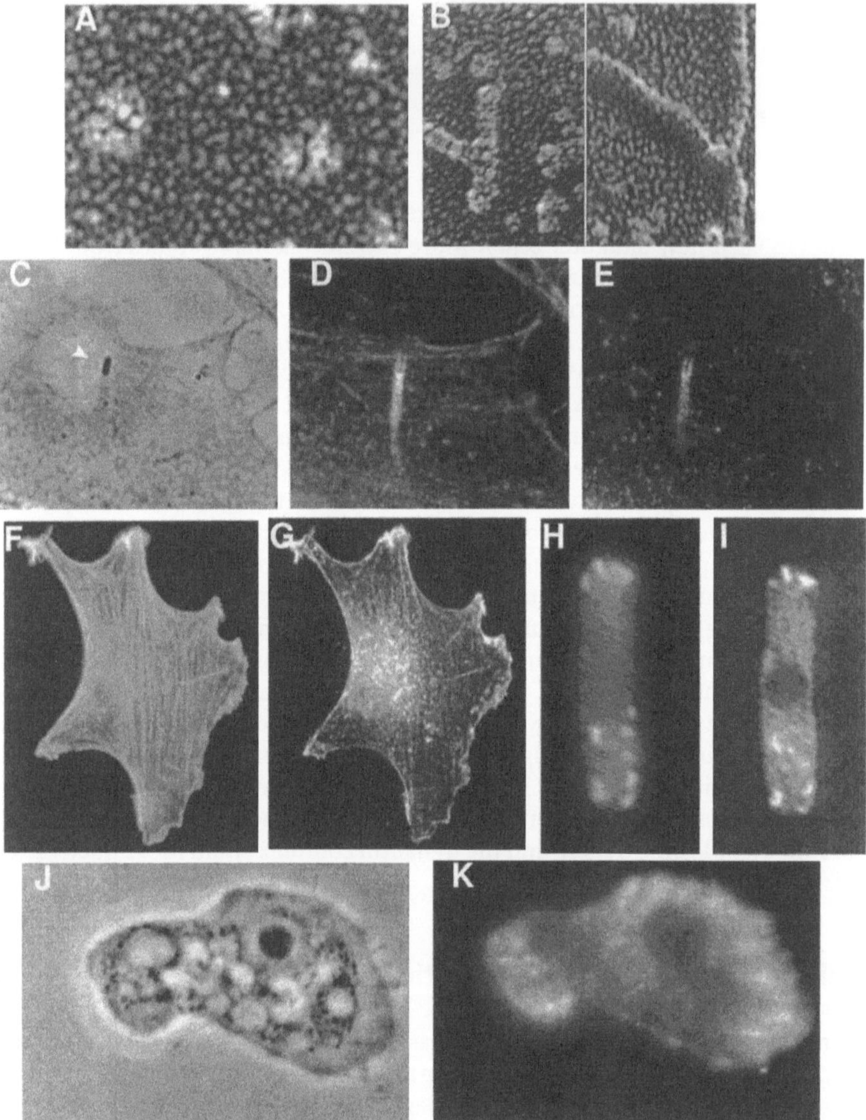

Fig. 3A–K. The Arp2/3 complex in vitro and in vivo **A** Rotary shadowing electron microscopy of purified Arp2/3 complex reveals horseshoe-shaped molecules. **B** When Arp2/3 complex is mixed with filamentous actin, branched structures are observed, presumably containing one Arp2/3 complex at the junction of each branch. These are similar to actual actin branches observed in motile cells by Svitkina et al. (Svitkina and Borisy 1998). **C** Phase contrast showing a *Listeria monocytogenes* bacterium making an actin comet in an infected cell (*white arrow* points to bacterium). **D** Filamentous actin staining using rhodamine phalloidin, showing the actin comet tail behind the bacterium. **E** Anti-p21-Arc monoclonal antibody labeling of the *Listeria* actin tail, showing that Arp2/3 complex is found along the length of the tail. **F** Rhodamine phalloidin labeling of filamentous actin in a mouse fibroblast. **G** Anti-p34-Arc staining showing the lamellipodia and punctate distribution of the Arp2/3 complex. **H** Rhodamine phalloidin labeling of filamentous actin in an *S. pombe* cell showing cortical filamentous actin spots. **I** Anti-Arp3 polyclonal antibody labelling showing the Arp2/3 complex distribution in cortical actin spots and in the cytoplasm of *S. pombe*. **J** Phase contrast showing an *Acanthamoeba castellanii* that was fixed while migrating across a glass surface. Notice the thick cortical lamellipodium at the front edge of the cell (*right side*). **K** Anti-p40-Arc polyclonal antibody labeling showing that Arp2/3 complex is found in the actin-rich cortex of leading and trailing edges of the amoeba. (Photos. A,B,J,K- Dyche Mullins, John Heuser, Tom Pollard; C-E- Lothar Groebe, Susanne Pistor, Juergen Wehland; H, I -Dan McCollum, Kathy Gould)

divisions (Hyman 1989; Hyman and White 1987; Waddle et al. 1994). Dynactin complex also localizes to cortical patches (Shrimankar et al. 1994), supporting this model.

Arp2 and Arp3: the Arp2/3 Complex

The actin cytoskeleton provides both the force and framework for much of cell motility, shape change and intracellular organization. Actin polymerization is thought to drive protrusion of the leading edge of a migrating cell, and is analogously able to drive the intracellular motility of many pathogenic organisms. Actin also has a central role in the assembly and function of myosin-containing contractile structures in the cell, such as the cleavage furrow during cytokinesis. Recent studies have shown that the assembly and maintenance of many actin-based structures in the cell are likely to depend on a protein complex containing the actin-related proteins Arp2 and Arp3, called the Arp2/3 complex. It is likely that this complex functions both as a coordinator of signals to the actin cytoskeleton and an initiator of actin assembly in response to these signals.

Structure of the Arp2/3 Complex

The Arp2/3 complex has seven subunits, present in a 1:1 stoichiometry with each other. In mammalian cells, these subunits are named Arp3, Arp2, p41-Arc, p34-Arc, p21-Arc, p20-Arc and p16-Arc (Welch et al. 1997). The names are roughly consistent in *S. cerevisiae, S. pombe* and *Acanthamoeba* except that some of the smaller subunits vary in size. The five non-Arp subunits of the complex have no significant homology to any previously well-characterized proteins. p41-Arc is homologous to SOP2Hs, the human counterpart of the *S. pombe* suppressor of profilin (Balasubramanian et al. 1996).

It is unclear whether SOP2Hs associates with human Arp2/3 complex, but so far, it has not been found in Arp2/3 complex purified from platelets and neutrophils (Machesky et al. 1997; Welch et al. 1997). Both p41-Arc and SOP2Hs are weakly homologous to the beta subunit of heterotrimeric G-proteins and contain WD repeats. The overall organization of the Arp2/3 complex is beginning to be understood, but still awaits further structural and biochemical studies. Interactions between subunits of the amoeba Arp2/3 complex have been characterized using chemical cross-linking. Arp3 and two of the smaller subunits also can cross-link to actin filaments (Mullins et al. 1997). Rotary shadowing revealed that purified complex takes on the shape of a horseshoe (Fig. 3A) (Mullins et al. 1997).

Arp2 and Arp3 are highly conserved in regions that bind to nucleotide and divalent cation, but are sufficiently divergent from actin and from each other to suggest that each has unique properties (Kelleher et al. 1995). Overall, Arp2 and Arp3 both closely fit a model based on the 3D structure of actin, although Arp3 contains several insertions (Kelleher et al. 1995). Arp2 binds to profilin in vitro and shows sequence conservation in the profilin-binding region of actin (Mullins et al. 1998). Divergence in the regions of actin intersubunit contacts, together with biochemical evidence, suggests that neither Arp2 nor Arp3 is likely to be able to assemble into filaments as actin does. However, Arp2 and Arp3 may associate in the complex to resemble an actin dimer (Kelleher et al. 1995).

Actin Nucleation and Cross-linking

In vitro studies, cellular concentration and immunolocalization have suggested that the Arp2/3 complex is a stable complex in cells and that it is likely to have an organizational role. There is a 20–40-fold lower molecular concentration of Arp2/3 complex than actin in amoebas (Mullins et al. 1997). However, amoeba Arp2 and Arp3 are more abundant than the other subunits, suggesting that in amoebas there may be Arp2 and Arp3 which are not associated with the complex (Mullins et al. 1997). In budding yeast, all of the subunits co-purify together, except for the p41-Arc subunit, which seems to be less tightly associated with the complex (Winter et al. 1997).

The Arp2/3 complex enhances the rate of actin filament nucleus formation and cross-links actin filaments (Mullins et al. 1998; Welch et al. 1998). The first physiological evidence for nucleating activity was the observation that Arp2/3 complex could induce actin polymerization on the surface of the bacterium *Listeria monocytogenes* (Welch et al. 1997). Enhancement of nucleation was directly demonstrated by in vitro studies showing that the Arp2/3 complex modestly reduced the lag phase during spontaneous actin polymerization (Mullins et al. 1998; Welch et al. 1998). As more than one subunit in the Arp2/3 complex can bind to actin filaments, it is perhaps not surprising that the complex also cross-links filaments in vitro (Mullins et al. 1998). In addition, purified Arp2/3 complex can promote branching of actin filaments (Fig. 3B), which suggests an attractive model for how cross-linking and nucleating activities could work together in vivo (Mullins et al. 1998).

In support of this model, branched structures similar to those promoted by Arp2/3 complex in vitro have been observed in lamellipodia of motile cells by electron microscopy (Svitkina et al. 1998). In *Listeria monocytogenes*, the Arp2/3 complex localizes along the whole actin tail (Fig. 3C-E; Welch et al. 1997), suggesting that the cross-linking and branching activity might be important in bacterial motility as well. The distribution of Arp2/3 complex in mammalian cells and amoebas also suggests a role in motility, as it is mainly concentrated in lamellipodia and punctate actin-rich structures (Fig. 3F,G,J,K).

In addition to its role in filament nucleation, Mullins et al. (1998) found that the Arp2/3 complex caps the pointed ends of actin filaments with high affinity. This has potentially significant implications for our understanding of actin filament turnover in cells. Previously, it was thought that pointed ends of actin filaments were free and that actin-depolymerizing proteins such as cofilin could speed up the spontaneous disassembly of ADP-bound actin monomers. Now it appears that disassembly must be a tightly regulated event, with dissociation of the Arp2/3 complex perhaps being a prerequisite. This could give the cell the ability to disassemble large networks of actin filaments rapidly, and could also provide a measure of stability under certain conditions.

Taking into account all of these activities, Mullins et al. (1998) proposed that the Arp2/3 complex induces dendritic nucleation of actin filaments at the leading edge of a cell (Fig. 4). This network could remain in place, with new actin branches pushing out the leading edge of the cell, while disassembly slowly takes place in the older regions of the actin network nearer the cell body. Although it is not known how actin assembly by the Arp2/3 complex is initiated or regulated, good candidates for targeting the Arp2/3 complex to lamellipodia are proteins of the WASP/Scar family. The absence of normal WASP, Wiskott-Aldrich syndrome protein, causes cytoskeletal defects leading to severe immune deficiencies (Derry et al. 1995). Both WASP and its relative, Scar, bind directly to the Arp2/3 complex in a manner thought to regulate lamellipodia assembly in cells (Machesky and Insall 1998). The *S. cerevisiae* WASP homologue, Las17p, has also been implicated as a suppressor of mutations in Arp2 and Arp3 (B. Winsor, personal comm.). These proteins lie downstream in signaling pathways from tyrosine kinase (Rivero-Lezcano et al. 1995) and G-protein-coupled receptors (Bear et al. 1998), and this may be how they connect signaling to actin polymerization.

Endocytosis and Cortical Actin Organization

The actin-related proteins Arp2 and Arp3 were originally found to be essential in yeast for growth and division (Lees-Miller et al. 1992; Schwob and Martin 1992). Further genetic studies have suggested more specific biological functions of the yeast Arp2/3 complex. Since yeasts do not move and do not contain lamellipodia, these studies pose an interesting contrast to studies in mammalian systems, where movement and protrusion have been the focus of most Arp2/3 complex studies to date.

The Arp2/3 complex has not yet been purified from *S. pombe*, but genes encoding several of the subunits have been identified, and both SOP2p (the p41-Arc homologous subunit) and Arp3p cosediment as a part of a large macromolecular complex (Balasubramanian et al. 1996; McCollum et al. 1996). Arp3p and SOP2p localize to cortical actin patches (Fig. 3H,I) and are required for reorganization of the actin cytoskeleton during the cell cycle (Balasubramanian et al. 1996; McCollum et al. 1996). Arp3p is not required for formation of the cytokinetic actin ring, but is required for normal actin patch movement and organization (McCollum et al. 1996). Actin patches have a role in polar-

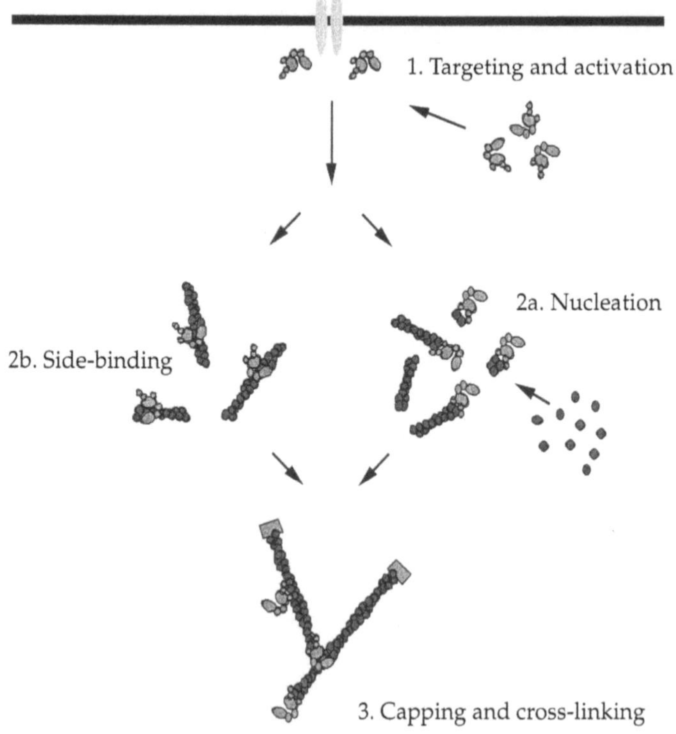

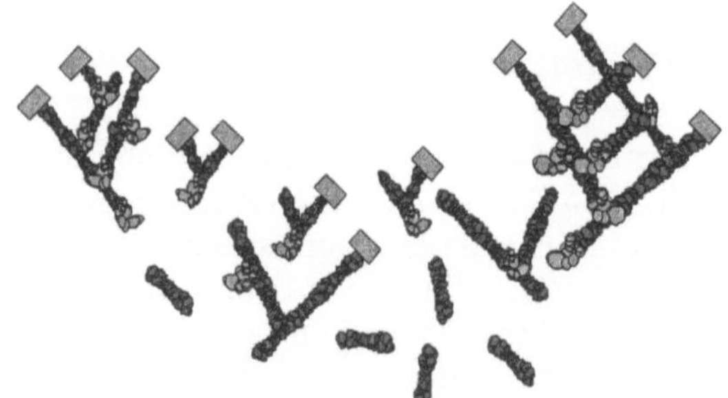

Fig. 4. Dendritic nucleation model for Arp2/3 complex mediated actin polymerization. Mullins et al. (1981) proposed that the Arp2/3 complex could promote the formation of dendritic networks of actin filaments at the leading edge of a cell. *1* The Arp2/3 complex may be targeted to or activated at the leading edge of the cell specifically. The mechanism for this recruitment or activation is unknown but could involve binding to WASP-related proteins (Machesky and Insall 1998). *2a* the Arp2/3 complex could nucleate new actin filaments from actin monomer in the cytoplasm and then stay attached *2b* to the new filaments to form branches or bind to filaments to form branches. In *3* the newly branched filaments are presumably capped by capping protein which may inhibit or regulate further elongation. In *4* the dendritic network would eventually stop growing as all of the new filaments became capped unless a signal for elongation or additional nucleation was present

ized septum deposition and formation of the new cell wall. Arp3p interacts genetically with profilin, as a specific mutation in Arp3 can suppress actin ring defects in a profilin mutant strain (McCollum et al. 1996). Similarly, a mutation in SOP2p can rescue temperature-sensitive lethality of a profilin mutation. SOP2p localizes to both actin patches and cables, suggesting that it might have additional roles in the cell to Arp2/3 complex function (Balasubramanian et al. 1996).

In *S. cerevisiae*, the Arp2/3 complex is important for endocytosis, actin patch localization and motility, and membrane growth and polarity. Arp2 (originally called Act2) was first described as an actin-like gene that was required late in the cell cycle (Schwob and Martin 1992). Further studies showed that Arp2p localizes to actin patches, similarly to *S. pombe* Arp3p, and is required for patch localization, polarity of budding, cell surface growth and endocytosis (Moreau et al. 1996, 1997).

The actin cytoskeleton is generally defective in Arp2 mutants, resulting in osmosensitivity. Polarity of actin distrubution is also aberrant in Arp2 mutants. The defects in endocytosis in Arp2 mutants are specific to the internalization step and do not affect the rate of constitutive secretion. Endocytosis generally requires proper functioning of the actin cytoskeleton, but the specificity of this defect suggests perhaps a previously undescribed requirement for actin nucleation or cross-linking during the internalization of vesicles. Arp3 is essential for viability (Huang et al. 1996). Additionally, mutations in Arp3 inhibit the motility of actin patches in yeast, suggesting some active role in intracellular actin-based motility for the Arp2/3 complex (Winter et al. 1997). This may be linked with endocytosis or polarized deposition of cell wall components.

Arp4 andArp7/9

The SWI/SNF Complex

In *S. cerevisiae*, the actin-related proteins, Arp4 (Weber et al. 1995), Arp7 and Arp9 (Peterson et al. 1998), have been implicated in chromatin structure and function in the nucleus. It is interesting to speculate that the localization of these proteins in the nucleus may account for some previous reports of actin in the nucleus. Arp4 (also called Act3p) is essential in *S. cerevisiae* (Harata et al. 1994). It contains a nuclear localization signal and is found in the nucleus, where it is postulated to be an essential component of chromatin. It is highly divergent from actin and other Arps, but is likely to have a similar overall structure to actin based on conserved regions in the primary sequence. Arp7 and Arp9 are also found in the nucleus and are components of the large SWI/SNF complex (Peterson et al. 1998). This complex is composed of 11 different protein subunits and is required for the transcription of a subset of yeast genes. The complex is thought to alleviate transcriptional repression by modulating chromatin structure. Arp7 and Arp9 are nonessential, but mutants grow poorly on media containing galactose, glycerol or sucrose as the sole carbon source, similarly to other *swi/snf* mutants (Peterson et al. 1998).

The *Drosophila* SWI/SNF complex, called the BRM complex, also contains an Arp subunit (Papoulas et al. 1998). This indicates that the function of Arps in transcriptional complexes is evolutionarily conserved. The actin-related protein in the BRM complex, BAP55, has only 38% identity to actin and shows a high degree of divergence from other Arps as well. The gene brahma (brm) is a trithorax group gene whose product activates homeotic gene transcription as a part of this complex. Brahma also has ATPase activity that is functional in the complex. The role of BAP55 is unknown, but it is possible that it has an ATPase-mediated conformational change which may contribute to the function of the complex.

Other Arps

Several Arp sequences are known, but have not yet been characterized with respect to function or localization in cells (Fig. 1). *Drosophila* Arp53D is a novel Arp found only in testes (Fyrberg et al. 1994). *Drosophila* Arp13E is abundant during embryogenesis, but present at all stages of development (Frankel et al. 1994). *S. cerevisiae* has Arp5, Arp8 and Arp10 (Poch and Winsor 1997). *Drosophila* and *C. elegans* have ArpX genes, and *S. pombe* has a sequence called Arp23D3. There is no known function yet for any of these genes or their products, but it will be interesting to see how their function compares to the more well-characterized Arps. We might predict that they will be involved in large multimolecular complexes and may be either nuclear or cytoskeletal, but only future research will tell.

Why Does the Cell Have Arps?

Actin is clearly a very old protein in evolution (also see chapter by Sheedy and Clarke this Vol.), and the actin fold appears to be a useful motif for cells to effect structural or enzymatic activities. Nucleotide hydrolysis is used by many different kinds of proteins to do work in the cell. Actin and tubulin use ATP and GTP (respectively) in an analogous fashion to form polarized filaments. Small GTPases and heterotrimeric G-proteins use GTP to effect a conformational switch that transmits a signal from outside the cell to the inside. Molecular motor proteins use nucleotide hydrolysis to change conformation and affinity of binding interactions, generating movement along filaments. The function of nucleotide binding to Arps in the dynactin complex, Arp2/3 complex or SWI/SNF complex is presently not known, but it is likely to be significant, given the high conservation in those regions of the sequence that bind to nucleotide.

Acknowledgements. We would like to thank Dr. Erika L. F. Holzbaur for sending a collection of reprints, as well as all those who sent reprints and copies of papers. We also are indebted to Dr. Barbara Winsor for helpful discussions and for the actin-related protein tree.

References

Balasubramanian MK, Feoktistova A, McCollum D, Gould KL (1996) Fission yeast SOP2p: a novel and evolutionally conserved protein that interacts with Arp3p and modulates profilin function. EMBO J 15:6426–6437

Bear JE, Rawles JF, Saxe CL III (1998) Scarl a WASP-related protein isolated as a suppressor of receptor defects late in development. J Cell Biol 142:1325–1335

Burkhardt JK, Echeverri CJ, Vallee RB (1997) Over-expression of the dynamitin (p50) subunit of the dynactin complex disrupts the dynein-dependent maintenance of membrane organelle distribution. J Cell Biol 139:469–484

Christopher LJ, Fletcher LD, Dykstra CC (1995) Cloning and identification of Arpl an actin-related protein from *Pneumocystis carinii.* J Eukaryot Microbiol 42:142–149

Clark SW, Meyer DI (1994) ACT3: A putative centractin homologue in *S. cerevisiae* is required for proper orientation of the mitotic spindle. J Cell Biol 127:129–138

Clark SW, Staub O, Clark IB, Holzbaur ELF, Paschal BM, Vallee RB, Meyer DI (1994) β-centractin: characterization and distribution of a new member of the centractin family of actin-related proteins. Mol Biol Cell 5:1301–1310

Derry JMJ, Kerns JA, Weinberg KI, Ochs HD, Volpini V, Estivill X, Walker AP, Francke U (1995) Wasp Gene-Mutations in Wiskott-Aldrich syndrome and X-linked thrombocytopenia. Hum Mol Genet 4:1127–1135

Echeverri CJ, Paschal BM, Vaughn KT, Vallee RB (1996) Molecular characterization of the 50-kD subunit of dynactin reveals function for the complex in chromosome alignment and spindle organization during mitosis. J Cell Biol 132:617–633

Frankel S, Heintzelman MB, Artavanis-Tsakonas S, Mooseker MS (1994) Identification of a divergent actin-related protein in *Drosophila.* J Mol Biol 235:1351–1356

Fyrberg C, Ryan L, Kenton M, Fyrberg E (1994) Genes encoding actin-related proteins of *Drosophila melanogaster.* J Mol Biol 241:498–503

Gaglio T, Saredi A, Bingham JB, Hasbani MJ, Gill SR, Schroer TA Compton DA (1996) Opposing motor activities are required for the organization of the mammalian mitotic spindle pole. J Cell Biol 135:399–414

Gill SR, Schroer TA, Szilak I, Steuer ER, Sheetz MP Cleveland DW (1991) Dynactin a conserved ubiquitously expressed component of an activator of vesicle motility mediated by cytoplasmic dynein. J Cell Biol 11:1639–1650

Harata M, Karwan A, Wintersberger U (1994) An essential gene of *Saccharomyces cerevisiae* coding for an actin-related protein. Proc Natl Acad Sci USA 91:8258–8262

Holleran EA, Karki S, Holzbaur ELF (1998) The role of the dynactin complex in intracellular motility. Int Rev Cytol 182:69–109

Holleran EA, Tokito MK, Karki S, Holzbaur ELF (1996) Centractin (ARP1) associates with spectrin revealing a potential mechanism to link dynactin to intracellular organelles. J Cell Biol 135:1815–1829

Huang M-E, Souciet J-L, Chaut J-C, Galibert F (1996) Identification of ACT4 a novel essential actin-related gene in the yeast *Saccharomyces cerevisiae.* Yeast 1:839–848

Hyman AA (1989) Centrosome movement in the early divisions of *Caenorhabditis elegans*: A cortical site determining centrosome position. J Cell Biol 109:1185–1193

Hyman AA, White JG (1987) Determination of cell division axes in the early embryogenesis of *Caenorhabditis elegans.* J Cell Biol 105:2123–2135

Karki S, Holzbaur ELF (1995) Affinity chromatography demonstrates a direct binding between cytoplasmic dynein and the dynactin complex. J Biol Chem 270:28806–28811

Kelleher JF, Atkinson SJ, Pollard TD (1995) Sequences structural models and cellular-localization of the actin-related proteins Arp2 and Arp3 From *Acanthamoeba.* J Cell Biol 131:385–397
Lees-Miller JP, Henry G, Helfman DM (1992) Identification of act2 an essential gene in the fission yeast *Schizosaccharomyces pombe* that encodes a protein related to actin. Proc Natl Acad Sci USA 89:80–83
Machesky, Reeves E, Wientjes F, Mattheyse F, Grogan A, Totty N, Burlingame AL, Hsuan JJ, Segal AW (1997) The mammalian Arp2/3 complex localizes to regions of lamellipodial protrusion and is composed of evolutionarily conserved proteins. Biochem J 328:105–112
Machesky LM, Insall RH (1998) Scar1 and the related Wiskott-Aldrich syndrome protein WASP regulate the actin cytoskeleton through the Arp2/3 complex. Curr Biol (in press)
McCollum D, Feoktistova A, Morphew M, Balasubramanian M, Gould KL (1996) The *Schizosaccharomyces pombe* actin-related protein Arp3 is a component of the cortical actin cytoskeleton and interacts with profilin. EMBO J 15:6438–6446
McGrail M, Gepner J, Silvanovich A, Ludman S, Serr M, Hays TS (1995) Regulation of cytoplasmic dynein function in vivo by the *Drosophila* glued complex. J Cell Biol 131:411–425
Moreau V, Galan J-M, Devilliers G, Haguenauer-Tsapis R, Winsor B (1997) The yeast actin-related protein Arp2p is required for the internalization step of endocytosis. Mol Biol Cell 8:1361–1375
Moreau V, Madania R, Martin P, Winsor B (1996) The *Saccharomyces cerevisiae* actin-related protein Arp2 is involved in the actin cytoskeleton. J Cell Biol 134:117–132
Muhua L, Karpova T, Cooper JA (1994) A yeast actin-related protein homologous to that in vertebrate dynactin complex is important for spindle orientation and nuclear migration. Cell 78:669–679
Mullins RD, Stafford WB, Pollard TD (1997) Structure subunit topology and actin-binding activity of the Arp2/3 complex from *Acanthamoeba.* J Cell Biol 136:331–343
Mullins RD, Heuser JA, Pollard TD (1998) The interaction of Arp2/3 complex with actin: nucleation high affinity pointed end capping and formation of branching networks of filaments. Proc Natl Acad Sci USA 95:6181–6186
Mullins RD, Kelleher JF, Xu J, Pollard TD (1998) Arp2/3 complex from *Acanthamoeba* binds profilin and cross-links actin filaments. Mol Biol Cell 9:841–852
Papoulas O, Beek SJ, Moseley SL, McCallum CM, Sarte M, Shearn A, Tamkun JW (1998) The *Drosophila* trithorax group proteins BRM ASH1 and ASH2 are subunits of distinct protein complexes. Development 125:3955–3966
Paschal BM, Holzbaur ELF, Pfister KK, Clark S, Meyer DI, Vallee RB (1993) Characterization of a 50-kDa polypeptide in cytoplasmic dynein preparations reveals a complex with p150 GLUED and a novel actin. J Biol Chem 268:15318–15323
Peterson CL, Zhao Y, Chait BT (1998) Subunits of the yeast SWI/SNF complex are members of the actin-related protein (ARP) family. J Biol Chem 273:23641–23644
Planmann M, Minke PF, Tinsley JH, Bruno KS (1994) Cytoplasmic dynein and actin-related protein Arp1 are required for normal nuclear distribution in filamentous fungi. J Cell Biol 127:139–149
Plough HH, Ives PT (1935) Induction of mutations by high temperature in *Drosophila.* Genetics 20:42–69
Poch O, Winsor B (1997) Who's who among the *Saccharomyces cerevisiae* actin-related proteins? a classification and nomenclature proposal for a large family. Yeast 13:••–••
Reinsch S, Karsenti E (1997) Movement of nuclei along microtubules in *Xenopus* egg extracts. Curr Biol 7:211–214
Schafer DA, Gill SR, Cooper JA, Heuser JE, Schroer TA (1994a) Ultrastructural analysis of the dynactin complex: an actin-related protein is a component of a filament that resembles F-actin. J Cell Biol 126:403–412
Schafer DA, Korshunova YO, Schroer TA, Cooper JA (1994b) Differential localization and sequence analysis of capping protein B-subunit isoforms of vertebrates. J Cell Biol 127:453–465
Schroer TA, Fyrberg E, Cooper JA, Waterston RH, Helfman D, Pollard, Meyer DI (1994) Actin-related protein nomenclature and classification. J Cell Biol 127:1777–1778

Schwob E, Martin P (1992) New yeast actin-like gene required late in the cell cycle. Nature 355:179–181

Shrimankar PV, Schriefer LA, Waterson RA (1994) Actin-related proteins *in C elegans.* Worm Breed Gaz 13:56

Svitkina TM, Borisy GG (1998) Correlative light and electron microscopy of the cytoskeleton of cultured cells. Methods Enzymol 298 (in press)

Vaughan KT, Vallee RB (1995) Cytoplasmic dynein binds dynactin through a direct interaction between the intermediate chains and p150 GLUED. J Cell Biol 131:1507–1516

Waddle JA, Cooper J, Waterson RH (1994) Transient localized accumulation of actin in *Caenorhabditis elegans* blastomeres with oriented asymmetric divisions. Development 120:2317–2328

Waterman-Storer CM, Holzbaur ELF (1996) The product of the *Drosophila* gene *glued* is the functional homologue of the p150glued component of the vertebrate dynactin complex. Proc Natl Acad Sci USA 271:1153–1159

Waterman-Storer CM, Karki S, Holzbaur ELF (1995) The p150 GLUED component of the dynactin complex binds to both microtubules and the actin-related protein centractin. Proc Natl Acad Sci USA 92:1634–1638

Weber V, Harata M, Hauser H, Wintersberger U (1995) The actin-related protein Act3p of *Saccharomyces cerevisiae* is located in the nucleus. Mol Biol Cell 6:1263–1270

Welch MD, DePace AH, Verma S, Iwamatsu A, Mitchison TJ (1997a) The human Arp2/3 complex is composed of evolutionarily conserved subunits and is localized to cellular regions of dynamic actin filament assembly J Cell Biol 138:375–384

Welch MD, Iwamatsu A, Mitchison TJ (1997) Actin polymerization is induced by the Arp2/3 protein complex at the surface of *Listeria* monocytogenes Nature 385:265–269

Welch MD, Rosenblatt J, Skoble J, Portnoy DA, Mitchison TJ (1998) Interaction of Arp2/3 complex and the *Listeria* monocytogenes ActA protein in actin filament nucleation Science 281:105–108

Winter D, Podtelejnikov AV, Mann M, Li R (1997) The complex containing actin-related proteins Arp2 and Arp3 is required for the motility and integrity of yeast actin patches Curr Biol 7:519–529

Control of the Actin Cytoskeleton by Extracellular Signals

Thomas Beck, Pierre-Alain Delley, and Michael N. Hall[1]

Introduction

Rho-type GTPases play a central role in linking extracellular signals to the organization of the actin cytoskeleton, from yeast to mammals (Van Aelst and D'Souza-Schorey 1997; Hall 1998b; Schmidt and Hall 1998). In addition to their role in actin organization, Rho-type GTPases also control signalling pathways that activate transcription factors. The Rho proteins thus regulate a broad variety of cellular events, such as shape changes, migration, attachment, proliferation, differentiation, and survival. The Rho family is a subclass of the Ras superfamily of GTPases. These small GTPases function as binary switches, cycling between an active, GTP-bound form and an inactive, GDP-bound form.

The localization and/or activity of Rho-type GTPases is regulated by three classes of proteins, GDP/GTP exchange factors (GEFs), GTPase-activating proteins (GAPs), and guanine nucleotide dissociation inhibitors (GDIs). GEFs convert a GTPase to the active form by facilitating the binding of GTP. Rho-GEFs display a tandem motif of a Dbl-homology (DH) domain followed by a pleckstrin homology (PH) domain. The DH domain is necessary and sufficient for guanine nucleotide exchange activity. PH domains are involved in protein-protein and protein-lipid interactions, and appear to be required for localization or full exchange activity of GEFs. Many GEFs possess additional protein-protein interaction motifs or catalytic domains, and the N-terminus of several GEFs appears to be autoinhibitory.

GAPs promote conversion of an active GTPase to the inactive form, by stimulating the intrinsic GTPase activity. Rho-GAPs share a catalytic domain of approximately 140 amino acids. Another class of potentially negative regulators are GDIs, which can associate with both the GDP- or GTP-bound form of a GTPase, thereby preventing nucleotide exchange or GTP hydrolysis, respectively. GDIs are able to extract a GTPase from the membrane, presumably by binding to and masking a membrane-anchoring prenyl group at the C-terminus of the GTPase. Phosphorylation of Rho GTPases may provide a mechanism to modulate the GDI binding affinity, as suggested for mammalian RhoA (Lang et al. 1996). It remains to be determined if phosphorylation is a general

[1] Department of Biochemistry, Biozentrum, University of Basel, CH-4056 Basel, Switzerland

Results and Problems in Cell Differentiation, Vol. 32
C. dos Remedios (Ed.): Molecular Interactions of Actin

mechanism to alter the activity, localization or protein-protein interactions of small GTPases.

How extracellular signals activate Rho-type GTPases and how the GTPases, in turn, elicit a particular actin-based cellular response are poorly understood. However, significant progress has been made in identifying pathways that couple extracellular signals to the actin cytoskeleton. Here we describe some of these pathways from yeast, fruit fly, and mammalian cells.

Saccharomyces Cerevisiae

The budding yeast *Saccharomyces cerevisiae* is a unicellular eukaryote. Cell growth (budding) is polarized, occurring at a defined position on the cell surface. This oriented growth is mediated by a polarized actin cytoskeleton (Li et al. 1995; Chant 1996; Madden and Snyder 1998). The yeast actin cytoskeleton is composed of two morphologically distinct pools of actin polymers, patches and cables. Patches are foci of actin and actin-binding proteins that are associated with plasma membrane invaginations (Mulholland et al. 1994), and move rapidly over long distances (Waddle et al. 1996). Cables are bundles of overlapping actin filaments (Karpova et al. 1998).

Recent studies suggest that patches and cables are associated and form an integrated system (Amberg 1998; Karpova et al. 1998). The distribution of actin patches and cables is polarized in a cell cycle-dependent manner (Adams and Pringle 1984; Schmidt and Hall 1998). Upon exposure to nutrients, a resting yeast cell enters the cell cycle. During early G1 phase, the actin cytoskeleton is randomly distributed, and isotropic growth occurs until a critical cell size is reached. Late in G1, at a point termed START, the cell is committed to undergo a new round of division, and the actin cytoskeleton becomes polarized toward a preselected bud site.

Actin cables orient toward the bud site where patches become concentrated and form a ring structure. This polarization mediates bud emergence and bud growth by directing secretory vesicle transport to the bud site and into the bud (Li et al. 1995; Chant 1996; Madden and Snyder 1998). During bud maturation, actin patches are found exclusively in the bud, and cables are aligned longitudinally from the mother cell into the bud. The polarization of the actin cytoskeleton and growth is maintained during S phase and early G2 phase, until the daughter cell reaches a critical size in late G2, and the asymmetric actin distribution is lost. Prior to cytokinesis, the actin cytoskeleton repolarizes. Patches accumulate at, and cables orient toward the mother-bud neck. The cell cycle-dependent organization of the actin cytoskeleton is controlled by five Rho-type GTPases (CDC42 and RHO1 through 4) (Schmidt and Hall 1998). The role of a newly identified sixth member of the Rho family, RHO5, is yet to be determined.

As in higher eukaryotes, reorganization of the yeast actin cytoskeleton is not only regulated by the cell cycle, but also in response to extracellular signals.

Changes in organization of the actin cytoskeleton are observed upon exposure of yeast cells to mating pheromone, upon nutrient limitation, and in response to environmental stress that affects cell integrity. Below we review signalling pathways that regulate the actin cytoskeleton in response to these different extracellular cues.

Pheromone-Induced Actin Reorganization

The mating process of yeast cells is a well-characterized example of extracellular signals controlling organization of the actin cytoskeleton (for recent reviews see Leberer et al. 1997; Schmidt and Hall 1998; Johnson 1999). Yeast cells of the α mating type produce a small peptide pheromone, α factor, which is secreted and bound to the α-factor receptor (STE2) on the surface of *a* cells. Cells of the *a* mating type produce and secrete *a* factor which is bound to the *a* factor receptor (STE3) on α cells.

The pheromone receptors STE2 and STE3 belong to the family of heptahelical G protein-coupled receptors. Pheromone binding induces a mating response characterized by a cell cycle arrest at START, polarization of the actin cytoskeleton toward the source of mating pheromone, and formation of a mating projection (shmoo). Upon cell-cell interaction at shmoo tips, the cell walls dissolve at the contact site and cells and nuclei fuse. Binding of mating pheromones to their cognate receptors activates the coupled heterotrimeric G protein.

The Gβγ dimer (STE4-STE18) binds the scaffold protein STE5 and the kinase STE20, which in turn activates a mitogen-activated protein kinase (MAPK) cascade tethered to STE5. The MAPK cascade consists of the MAPK kinase kinase (MAPKKK) STE11, the MAPK kinase (MAPKK) STE7 and the MAPK FUS3. FUS3 stimulates transcription of specific pheromone-induced genes by activating the transcription factor STE12 (Oehlen et al. 1996). FUS3 also induces cell cycle arrest via FAR1. FUS3-phosphorylated FAR1 binds and inhibits the G1-specific cyclin-dependent protein kinase (CDK) complex CLN/CDC28 (Peter and Herskowitz 1994; Jeoung et al. 1998).

Several proteins involved in polarization of the actin cytoskeleton toward the source of mating pheromone have been identified. A central function in this process is associated with the Rho-type GTPase CDC42 and the CDC42-specific GEF CDC24 (for a comprehensive review, see Johnson 1999). During mating, CDC24 and CDC42 are localized to the tips of mating projections, and mating-deficient mutants have been isolated for both. Among these, specific *cdc24* mutants are deficient in orienting the mating projection toward the source of pheromone. CDC24 associates with BEM1 and the STE4-STE18 Gβγ dimer.

The src homology 3 (SH3) domain-containing protein BEM1 is required for mating, and was isolated as a multicopy suppressor of a *ste4* mutant deficient in pheromone signalling. BEM1 associates with actin (Leeuw et al. 1995), and colocalizes in growing cells with CDC24 and CDC42 to a preselected bud site and to tips of growing buds. Interestingly, this localization is independent of an intact actin cytoskeleton (Ayscough et al. 1997). BEM1 also interacts with STE5, STE20 and FAR1. In addition to its role in cell cycle arrest, FAR1 is required for polarization of the actin cytoskeleton toward the source of mating pheromone (Valtz et al. 1995).

Triggering of the mating response induces export of FAR1 from the nucleus. Cytoplasmic FAR1 binds to the STE4-STE18 Gβγ dimer and is thus recruited to the site of pheromone receptor activation (Butty et al. 1998; Nern and Arkowitz 1999). Here, FAR1 is possibly required to recruit the CDC42-CDC24 complex. FAR1 interacts directly with CDC24 and the binding of CDC24 to STE4 is FAR1-dependent (Butty et al. 1998; Nern and Arkowitz 1998, 1999). Thus, FAR1 may provide a platform to assemble a complex containing the STE4-STE18 Gβγ dimer, the MAPKKK kinase STE20, STE5, the STE MAPK cascade, CDC24, CDC42, and BEM1.

CDC42 Effectors Involved in Actin Polarization

What are the links between CDC42 activation and polarity establishment in response to pheromone? Several proteins have been identified that are required for the mating process and appear to act as CDC42 effectors (Fig. 1).

STE20

STE20 is the founding member of a structurally related family of kinases, referred to as the p21-activated kinases (PAKs). This family includes the budding yeast kinases STE20, CLA4 and SKM1. PAKs contain a characteristic Cdc42/Rac interactive binding (CRIB) motif that mediates direct binding to activated Cdc42 and Rac (Burbelo et al. 1995). STE20 and CLA4 are required for organization of the actin cytoskeleton during vegetative growth (Eby et al. 1998). However, during mating, STE20 mediates transcription of pheromone-induced genes, and this function is independent of STE20 binding to CDC42 (Leberer et al. 1992; Peter et al. 1996).

To activate transcription, STE20 binds STE4 (Gβ) in a pheromone-dependent manner (Leeuw et al. 1998). STE4 also binds STE5 (Whiteway et al. 1995), and thereby possibly enables STE20 to activate the STE5-associated MAPK cascade that mediates pheromone-induced transcription and cell cycle arrest. STE20 is also required for formation of the mating projection, and localization of STE20 to the tips of emerging buds in growing cells as well as to tips of mating projections depends on binding to CDC42 (Peter et al. 1996). Thus, STE20 may have a CDC42-dependent and transcription-independent function in organizing the actin cytoskeleton during vegetative growth and during mating.

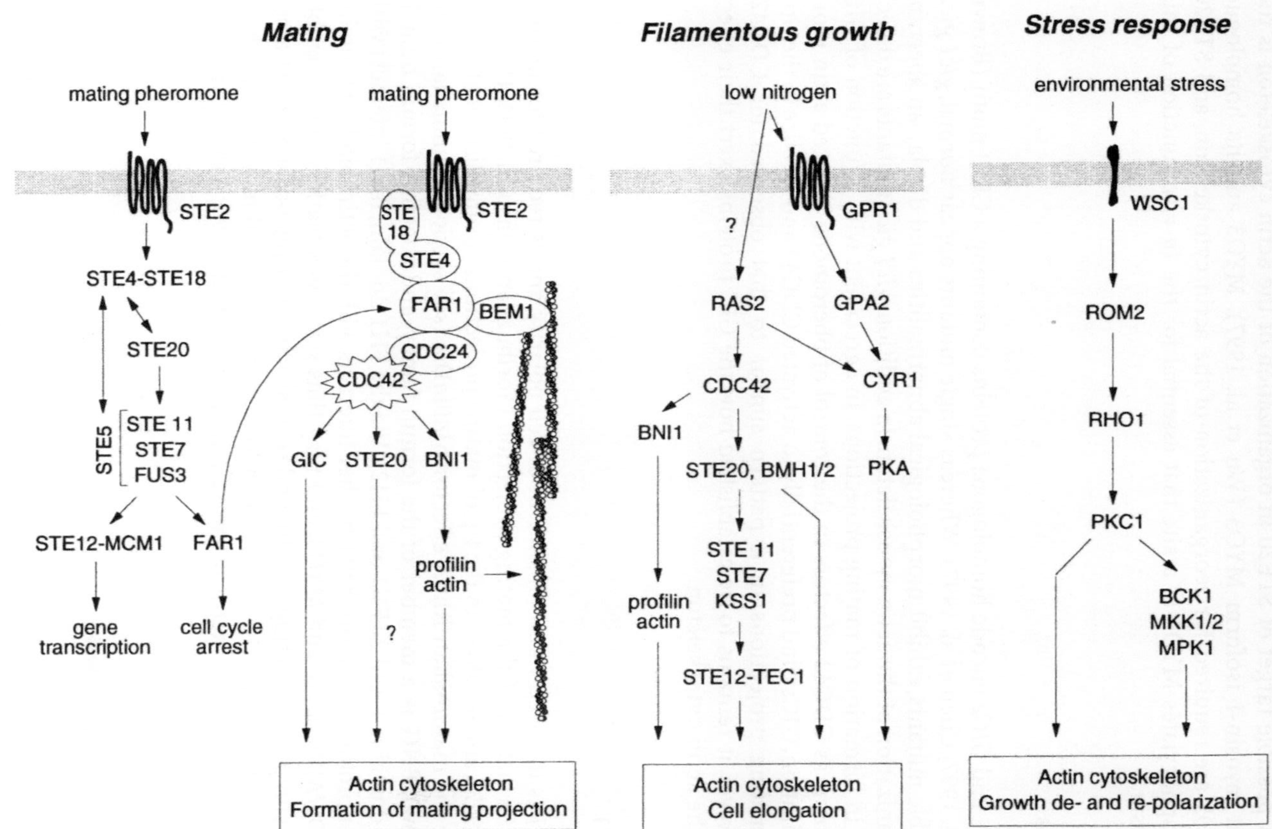

Fig. 1. Yeast pathways that control the actin cytoskeleton in response to mating pheromone, nutrient limitation and environmental stress

A possible target of STE20 in organization of the actin cytoskeleton is the yeast myosin-I isoform MYO3 (Wu et al. 1997). MYO3 and its homologue MYO5 are required for reorganization of the actin cytoskeleton, and STE20 phosphorylates MYO3 at a site that essential for the *in vivo* function of this myosin.

GICs

GIC1 and *GIC2* encode homologous proteins containing a CRIB motif (Brown et al. 1997; Chen et al. 1997). Whereas single mutants appear normal, *gic1 gic2* double mutants exhibit morphological abnormalities and display an aberrant organization of the actin cytoskeleton. In addition, *gic1 gic2* mutants are defective in formation of mating projections. In agreement with a function of GIC proteins as CDC42 effectors in the control of pheromone-induced actin reorganization, GICs bind preferentially to activated CDC42 and localize to the tips of mating projections in a pattern similar to that observed for CDC42. However, it remains to be established how the GIC proteins exert their effect on the actin cytoskeleton.

BNI1

BNI1 is involved in establishment of cell polarity and is required for mating, i.e., formation of the mating projection (Yorihuzi and Ohsumi 1994; Dorer et al. 1997), and localizes to the tip of mating projections (Evangelista et al. 1997). Like the mammalian RhoA effector p140mDia (see section Mammalian Cells below), BNI1 is a member of the formin family, and contains formin homology domains 1 and 2 (FH1 and FH2). Via the FH1 domain, BNI1 interacts with yeast profilin and actin. BNI1 further interacts with the actin-binding proteins BUD6/AIP3, SPA2 and MYO3, and may thus serve as a platform for assembly of a complex consisting of actin and several actin-binding proteins. BNI1 associates with other RHO proteins in addition to CDC42 and may therefore have additional roles in actin organization (Kohno et al. 1996; Evangelista et al. 1997).

Filamentous Growth

Yeast cells respond to nutritional stress by undergoing changes in cell shape and budding pattern, possibly as a foraging response (Madhani and Fink 1998). These changes are presumably mediated by changes in the actin cytoskeleton. Upon nitrogen limitation, diploid cells alter their growth mode, budding pattern and morphology, to switch from the unicellular yeast form to a multicellular, filamentous (pseudohyphal) form. Haploid cells can undergo a similar morphological transition that is manifested by invasive filamentous growth.

It is not known how nitrogen limitation is sensed but, similar to the pheromone pathway, a G protein-coupled receptor may be involved. A newly identified heptahelical receptor (GPR1) is required for pseudohyphal differentiation (cited in Pan and Heitman 1999). GPR1 associates with GPA2, one of two Gα subunits encoded in the yeast genome (Xue et al. 1998). GPA2 is required for filamentous growth, and appears to signal in concert with RAS2 to adenylate cyclase (CYR1) and the cAMP-dependent protein kinase PKA (Kubler et al. 1997; Pan and Heitman 1999).

Activated CDC42, like activated RAS2, induces filamentous growth, and dominant negative CDC42 blocks this morphogenic switch in response to nitrogen limitation or RAS2 activation (Mosch et al. 1996). Furthermore, CDC42 requires STE20 and BMH1/BMH2 (14-3-3 proteins) to induce filamentous growth (Mosch et al. 1996; Roberts et al. 1997). STE20 and 14-3-3 proteins physically interact and appear to function at the same level in a signalling pathway (Roberts et al. 1997).

Besides STE20, several other signalling components of the pheromone MAPK pathway, including the MAPKKK STE11, the MAPKK STE7 and the transcription factor STE12, are required for filamentous growth. However, in the pathway signalling filamentous growth, the pheromone pathway-specific MAPK FUS3 is replaced by the MAPK KSS1, and STE12 forms a heterodimer with the transcription factor TEC1 (Madhani and Fink 1998).

The STE12-TEC1 dimer activates transcription of filamentation-specific genes such as FLO11 (Rupp et al. 1999). Taken together, these results suggest a signalling pathway possibly initiated by the heptahelical receptor GPR1 and the associated Gα subunit GPA2 that converges with RAS2 to control the PKA pathway. In addition, RAS2 signals to the STE20-linked MAPK cascade via CDC42 (Fig. 1). An important yet unsolved question is how these signalling pathways may control organization of the actin cytoskeleton. CDC42 may play a central role in this process, as the CDC42 effector BNI1 is involved in polarization of growth during pseudohyphal development (Mosch and Fink 1997). It remains to be determined whether other CDC42 effectors, like the GIC proteins, or CDC42-independent pathways are also involved in the control of actin reorganization that underlies the change between the unicellular budding yeast form and the multicellular filamentous form.

Growth De- and Re-Polarization in Response to Environmental Stress

Like mammalian cells, yeast reorganize their actin cytoskeleton in response to environmental stress. Hypertonic or cell wall stress induces a transient depolarization of the actin cytoskeleton (Chowdhury et al. 1992; Gustin et al. 1998; Delley and Hall 1999). Cell wall stress caused by heat shock induces a rapid and severe, but transient, depolarization of the actin cytoskeleton. This is necessary for depolarization of the cell wall biosynthetic enzyme glucan synthase from the growth site to the overall cell periphery, suggesting that the depolarization is a homeostasis mechanism to repair general cell wall damage (Delley and Hall 1999).

Depolarization of the actin cytoskeleton and glucan synthase in response to cell wall damage is mediated by the integrin-like plasma membrane protein WSC1, the RHO1 GTPase switch, PKC1 and a yet-to-be identified PKC1 effector branch (Bickle et al. 1998; Delley and Hall 1999). The PKC1-activated MAP kinase cascade consisting of BCK1 (MAPKKK), MKK1 and MKK2 (MAPKK) and MPK1 (MAPK) (Banuett 1998; Gustin et al. 1998) is not required for depolarization, but rather for repolarization of the actin cytoskeleton and glucan synthase (Delley and Hall 1999). Thus, activated RHO1 can mediate both polarized and depolarized cell growth via the same effector, PKC1, suggesting that RHO1 may function as a rheostat rather than as a simple on-off switch. The mechanism by which a yet-to-be determined PKC1 effector pathway and the PKC1-activated MAP kinase cascade control depolarization and repolarization of the actin cytoskeleton remains to be determined.

Drosophila Melanogaster

Studies with the fruit fly *Drosophila melanogaster* have implicated actin-controlling signalling pathways in several developmental processes, such as oogenesis (Robinson and Cooley 1997), gastrulation (Barrett et al. 1997; Hacker and Perrimon 1998), and dorsal closure (Noselli 1998).

Dorsal Closure

Dorsal closure (DC) is a morphogenetic event in which two symmetric sheets of lateral ectodermal cells close a hole in the dorsal epithelium of the embryo. This process, which takes place without cell recruitment or proliferation, is mediated by cell elongation causing migration of the cell sheets toward each other (Campos-Ortega and Hartenstein 1985; Noselli 1998). DC is initiated by the dorsoventral elongation of a row of cells at the dorsal most end of the lateral epithelium, the leading edge (LE). These LE cells show a strong accumulation of filamentous actin and nonmuscle myosin, which may drive cell elongation by forming a contractile actin-myosin apparatus (Young et al. 1993). Subsequently the elongation of more lateral cells and movement of the epithelia toward the dorsal midline occurs. DC is completed when the LE cells of the two symmetric ectodermal sheets meet. Analysis of mutants with a dorsal open phenotype have implicated the actin cytoskeleton and two sequentially acting signalling pathways, the Jun-N-terminal kinase (JNK) pathway and the transforming growth factor-β (TGF-β) pathway, in DC (Fig. 2).

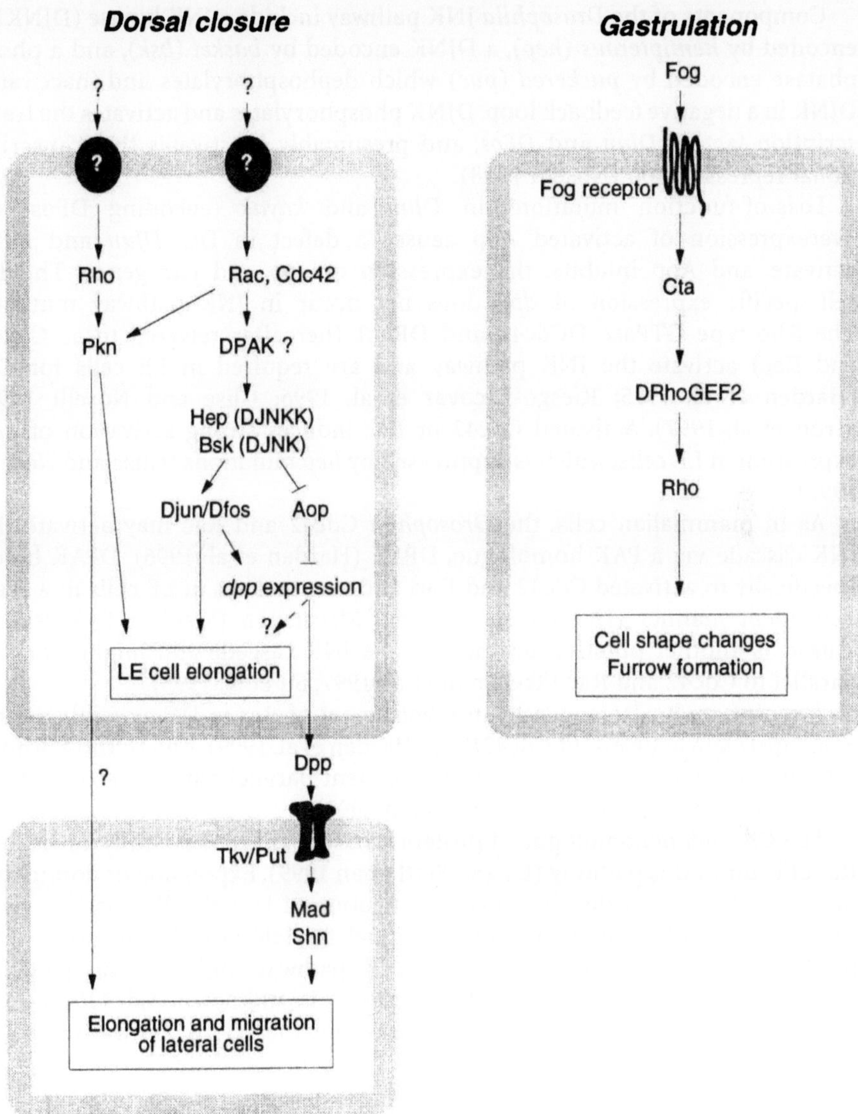

Fig. 2. Signalling pathways that control dorsal closure and furrow formation in *Drosophila*

Function of the JNK Pathway in Dorsal Closure

Activation of the JNK pathway in LE cells, by an unknown developmental signal, leads to elongation of LE cells and expression of the secreted TGF-β family member Decapentaplegic (Dpp). Mutants defective in different components of the JNK cascade exhibit similar defects in actin cytoskeleton assembly and *dpp* expression in LE cells (Ricos et al. 1999).

Components of the *Drosophila* JNK pathway include a JNK kinase (DJNKK) encoded by *hemipterous* (*hep*), a DJNK encoded by *basket* (*bsk*), and a phosphatase encoded by *puckered* (*puc*) which dephosphorylates and inactivates DJNK in a negative feedback loop. DJNK phosphorylates and activates the transcription factors *DJun* and *DFos*, and presumably inactivates the transcriptional repressor Aop (Noselli 1998).

Loss-of-function mutations in *DJun* and *kayak* (encoding DFos) or overexpression of activated Aop causes a defect in DC. *DJun* and *DFos* activate, and Aop inhibits, the expression of *dpp* and *puc* genes. The LE cell-specific expression of *dpp* does not occur in JNK pathway mutants. The Rho-type GTPase DCdc42 and DRac1 (hereafter referred to as Cdc42 and Rac) activate the JNK pathway and are required in LE cells for DC (Harden et al. 1995; Riesgo-Escovar et al. 1996; Glise and Noselli 1997; Strutt et al. 1997). Activated Cdc42 or Rac induces strong activation of *dpp* expression in LE cells, which is suppressed by *hep* mutations (Glise and Noselli 1997).

As in mammalian cells, the *Drosophila* Cdc42 and Rac may activate the JNK cascade via a PAK homologue, DPAK (Harden et al. 1996). DPAK binds specifically to activated Cdc42 and Rac, and accumulates in LE cells in a Rac-dependent manner (Harden et al. 1996). Misshapen (Msn), a PAK-related kinase, is another putative activator of the JNK cascade and might work in parallel to Cdc42 and Rac (Treisman et al. 1997; Su et al. 1998).

A recent study also suggests an involvement of Ras in DC, possibly acting as an upstream activator of Cdc42/Rac (Harden et al. 1999). Rho (DrhoA/Rho1) acts in a Cdc42/Rac-JNK cascade-independent parallel pathway required for DC (Harden et al. 1999; Lu and Settleman 1999).

The *Drosophila* homologue of protein kinase N (Pkn) is possibly a Rho and Rac effector in this pathway (Lu and Settleman 1999). Expression of dominant-negative Rho in LE cells also affects elongation of lateral cells possibly via a second, Dpp-independent instructive signal for lateral cell elongation. The mechanism by which the Cdc42/Rac-JNK pathway and the Rho pathway mediate actin-dependent LE cell elongation is unknown. Cdc42/Rac-JNK pathway-induced Dpp mediates DC by activating the TGF-β pathway in neighboring lateral cells.

Function of the TGF-β Pathway in Dorsal Closure

The TGF-β pathway is a second signalling pathway required for DC. The JNK kinase pathway controls expression of *dpp*. Secreted Dpp is the ligand for TGF-β receptors on lateral cells. The TGF-β receptor is encoded by *thick veins* (*tkv*) and *punt* (*put*), mutations which cause a DC defect (Affolter et al. 1994; Brummel et al. 1994; Nellen et al. 1994; Penton et al. 1994; Letsou et al. 1995; Ruberte et al. 1995). Ectopic expression of Dpp or expression of an activated Dpp receptor, Tkv, suppresses mutations in the JNK pathway, and mutations in *tkv* compromise elongation and migration of lateral ectodermal cells but not of LE cells (Hou et al. 1997; Riesgo-Escovar and Hafen 1997).

Based on genetic interactions and a DC phenotype, the *Drosophila* SMAD homologue Mothers-against-Dpp (Mad) and a nuclear zinc finger protein (Shn), encoded by *schnurri*, are likely downstream components of the TGF-β receptor (Arora et al. 1995; Grieder et al. 1995; Staehling-Hampton et al. 1995; Hudson et al. 1998). A recent study suggests a role also for Cdc42 as a component downstream of the TGF-β receptor (Ricos et al. 1999). Inhibition of Cdc42 results in defects typically observed for TGF-β pathway mutants.

In addition, expression of activated Cdc42, but not Rac, rescues the DC defect of a *tkv* mutant, and activated Tkv does not rescue the DC defect caused by expression of dominant negative Cdc42. Further genetic interactions should provide more detailed information about the signalling steps downstream of the TGF-β receptor. The mechanism by which the TGF-β pathway may impinge on the actin cytoskeleton to mediate lateral cell elongation is, again, unknown.

Upstream of Cdc42, Rac and Rho

Whereas Dpp clearly has a role as an extracellular stimulus for the TGF-β pathway in lateral cells, little is known about the signal(s) and pathway(s) that initiate and maintain activation of Cdc42, Rac and Rho in LE cells. A potential Rac-specific activator is encoded by *myoblast city* (*mbc*), a gene required for myoblast fusion, cytoskeletal organization, and DC (Rushton et al. 1995; Erickson et al. 1997; Nolan et al. 1998). Mbc associates specifically with Rac in a nucleotide-independent manner, and is a homologue of mammalian DOCK180 which has been implicated in the control of cell morphology and Rac1-dependent JNK activation (Hasegawa et al. 1996; Nolan et al. 1998). Since *mbc* mutations have no apparent effect on JNK activation in *Drosophila*, the functional role of Mbc in the control of cytoskeletal changes during DC remains elusive. Genetic screens may provide additional players in these signalling cascades.

Gastrulation

Both Rho and DRhoGEF2, a predicted Rho-specific GEF, are required for cell shape changes driving tissue invagination during gastrulation (Barrett et al. 1997; Hacker and Perrimon 1998). Two more signalling molecules with a function in this morphogenetic event are known, the secreted protein Folded gastrulation (Fog) and a Gα subunit of a heterotrimeric G protein, Concertina (Cta) (Parks and Wieschaus 1991; Costa et al. 1994). *cta* has genetically been placed downstream of *fog* (Morize et al. 1998).

In mammalian cells, the Gα subunit of G_{13} binds and activates the Lsc/p115RhoGEF (Hart et al. 1998; Kozasa et al. 1998). Like Lsc, DRhoGEF2 possesses a RGS (regulators of G protein signalling) domain for interaction with activated Gα. By analogy, DRhoGEF2 may be activated by Cta to control ventral furrow formation during gastrulation (Hall 1998a). The gene encoding the Fog receptor has not yet been cloned, but would, according to the model depicted in Fig. 2, belong to the family of heptahelical G protein-coupled receptors. Unlike *DRhoGEF2*, *fog* and *cta* are not essential for ventral furrow formation, suggesting that a second, Fog- and Cta-independent pathway can activate DRhoGEF2.

Mammalian Cells

As in lower eukaryotes and in *Drosophila*, Rho-type GTPases play a central role in mammalian cells in coordinating actin organization and transcriptional regulation in response to extracellular and intracellular stimuli. The mammalian Rho family currently comprises at least 14 members (Zohn et al. 1998; Aspenstrom 1999), among which Cdc42, Rac1 and RhoA are the best characterized.

More than 20 GEFs and more than 10 GAPs are known (Van Aelst and D'Souza-Schorey 1997; Zohn et al. 1998). Some of these regulators display *in vitro* activity towards several Rho-type GTPases, whereas others seem to be more specific (Van Aelst and D'Souza-Schorey 1997). In addition, there are cell type-specific differences in expression, and in most cases the contribution of a given GEF or GAP to the regulation of a specific Rho protein *in vivo* is not established. Pathways in mammalian cells leading from extracellular signals to the actin cytoskeleton are summarized in Fig. 3. In general, the mechanisms by which these pathways impinge on the actin cytoskeleton are poorly understood.

Control of Actin Organization by Rho-type GTPases

The role of Rho-type GTPases in the regulation of actin structures was initially explored in fibroblasts (Ridley and Hall 1992; Ridley et al. 1992). Activation of Cdc42 in response to bradykinin causes formation of filopodia, finger-like protrusions at the leading edge of the cell that consist of bundles of thin actin filaments attached to small focal complexes. Rac1 activation by growth factors (EGF, PDGF) or insulin causes formation of lamellipodia, protrusive sheets of cross-linked thin actin filaments at the edges of the cell, which are also in contact with small focal complexes.

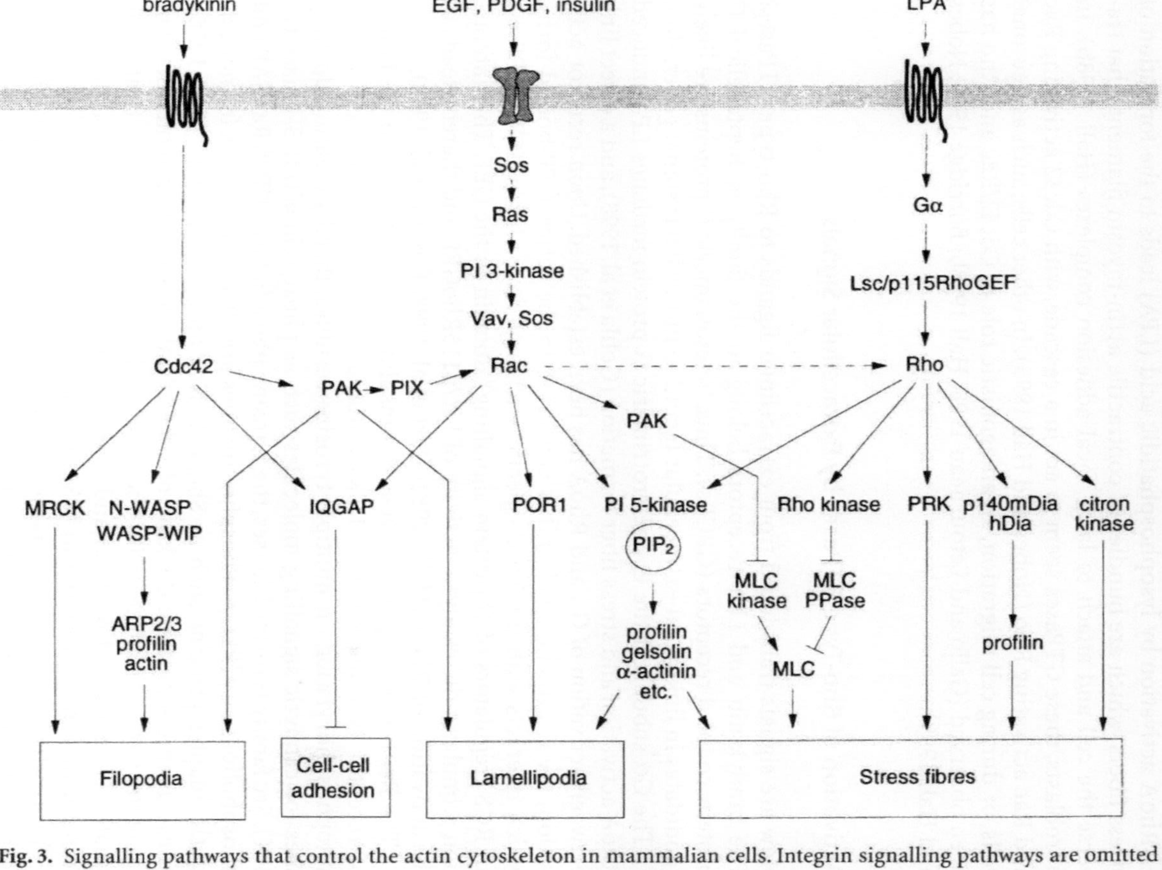

Fig. 3. Signalling pathways that control the actin cytoskeleton in mammalian cells. Integrin signalling pathways are omitted

RhoA activation by lysophosphatidic acid (LPA) leads to the formation of stress fibers, which are bundles of contractile actin-myosin filaments that traverse the cell and attach to large focal adhesion complexes (Hall 1998b). In fibroblasts, these GTPases seem to act in a cascade, with Cdc42 activating Rac and Rac activating Rho (Nobes and Hall 1995). In other cells, such as neuronal cells, or during cell migration, an antagonistic role of Cdc42/Rac and Rho has been observed (Gallo and Letourneau 1998; Hall 1998b; Burridge 1999; Nobes and Hall 1999).

Activation of Rho-Type GTPases by Extracellular Signals

How are signals transduced from extracellular ligands to Rho-type GTPases? The bradykinin and LPA receptors belong to the family of heptahelical G protein-coupled receptors (GPCRs). Thus, heterotrimeric G proteins are likely candidates in linking an extracellular ligand to intracellular signal generation.

The Gα subunit of the G_{13} heterotrimeric G protein mediates LPA-induced RhoA activation and stress fiber formation (Gohla et al. 1998), and a direct link between activation of G_{13} and RhoA has been established. Upon receptor activation, the G protein α subunit is converted to the active, GTP-bound form, which interacts with the RhoA-specific exchange factor Lsc/p115RhoGEF via an RGS (regulators of G protein signalling) domain in the GEF. This interaction stimulates the catalytic activity of Lsc/p115RhoGEF, and thereby leads to the activation of RhoA (Hall 1998a; Hart et al. 1998; Kozasa et al. 1998).

The Rac1-activating EGF, PDGF and insulin receptors belong to the family of receptor tyrosine kinases (RTKs). Stimulation of these receptors causes autophosphorylation on multiple tyrosine residues, thereby creating docking sites for catalytic signalling molecules such as phosphoinositide 3-kinase (PI 3-K), Src family tyrosine kinases, the tyrosine phosphatase SHP-2, RasGAP and phospholipase Cγ (PLCγ). Phosphotyrosines are also docking sites for noncatalytic adapter proteins such as Shc, Crk, Nck and Grb2 (Heldin et al. 1998). These adapters mediate docking of other catalytic signalling molecules. For example, Grb2 complexes with the RasGEF Sos, which in turn activates the small GTPase Ras. Cdc42, Rac1 and RhoA all seem to play a role in Ras-induced transformation (Van Aelst and D'Souza-Schorey 1997).

In T-cells, activation of Ras- and Rho-type GTPases is required for Cap formation (polarized clustering of receptors and signalling molecules at the T-cell/APC contact site), for cytokine expression and secretion, and for proliferation and maturation. These cellular responses also require an intact actin cytoskeleton (Penninger and Crabtree 1999). Contact between a T-cell and an antigen-presenting cell (APC) leads to activation of the T-cell receptor complex, recruitment of the Src tyrosine kinase $p56^{lck}$ via CD4 and CD8 co-receptors, and subsequent recruitment and activation of the tyrosine kinases ZAP-70 and SYK. These kinases phosphorylate immune cell-specific adapter molecules, such as LAT and SLP-76, leading to recruitment of PLCγ1, Grb2-Sos,

PI 3-K and the Rac1-specific GEF Vav1 (Rudd 1999). Sos activates Ras which subsequently activates Rac1 via PI 3-K and Vav1 (Rodriguez-Viciana et al. 1997; Han et al. 1997, 1998; Nimnual et al. 1998).

Two mechanisms have been proposed for Ras-mediated activation of Rac1. First, the product of the Ras-effector PI 3-K, phosphatidylinositol-3,4,5-trisphosphate (PIP_3), binds to the Vav1 PH domain which results in enhanced phosphorylation and activation of Vav1 by $p56^{lck}$ (Han et al. 1998). A second mechanism for Rac1 activation is provided by the Rac1-specific exchange activity of Sos itself. Besides the RasGEF domain, Sos also contains a DH domain with Rac1-specific GEF activity. A DH-adjacent PH domain inhibits exchange activity of Sos toward Rac1 unless Ras is activated. The Sos PH domain binds PIP_3, and the effect of Ras is mediated by PI 3-K, suggesting that PIP_3 binding abolishes the inhibitory effect of the Sos PH domain (Nimnual et al. 1998). In addition to activating Rac1, Vav1 directly associates with the actin-nucleating protein talin and with vinculin (Fischer et al. 1998). A regulatory significance of these interactions, however, remains to be established.

Effectors of Cdc42, Rac1 and RhoA Involved in Actin Organization

WASP and N-WASP

A Cdc42 effector specifically expressed in hematopoietic cells is the Wiskott-Aldrich syndrome protein (WASP). WASP may be a key component in Cdc42-dependent actin regulation, as lymphocytes from WAS patients have an abnormal cytoskeleton and distorted microvilli (Ramesh et al. 1999). WASP binds specifically to activated Cdc42, and WASP-induced actin polymerization is blocked by coexpression of dominant-negative Cdc42 (Symons et al. 1996).

The binding of WASP to Cdc42 is mediated by a CRIB motif. Cdc42 is essential for polarity establishment of T-cells toward APCs (Stowers et al. 1995), and WASP-deficient cells display a failure in cell polarization and polarized migration (Badolato et al. 1998; Snapper et al. 1998; Zicha et al. 1998). Proline-rich domains in WASP are bound by adapter proteins (Anton et al. 1998) and by several protein tyrosine kinases such as Abl and the Src kinase Fgr (Ramesh et al. 1999). In addition, the WASP-interacting protein (WIP) coprecipitates with profilin, and might bind directly to actin (Ramesh et al. 1997). WASP also associates with and possibly acts through the actin filament nucleating ARP2/3 complex (Machesky and Insall 1998) (See also Chapter 15).

A ubiquitously expressed isoform of WASP, N-WASP (Miki et al. 1996) and WASP-related proteins (Machesky and Insall 1998; Miki et al. 1998b) may replace WASP in other cell types. In agreement with this, overexpression of N-WASP causes filopodia formation (Miki et al. 1996). Further studies indicate a role for N-WASP as a Cdc42 effector in EGF- and bradykinin-induced filopodia formation (Miki et al. 1998a). N-WASP binds to the ARP2/3 complex, and thereby stimulates the activity of the ARP2/3 complex in nucleation of actin polymerization (Rohatgi et al. 1999). In addition, N-WASP directly binds to filamentous actin and to profilin, and both binding activities are required for filopodia formation (Miki and Takenawa 1998; Suetsugu et al. 1998).

Like WASP, N-WASP has a PH domain, and binding of phosphatidylinositol 4,5-bisphosphate (PIP_2) appears to be required for N-WASP function and localization (Miki et al. 1996; Rohatgi et al. 1999). The function of WASP family proteins may not be restricted to Cdc42 signalling, as a novel member has been implicated in Rac1-dependent membrane ruffling (Miki et al. 1998b).

PAK and PIX

p21-activated kinases (PAK1, 2 and 3) have been isolated as Cdc42/Rac1 effectors with a role in regulation of the actin cytoskeleton and JNK/SAPK activation (Manser et al. 1994; Sells and Chernoff 1997; Sells et al. 1997). Mammalian PAKs are homologous to yeast STE20 and to *Drosophila* DPAK. PAKs possess a CRIB motif, and their kinase activity is stimulated upon binding to activated Cdc42 or Rac1.

PAK1 phosphorylates and inhibits myosin light chain kinase (MLCK), a key regulator of myosin activity and actin-myosin filament contractility (Sanders et al. 1999). Expression of activated PAK causes changes in actin organization similar to those observed upon expression of activated Cdc42 and Rac1 (Manser et al. 1997; Sells et al. 1997). Expression of dominant-negative PAK-derived polypeptides inhibits several aspects of Cdc42 and Rac1 function in actin reorganization (Obermeier et al. 1998; Zhao et al. 1998).

PAK3-induced actin reorganization can be blocked by expression of dominant-negative Rac1 in PC12 cells, whereas PAK1 can affect actin structures in a Rac1-independent and kinase activity-independent manner in fibroblasts (Obermeier et al. 1998; Sells et al. 1999). Effector site mutants of Cdc42 and Rac1 that are unable to bind PAK can still induce filopodia and membrane ruffles, arguing against a requirement of PAK in these processes (Joneson et al. 1996; Lamarche et al. 1996; Westwick et al. 1997). In addition, GTPase binding may not be required for PAK localization (Zhao et al. 1998), and GTPase binding-deficient PAK mutants are still able to induce actin reorganization (Obermeier et al. 1998). This illustrates the possibility of cell type- and isoform-specific differences in PAK signalling, and of GTPase- and kinase activity-dependent and independent mechanisms. A possible clarification of the apparently conflicting data on PAK function is provided by a newly identified family of PAK-interacting exchange factors (PIX) (Manser et al. 1998).

PIX provides a GTPase-independent targeting mechanism for PAK, is able to induce membrane ruffling, and acts as a Rac1-specific GEF *in vivo* (Manser et al. 1998; Obermeier et al. 1998). PIX interacts with the noncatalytic N-terminus of PAK via a SH3 domain, and the PAK-PIX interaction, but not PAK kinase activity, is required for induction of lamellipodia (Obermeier et al. 1998; Sells et al. 1999). Thus PAK appears to be able to activate Rac1 through PIX via a noncatalytic mechanism.

MRCK

Myotonic dystrophy kinase-related Cdc42-binding kinase (MRCK) phosphorylates nonmuscle myosin light chain and possesses a PH domain and a CRIB motif (Leung et al. 1998). Overexpression of MRCK results in actin reorganization and enhanced Cdc42 effects, and kinase-dead MRCK blocks Cdc42-dependent formation of focal complexes and filopodia. This suggests a role for MRCK as a Cdc42 effector in cytoskeletal reorganization.

IQGAP

IQGAP binds and cross-links actin filaments, and activated Cdc42 potentiates the cross-linking activity of IQGAP *in vitro* (Bashour et al. 1997; Fukata et al. 1997). The binding of IQGAP to E-cadherin and β-catenin, important players in cell-cell adhesion, and inhibitory effects of IQGAP overexpression on cell-cell adhesion suggest a role for IQGAP in Cdc42- and Rac1-regulated, E-cadherin-mediated cell-cell adhesion (Kuroda et al. 1998). IQGAP possesses a calmodulin-binding (IQ) domain and several protein-protein interaction domains, including a putative RasGAP domain, an actin-binding domain, and a Cdc42/Rac1-interacting domain (Johnson 1999).

PI 5-Kinase

Phosphoinositide 5-kinase (PI 5-K) converts phosphatidylinositol 4-phosphate to phosphatidylinositol 4,5-bisphosphate (PIP_2). PIP_2 binds and alters the activity of many actin-binding proteins, such as profilin, gelsolin, α-actinin, CapZ, N-WASP and ezrin/radixin/moesin (ERM) proteins (Schmidt and Hall 1998; Toker 1998, Rohatgi et al. 1999). For example, PIP_2 binding triggers the dissociation of actin monomers from profilin, promotes uncapping of actin filaments, and enhances the actin cross-linking activity of α-actinin. Furthermore, PI 5-K provides the precursor for generation of other second messengers, such as PIP_3 (via PI 3-K), DAG and IP_3 (via PLC).

IP_3 induces the release of Ca^{2+} from intracellular stores into the cytoplasm. Changes in the intracellular Ca^{2+} concentration can greatly influence actin organization such as in the dynamics of neuronal growth cones (Goldberg and Grabham 1999). DAG activates conventional protein kinase C (PKC), and several PKC isoforms promote actin rearrangements, possibly by direct phosphorylation of cytoskeletal proteins (Keenan and Kelleher 1998). PI 5-K is reported to interact with Rac1 and RhoA, and appears to be regulated by these GTPases (Chong et al. 1994; Hartwig et al. 1995; Tolias et al. 1995; Ren et al. 1996).

Rho Kinases

The homologous serine/threonine protein kinases p160ROCK (also called ROKβ or ROCK-I) and ROKα (also called Rho kinase or ROCK-II) are direct downstream effectors of RhoA and can induce the formation of actin stress fibers (Leung et al. 1995; Ishizaki et al. 1996; Matsui et al. 1996). The induction of stress fibers and focal adhesions by Rho kinases requires kinase activity, but not the Rho-binding domain, and expression of kinase-dead variants inhibits stress fiber assembly or induces their disassembly (Leung et al. 1996; Amano et al. 1997).

Myosin light chain (MLC), the myosin-binding subunit (MBS) of MLC phosphatase, and ERM proteins are substrates of Rho kinases (Amano et al. 1996a; Kimura et al. 1996; Matsui et al. 1996, 1998; Fukata et al. 1998). Phosphorylation of MBS reduces MLC phosphatase activity, and phosphorylated MLC shows increased binding to actin filaments and triggers stress fiber formation. In agreement with a role of MLC phosphorylation in induction of stress fibers *in vivo*, inhibition of MLC kinase promotes loss of stress fibers and focal adhesions (Chrzanowska-Wodnicka and Burridge 1996).

ERM proteins anchor actin filaments to membranes. Phosphorylation of specific serine and threonine residues in ERM proteins in response to LPA treatment correlates with ERM translocation to a membrane-cytoskeleton pool (Mangeat et al. 1999). Thus, ERM proteins may be important players in the control of RhoA-induced focal adhesion and stress fiber formation.

PRKs

The kinase domains of protein kinase N (PKN or PRK1) and its homologue PRK2 are highly related to that of PKC. PKN was identified as a putative RhoA effector (Amano et al. 1996b; Watanabe et al. 1996). α-actinin was isolated as a PKN-interactor and proposed to link PKN to the actin cytoskeleton (Mukai et al. 1997). PRK2 was initially reported to associate specifically with RhoA in a GTP-dependent manner (Quilliam et al. 1996), but the specificity has been extended to Rac1 (Vincent and Settleman 1997).

The expression of a kinase-dead PRK2 causes disruption of actin stress fibers, suggesting that PRK2 might function as Rac1/RhoA effector to promote rearrangements of the actin cytoskeleton (Vincent and Settleman 1997). Rhotekin and Rhophilin are two additional putative RhoA effectors that share homology to PRKs in the Rho-binding domain (Reid et al. 1996; Watanabe et al. 1996). The cellular function of these proteins is unknown.

Formins

Like the yeast BNI1 and the *Drosophila* Diaphanous (Dia) proteins, the mouse and human Dia homologues p140mDia and hDia, respectively, belong to the FH domain-containing formin family. These proteins are effectors of Rho-type GTPases in cytoskeletal organization (Wasserman 1998). p140mDia was isolated as a RhoA effector, selectively binding to the GTP-bound form of RhoA (Watanabe et al. 1997). Like yeast BNI1, p140mDia binds to profilin, and p140mDia, profilin and RhoA colocalize to RhoA-dependent actin structures. Overexpression of p140mDia induces the formation of actin filaments. hDia is encoded by *DFNA1*, mutation of which causes deafness possibly due to a defect in actin organization in hair cells of the inner ear (Lynch et al. 1997).

. . . and more

Citron was identified as a protein that interacts with activated Rac1 and RhoA (Madaule et al. 1995). Subsequently, a splice variant of citron that includes a kinase domain was identified and termed Citron kinase (Madaule et al. 1998). Citron kinase is a RhoA effector that possibly controls contraction of actin-myosin filaments during cytokinesis. POR (partner of Rac) specifically binds to activated Rac1, and mutant versions of POR inhibit Rac1-induced membrane ruffling (Van Aelst et al. 1996).

Another Rac1-specific effector with a possible role in lamellipodia formation is Sra-1 (specifically Rac1-associated protein) (Kobayashi et al. 1998). Sra-1 colocalizes with activated Rac1 and actin filaments at membrane ruffles. Sra-1 also co-sediments with filamentous actin, and may thus interact directly with F-actin.

Crosstalk: Signalling Between Rho GTPases

Crosstalk between Rho signalling pathways has been observed repeatedly. This may occur at different levels in signalling cascades. An emerging theme is crosstalk between growth factor receptors and integrins. Integrin-signaled cell adhesion can induce activation of G protein-coupled receptors (GPCRs) and receptor tyrosine kinases (RTKs), and GPCRs can activate RTKs (Luttrell et al. 1999; Schwartz and Baron 1999). On the level of the GTPases, crosstalk between Rac1 and Cdc42 may occur through PAK and PIX. A model has been proposed in which PAK, as a Cdc42 effector, regulates Rac1 via the GEF activity of PIX toward Rac1 (Manser et al. 1998; Obermeier et al. 1998). Alternatively, PI 3-K might, as a Cdc42 effector (Zheng et al. 1994), activate Rac1 via PIP_3-stimulated Rac1GEFs.

A model for opposing effects of Rac1 and RhoA on stress fiber assembly and contractility of actin-myosin filaments has been proposed recently

(Burridge 1999). Whereas RhoA inhibits MLC dephosphorylation via Rho kinase-dependent inactivation of MLC phosphatase, Rac1 inhibits MLC phosphorylation via PAK1-induced inactivation of MLC kinase. As a result, RhoA stimulates stress fiber assembly and contractility, whereas Rac1 promotes stress fiber disassembly.

Cell Migration

Cell migration is an essential aspect of many cellular functions. For example, leukocytes migrate in response to inflammation, and epithelial cells migrate during development or wound healing. Mammalian cells move by crawling, and migration is mediated by a dynamic remodeling of both the actin cytoskeleton and cell-substratum contact sites. This remodeling is controlled by Rho-type GTPases (Lauffenburger and Horwitz 1996).

Cell adhesion is established by binding of extracellular matrix (ECM) proteins like fibronectin to specific cell surface receptors, the integrins (Howe et al. 1998; Keely et al. 1998). Integrins are associated with actin filaments, and thus couple the ECM to the actin cytoskeleton. Integrin-mediated cell adhesion is controlled by Rho-type GTPases on the level of organization of focal adhesion complexes. These complexes are composed of integrins, actin-myosin filaments and associated components such as α-actinin and paxillin, and signalling molecules such as focal adhesion kinase (FAK), Src and the adapter proteins p130^{Cas} and Grb2. The affinity of integrins for ECM ligands is modulated by Ras and at least RhoA (Howe et al. 1998). However, the signalling between small GTPases and integrins is not unidirectional, as integrin-mediated adhesion also induces activation of Cdc42, Rac1 and RhoA (reviewed in Schoenwaelder and Burridge 1999).

During cell migration, e.g., the chemotaxis of macrophages (Allen et al. 1998), or the coordinated movement of fibroblasts in wound healing (Nobes and Hall 1999), Cdc42, Rac1 and Rho1 appear to have distinct functions. Cdc42 is nonessential for migration, but is required for maintenance of cell polarity during fibroblast movement and macrophage chemotaxis. Rac1, controlling lamellipodial activity, is essential for movement in both fibroblasts and macrophages. RhoA, like Rac1, is required for migration of macrophages induced by the chemoattractant colony-stimulating factor 1 (CSF-1).

In fibroblasts, activated RhoA has a migration-inhibitory effect, possibly due to inhibition of a dynamic turnover of focal adhesions and stress fibers. Basal RhoA activity, however, is required for maintenance of cell adhesion in fibroblasts. Inhibition of Ras in fibroblasts has no effect on cell polarity (controlled by Cdc42) or lamellipodia formation (controlled by Rac1), but severely affects migration velocity. This effect is neutralized by inhibition of RhoA or of the RhoA effector p160ROCK. As mentioned above, Ras can affect integrin-ECM interactions by modulating integrin affinity, and a concentration of activated Ras was observed at focal adhesion sites in fibroblasts (Nobes and Hall 1999).

Thus Ras and RhoA possibly have opposing roles in focal adhesion assembly and turnover.

In addition to the small GTPases, the Cdc42/Rac1 effector PAK1 is required for cell motility (Sells et al. 1999). PAK1 kinase activity is apparently only required for the directionality of movement, as a kinase-dead PAK1 mutant is able to induce lamellipodia and cell motility, but fails to trigger a sustained polarized migration.

Subversion of Host Cell Actin Cytoskeleton by Extracellularly-Bound Bacterial Pathogens

Pathogenic bacterial cells adhere to and enter into mammalian cells. As the target cells are often non-phagocytic, and thus normally incapable of internalizing a surface-bound bacterial cell, externally-bound invasive bacterial cells induce their own internalization. This induced internalization of surface-bound bacterial cells is mediated by morphological changes in the host plasma membrane and the underlying actin cytoskeleton. Bound pathogens induce these changes by subverting signalling pathways controlling the actin cytoskeleton. Furthermore, the nature of the induced morphological changes can vary dramatically depending on the particular type of pathogen (Ireton and Cossart 1998).

Yersinia and *Listeria* enter epithelial cells via a 'zippering' mechanism in which there is minor remodeling of the host membrane. *Salmonella* and *Shigella* enter epithelial cells and macrophages by inducing large membrane folds or ruffles, often referred to as a trigger mechanism or macropinocytosis. *Legionella* cells are entrapped by phagocytes and engulfed by a large pseudopod coil. Enteropathogenic *Escherichia coli* (EPEC) induce formation of an actin-rich pedestal or platform, and cause a localized degeneration of microvilli, on intestinal epithelial cells. Unlike other pathogens, *Bartonella henselae* cells form aggregates, which are internalized as single cells that are slowly engulfed by endothelial cells via a unique host structure referred to as the invasome. This diversity of morphological changes induced by bacterial pathogens underscores the malleability of the actin cytokeleton and the complexity of the signalling pathways that control it.

Here we review the strategies by which some of the above pathogens subvert signalling pathways controlling the actin cytoskeleton to enter a target cell. We omit discussion of many pathogens because this is a relatively new area of investigation and, in many cases, very little is known about the molecular events behind the morphological changes. We also do not review mechanisms by which some bacterial pathogens, once inside a host cell, further exploit the actin cytoskeleton for motility.

For more information, the reader is referred to several recent reviews stressing different pathogens or different aspects of pathogen entry and motility (Isberg and Van Nhieu 1994; Galan and Bliska 1996; Virji 1996; Finlay and

Cossart 1997; Cossart and Lecuit 1998; Dramsi and Cossart 1998; Ireton and Cossart 1998; Brumell et al. 1999).

In a pioneering study, Leong et al. (1990) showed that integrins bind the *Yersinia* surface protein invasin prior to penetration of a *Yersinia* cell into a mammalian cell. This was the first identification of a mammalian cell receptor that promotes entry of bacteria into non-phagocytic cells. Integrins are a superfamily of heterodimeric transmembrane proteins that contain a small cytoplasmic tail linked to the actin cytoskeleton via focal adhesion complexes.

Focal adhesion complexes are a mixture of structural (e.g., vinculin, talin, α-actinin) and signalling proteins (e.g., Rho, the $p130^{cas}$ SH2-docking protein, and the focal adhesion kinase FAK) (Howe et al. 1998). The events subsequent to the invasin-integrin interaction that mediate *Yersinia* internalization are poorly understood. Invasin-mediated internalization is inhibited by drugs that depolymerize actin filaments, indicating that observed local rearrangements of the actin cytoskeleton are required for bacterial uptake (Isberg and Tran Van Nhieu 1995). However, mutagenic analysis of an integrin cytoplasmic domain critical for uptake suggests that the association of integrins with the actin cytoskeleton interferes with integrin-mediated bacterial uptake (Tran Van Nhieu et al. 1996).

Additional insight into the mechanism of *Yersinia* internalization has come from the findings that tyrosine phosphorylation is necessary for uptake, and that surface-bound *Yersinia* cells can inhibit their own internalization by injecting a tyrosine phosphatase, YopH, into a potential host cell (Persson et al. 1995). YopH disrupts focal adhesions by dephosphorylating $p130^{cas}$ and FAK, suggesting that integrin-mediated internalization normally involves signalling by $p130^{cas}$ and FAK to the actin cytoskeleton (Black and Bliska 1997; Persson et al. 1997).

The internalization of *Listeria* cells occurs upon binding of the bacterial surface protein InlA to the mammalian protein E-cadherin (Mengaud et al. 1996). E-cadherin, like integrins, is also an adhesion receptor with a short non-catalytic cytoplasmic tail. Entry of *Listeria* into a target cell requires the actin cytoskeleton, and E-cadherin interacts with the actin cytoskeleton via its tail and at least the two cytoplasmic proteins α- and β-catenin. However, it remains to be determined if the E-cadherin tail or the catenins are necessary for uptake of *Listeria*. For entry into some cell lines, the *Listeria* surface protein InlB is necessary instead of InlA. InlB-mediated entry requires tyrosine phosphorylation and PI 3-kinase (p85–p110) (Ireton et al. 1996). It is not known how PI 3-kinase might be signalling to the actin cytoskeleton to mediate entry, but this might involve Rac, a Rho family GTPase and a PI 3-kinase effector known to control organization of the actin cytoskeleton.

Salmonella cells invade a host cell by injecting a number of virulence factors via a specialized (type III) secretion system (Brumell et al. 1999). These factors promote internalization of a *Salmonella* cell by, among other effects, controlling Rho family GTPase directly and, thereby, the host actin cytoskeleton and ruffling. Chen et al. (1996) demonstrated that dominant negative Cdc42 prevents *Salmonella*-induced actin cytoskeleton reorganization and subsequent bacterial internalization, whereas constitutively active Cdc42 triggers internalization of a *Salmonella* mutant unable to induce its own internalization. SopE, one of the injected virulence factors, was then shown to be a GDP/GTP exchange factor for Cdc42 and Rac (Hardt et al. 1998). These findings fit well with the previously known role of Cdc42 and Rac in ruffle formation, and provide the clearest example of how a bacterial pathogen impinges on a signalling pathway controlling the actin cytoskeleton to induce internalization. *Salmonella*, via SopE, also activates Cdc42- and Rac-controlled MAP kinase cascades which trigger a nuclear response, and it is now important to dissect the role of this nuclear response in the uptake process.

Shigella, like *Salmonella*, induces long bundles of actin filaments that cause large bacteria-engulfing membrane folds (Adam et al. 1995). A *Shigella* cell bound to the host cell surface recruits an actin nucleation site that includes many known focal adhesion complex proteins such as vinculin, paxillin, talin, α-actinin, and FAK (Adam et al. 1995; Tran Van Nhieu et al. 1997). *Shigella* cells secrete, via a contact-activated type III secretion system, at least three proteins, IpaB, IpaC and IpaD (invasins), which are necessary for bacterial entry.

The IpaB and IpaC proteins form the IpaA complex that, like the *Yersinia* YopH protein, quickly gains access to the interior of the host cell. The IpaA complex appears to interact with an integrin receptor on the host cell surface and, once inside the target cell, with vinculin (Tran Van Nhieu et al. 1997; Watarai et al. 1997). Although an *ipaA* mutant is still able to induce foci of actin polymerization, it differs from wildtype *Shigella* in its ability to recruit vinculin and α-actinin, and the IpaA-vinculin interaction has thus been proposed to initiate the formation of the focal adhesion-like structures required for efficient invasion (Tran Van Nhieu et al. 1997). Rho GTPase is also required for invasion by a *Shigella* cell (Adam et al. 1996; Watarai et al. 1997). Three Rho isoforms are recruited to entry sites, and treatment with the Rho-specific inhibitor C3 blocks *Shigella*-induced membrane folding and reduces bacterial entry. The relationship between IpaA and the Rho isoforms in mediating uptake is intriguing, and remains to be determined. Watarai et al. (1997) suggest that Rho is involved in a signalling pathway triggered by the IpaA-integrin interaction.

The field of molecular bacterial pathogenicity is an emerging field that has clear medical relevance and applications. A 'side' benefit is that it has provided a new set of reagents, the pathogens themselves, to probe further the signalling pathways that link the cell surface to the actin cytoskeleton.

Acknowledgements. We thank Markus Affolter and Matthias Peter for helpful discussions. The authors were supported by grants from the Swiss National Science Foundation and the Canton of Basel to MNH.

References

Adam T, Arpin M, Prevost MC, Gounon P, Sansonetti PJ (1995) Cytoskeletal rearrangements and the functional role of T-plastin during entry of *Shigella flexneri* into HeLa cells. J Cell Biol 129:367–381

Adam T, Giry M, Boquet P, Sansonetti P (1996) Rho-dependent membrane folding causes *Shigella* entry into epithelial cells. EMBO J 15:3315–3321

Adams AE, Pringle JR (1984) Relationship of actin tubulin distribution to bud growth in wildtype morphogenetic-mutant *Saccharomyces cerevisiae.* J Cell Biol 98:934–945

Affolter M, Nellen D, Nussbaumer U, Basler K (1994) Multiple requirements for the receptor serine/threonine kinase thick veins reveal novel functions of TGF beta homologs during *Drosophila* embryogenesis. Development 120:3105–3117

Allen WE, Zicha D, Ridley AJ, Jones GE (1998) A role for Cdc42 in macrophage chemotaxis. J Cell Biol 141:1147–1157

Amano M, Ito M, Kimura K, Fukata Y, Chihara K, Nakano T, Matsuura Y, Kaibuchi K (1996a) Phosphorylation and activation of myosin by Rho-associated kinase (Rho-kinase). J Biol Chem 271:20246–20249

Amano M, Mukai H, Ono Y, Chihara K, Matsui T, Hamajima Y Okawa K Iwamatsu A Kaibuchi K (1996b) Identification of a putative target for Rho as the serine-threonine kinase protein kinase N. Science 271:648–650

Amano M, Chihara K, Kimura K, Fukata Y, Nakamura N, Matsuura Y, Kaibuchi K (1997) Formation of actin stress fibers and focal adhesions enhanced by Rho-kinase. Science 275:1308–1311

Amberg DC (1998) Three-dimensional imaging of the yeast actin cytoskeleton through the budding cell cycle. Mol Biol Cell 9:3259–3262

Anton IM, Lu W, Mayer BJ, Ramesh N, Geha RS (1998) The Wiskott-Aldrich syndrome protein-interacting protein (WIP) binds to the adaptor protein Nck. J Biol Chem 273:20992–20995

Arora K, Dai H, Kazuko SG, Jamal J, O'Connor MB, Letsou A, Warrior R (1995) The *Drosophila schnurri* gene acts in the Dpp/TGF beta signalling pathway and encodes a transcription factor homologous to the human MBP family. Cell 81:781–790

Aspenstrom P (1999) Effectors for the Rho GTPases. Curr Opin Cell Biol 11:95–102

Ayscough KR, Stryker J, Pokala N, Sanders M, Crews P, Drubin DG (1997) High rates of actin filament turnover in budding yeast and roles for actin in establishment and maintenance of cell polarity revealed using the actin inhibitor latrunculin-A. J Cell Biol 137:399–416

Badolato R, Sozzani S, Malacarne F, Bresciani S, Fiorini M, Borsatti A, Albertini A, Mantovani A, Ugazio AG, Notarangelo LD (1998) Monocytes from Wiskott-Aldrich patients display reduced chemotaxis and lack of cell polarization in response to monocyte chemoattractant protein-1 and formyl-methionyl-leucyl-phenylalanine. J Immunol 161:1026–1033

Banuett F (1998) Signalling in the yeasts: an informational cascade with links to the filamentous fungi. Microbiol Mol Biol Rev 62:249–274

Barrett K, Leptin M, Settleman J (1997) The Rho GTPase and a putative RhoGEF mediate a signalling pathway for the cell shape changes in *Drosophila* gastrulation. Cell 91:905–915

Bashour AM, Fullerton AT, Hart MJ, Bloom GS (1997) IQGAP1 a Rac- and Cdc42-binding protein directly binds and cross-links microfilaments. J Cell Biol 137:1555–1566

Bickle M, Delley PA, Schmidt A, Hall MN (1998) Cell wall integrity modulates RHO1 activity via the exchange factor ROM2. EMBO J 17:2235–2245

Black DS, Bliska JB (1997) Identification of p130Cas as a substrate of *Yersinia* YopH (Yop51) a bacterial protein tyrosine phosphatase that translocates into mammalian cells and targets focal adhesions. EMBO J 16:2730–2744

Brown JL, Jaquenoud M, Gulli MP, Chant J, Peter M (1997) Novel Cdc42-binding proteins Gic1 and Gic2 control cell polarity in yeast. Genes Dev 11:2972–2982

Brumell JH, Steele-Mortimer O, Finlay BB (1999) Force feeding by *Salmonella.* Curr Biol 9: R277–280

Brummel TJ, Twombly V, Marques G, Wrana JL, Newfeld SJ, Attisano L, Massague J, O'Connor MB Gelbart WM (1994) Characterization and relationship of Dpp receptors encoded by the saxophone and thick veins genes in *Drosophila*. Cell 78:251–261

Burbelo PD, Drechsel D, Hall A (1995) A conserved binding motif defines numerous candidate target proteins for both Cdc42 and Rac GTPases. J Biol Chem 270:29071–29074

Burridge K (1999) Crosstalk between Rac and Rho. Science 283:2028–2029

Burridge K, Chrzanowska-Wodnicka M (1996) Focal adhesions contractility and signalling. Annu Rev Cell Dev Biol 12:463–518

Butty AC, Pryciak PM, Huang LS, Herskowitz I, Peter M (1998) The role of Far1p in linking the heterotrimeric G protein to polarity establishment proteins during yeast mating. Science 282:1511–1516

Campos-Ortega JA, Hartenstein V (1985) The embryonic development of Drosophila melanogaster. Springer Berlin, Heidelberg, new York

Chant J (1996) Generation of cell polarity in yeast. Curr Opin Cell Biol 8:557–565

Chen GC, Kim YJ, Chan CS (1997) The Cdc42 GTPase-associated proteins Gicl and Gic2 are required for polarized cell growth in *Saccharomyces cerevisiae*. Genes Dev 11:2958–2971

Chen LM, Hobbie S, Galan JE (1996) Requirement of CDC42 for *Salmonella*-induced cytoskeletal and nuclear responses. Science 274:2115–2118

Chong LD, Traynor-Kaplan A, Bokoch GM, Schwartz MA (1994) The small GTP-binding protein Rho regulates a phosphatidylinositol 4-phosphate 5-kinase in mammalian cells. Cell 79:507–513

Chowdhury S, Smith KW, Gustin MC (1992) Osmotic stress and the yeast cytoskeleton: phenotype-specific suppression of an actin mutation. J Cell Biol 118:561–571

Chrzanowska-Wodnicka M, Burridge K (1996) Rho-stimulated contractility drives the formation of stress fibers and focal adhesions. J Cell Biol 133:1403–1415

Cossart P, Lecuit M (1998) Interactions of *Listeria* monocytogenes with mammalian cells during entry and actin-based movement: bacterial factors cellular ligands and signalling. EMBO J 17:3797–3806

Costa M, Wilson ET, Wieschaus E (1994) A putative cell signal encoded by the *folded gastrulation* gene coordinates cell shape changes during *Drosophila* gastrulation. Cell 76:1075–1089

Delley P-A, Hall MN (1999) Cell wall stress depolarizes cell growth via hyperactivation of RHO1. J Cell Biol 147:163–174

Dorer R, Boone C, Kimbrough T, Kim J, Hartwell LH (1997) Genetic analysis of default mating behaviour in *Saccharomyces cerevisiae*. Genetics 146:39–55

Dramsi S, Cossart P (1998) Intracellular pathogens and the actin cytoskeleton. Annu Rev Cell Dev Biol 14:137–166

Eby JJ, Holly SP, van Drogen F, Grishin AV, Peter M, Drubin DG, Blumer KJ (1998) Actin cytoskeleton organization regulated by the PAK family of protein kinases. Curr Biol 8:967–970

Erickson MR, Galletta BJ, Abmayr SM (1997) *Drosophila myoblast city* encodes a conserved protein that is essential for myoblast fusion dorsal closure and cytoskeletal organization. J Cell Biol 138:589–603

Evangelista M, Blundell K, Longtine MS, Chow CJ, Adames N, Pringle JR, Peter M, Boone C (1997) Bni1p a yeast formin linking Cdc42p and the actin cytoskeleton during polarized morphogenesis. Science 276:118–122

Finlay BB, Cossart P (1997) Exploitation of mammalian host cell functions by bacterial pathogens. Science 276:718–725

Fischer KD, Kong YY, Nishina H, Tedford K, Marengere LE, Kozieradzki I, Sasaki T, Starr M, Chan G, Gardener S, Nghiem MP, Bouchard D, Barbacid M, Bernstein A, Penninger JM (1998) Vav is a regulator of cytoskeletal reorganization mediated by the T-cell receptor. Curr Biol 8:554–562

Fukata M, Kuroda S, Fujii K, Nakamura T, Shoji I, Matsuura Y, Okawa K, Iwamatsu A, Kikuchi A, Kaibuchi K (1997) Regulation of cross-linking of actin filament by IQGAP1 a target for Cdc42. J Biol Chem 272:29579–29583

Fukata Y, Kimura K, Oshiro N, Saya H, Matsuura Y, Kaibuchi K (1998) Association of the myosin-binding subunit of myosin phosphatase and moesin: dual regulation of moesin phosphorylation by Rho-associated kinase and myosin phosphatase. J Cell Biol 141:409–418

Galan JE, Bliska JB (1996) Cross-talk between bacterial pathogens and their host cells. Annu Rev Cell Dev Biol 12:221–255

Gallo G, Letourneau PC (1998) Axon guidance: GTPases help axons reach their targets. Curr Biol 8: R80–82

Glise B, Noselli S (1997) Coupling of Jun amino-terminal kinase and decapentaplegic signalling pathways in *Drosophila* morphogenesis. Genes Dev 11:1738–1747

Gohla A, Harhammer R, Schultz G (1998) The G-protein G13 but not G12 mediates signalling from lysophosphatidic acid receptor via epidermal growth factor receptor to Rho. J Biol Chem 273:4653–4659

Goldberg DJ, Grabham PW (1999) Braking news: calcium in the growth cone. Neuron 22:423–425

Grieder NC, Nellen D, Burke R, Basler K, Affolter M (1995) *Schnurri* is required for *Drosophila* Dpp signalling and encodes a zinc finger protein similar to the mammalian transcription factor PRDII-BF1. Cell 81:791–800

Gustin MC, Albertyn J, Alexander M, Davenport K (1998) MAP kinase pathways in the yeast *Saccharomyces cerevisiae*. Microbiol Mol Biol Rev 62:1264–1300

Hacker U, Perrimon N (1998) DRhoGEF2 encodes a member of the Dbl family of oncogenes and controls cell shape changes during gastrulation in *Drosophila*. Genes Dev 12:274–284

Hall A (1998a) G proteins and small GTPases: distant relatives keep in touch. Science 280:2074–2075

Hall A (1998b) Rho GTPases and the actin cytoskeleton. Science 279:509–514

Han J, Das B, Wei W, Van Aelst L, Mosteller RD, Khosravi-Far R, Westwick JK, Der CJ, Broek D (1997) Lck regulates Vav activation of members of the Rho family of GTPases. Mol Cell Biol 17:1346–1353

Han J, Luby-Phelps K, Das B, Shu X, Xia Y, Mosteller RD, Krishna UM, Falck JR, White MA, Broek D (1998) Role of substrates and products of PI 3-kinase in regulating activation of Rac-related guanosine triphosphatases by Vav. Science 279:558–560

Harden N, Loh HY, Chia W, Lim L (1995) A dominant inhibitory version of the small GTP-binding protein Rac disrupts cytoskeletal structures and inhibits developmental cell shape changes in *Drosophila*. Development 121:903–914

Harden N, Lee J, Loh HY, Ong YM, Tan I, Leung T, Manser E, Lim L (1996) A *Drosophila* homolog of the Rac- and Cdc42-activated serine/threonine kinase PAK is a potential focal adhesion and focal complex protein that colocalizes with dynamic actin structures. Mol Cell Biol 16:1896–1908

Harden N, Ricos M, Ong YM, Chia W, Lim L (1999) Participation of small GTPases in dorsal closure of the *Drosophila* embryo: distinct roles for Rho subfamily proteins in epithelial morphogenesis. J Cell Sci 112:273–284

Hardt WD, Chen LM, Schuebel KE, Bustelo XR, Galan JE (1998) *S. typhimurium* encodes an activator of Rho GTPases that induces membrane ruffling and nuclear responses in host cells. Cell 93:815–826

Hart MJ, Jiang X, Kozasa T, Roscoe W, Singer WD, Gilman AG, Sternweis PC, Bollag G (1998) Direct stimulation of the guanine nucleotide exchange activity of p115 RhoGEF by Galpha13. Science 280:2112–2114

Hartwig JH, Bokoch GM, Carpenter CL, Janmey PA, Taylor LA, Toker A, Stossel TP (1995) Thrombin receptor ligation and activated Rac uncap actin filament barbed ends through phosphoinositide synthesis in permeabilized human platelets. Cell 82:643–653

Hasegawa H, Kiyokawa E, Tanaka S, Nagashima K, Gotoh N, Shibuya M, Kurata T, Matsuda M (1996) DOCK180 a major CRK-binding protein alters cell morphology upon translocation to the cell membrane. Mol Cell Biol 16:1770–1776

Heldin CH, Ostman A, Ronnstrand L (1998) Signal transduction via platelet-derived growth factor receptors. Biochim Biophys Acta 1378:F79–113

Hou XS, Goldstein ES, Perrimon N (1997) *Drosophila* Jun relays the Jun amino-terminal kinase signal transduction pathway to the Decapentaplegic signal transduction pathway in regulating epithelial cell sheet movement. Genes Dev 11:1728–1737

Howe A, Aplin AE, Alahari SK, Juliano RL (1998) Integrin signalling and cell growth control. Curr Opin Cell Biol 10:220–231

Hudson JB, Podos SD, Keith K, Simpson SL, Ferguson EL (1998) The *Drosophila* Medea gene is required downstream of dpp and encodes a functional homolog of human Smad 4. Development 125:1407–1420

Ireton K, Cossart P (1998) Interaction of invasive bacteria with host signalling pathways. Curr Opin Cell Biol 10:276–283

Ireton K, Payrastre B, Chap H, Ogawa W, Sakaue H, Kasuga M, Cossart P (1996) A role for phosphoinositide 3-kinase in bacterial invasion. Science 274:780–782

Isberg RR, Tran Van Nhieu G (1995) The mechanism of pathogenic uptake promoted by invasin-integrin interaction. Trends Cell Biol 5:120–124

Isberg RR, Van Nhieu GT (1994) Two mammalian cell internalization strategies used by pathogenic bacteria. Annu Rev Genet 28:395–422

Ishizaki T, Maekawa M, Fujisawa K, Okawa K, Iwamatsu A, Fujita A, Watanabe N, Saito Y, Kakizuka A, Morii N, Narumiya S (1996) The small GTP-binding protein Rho binds to and activates a 160 kDa Ser/Thr protein kinase homologous to myotonic dystrophy kinase. EMBO J 15:1885–1893

Jeoung DI, Oehlen LJ, Cross FR (1998) Cln3-associated kinase activity in *Saccharomyces cerevisiae* is regulated by the mating factor pathway. Mol Cell Biol 18:433–441

Johnson DI (1999) Cdc42: An essential Rho-type GTPase controlling eukaryotic cell polarity. Microbiol Mol Biol Rev 63:54–105

Joneson T, McDonough M, Bar-Sagi D, Van Aelst L (1996) RAC regulation of actin polymerization and proliferation by a pathway distinct from Jun kinase. Science 274:1374–1376

Karpova TS, McNally JG, Moltz SL, Cooper JA (1998) Assembly and function of the actin cytoskeleton of yeast: relationships between cables and patches. J Cell Biol 142:1501–1517

Keely P, Parise L, Juliano R (1998) Integrins and GTPases in tumour cell growth motility and invasion. Trends Cell Biol 8:101–106

Keenan C, Kelleher D (1998) Protein kinase C and the cytoskeleton. Cell Signal 10:225–232

Kimura K, Ito M, Amano M, Chihara K, Fukata Y, Nakafuku M, Yamamori B, Feng J, Nakano T, Okawa K, Iwamatsu A, Kaibuchi K (1996) Regulation of myosin phosphatase by Rho and Rho-associated kinase (Rho-kinase). Science 273:245–248

Kobayashi K, Kuroda S, Fukata M, Nakamura T, Nagase T, Nomura N, Matsuura Y, Yoshida-Kubomura N, Iwamatsu A, Kaibuchi K (1998) p140Sra-1 (specifically Rac1-associated protein) is a novel specific target for Rac1 small GTPase. J Biol Chem 273:291–295

Kohno H, Tanaka K, Mino A, Umikawa M, Imamura H, Fujiwara T, Fujita Y, Hotta K, Qadota H, Watanabe T, Ohya Y, Takai Y (1996) Bni1p implicated in cytoskeletal control is a putative target of Rho1p small GTP binding protein in *Saccharomyces cerevisiae.* EMBO J 15:6060–6068

Kozasa T, Jiang X, Hart MJ, Sternweis PM, Singer WD, Gilman AG, Bollag G, Sternweis PC (1998) p115 RhoGEF a GTPase activating protein for Galpha12 and Galpha13. Science 280:2109–2111

Kubler E, Mosch HU, Rupp S, Lisanti MP (1997) Gpa2p a G-protein alpha-subunit regulates growth and pseudohyphal development in *Saccharomyces cerevisiae* via a cAMP-dependent mechanism. J Biol Chem 272:20321–20323

Kuroda S, Fukata M, Nakagawa M, Fujii K, Nakamura T, Ookubo T, Izawa I, Nagase T, Nomura N, Tani H, Shoji I, Matsuura Y, Yonehara S, Kaibuchi K (1998) Role of IQGAP1 a target of the small GTPases Cdc42 and Rac1 in regulation of E-cadherin-mediated cell-cell adhesion. Science 281:832–835

Lamarche N, Tapon N, Stowers L, Burbelo PD, Aspenstrom P, Bridges T, Chant J, Hall A (1996) Rac and Cdc42 induce actin polymerization and G1 cell cycle progression independently of p65PAK and the JNK/SAPK MAP kinase cascade. Cell 87:519–529

Lang P, Gesbert F, Delespine-Carmagnat M, Stancou R, Pouchelet M, Bertoglio J (1996) Protein kinase A phosphorylation of RhoA mediates the morphological and functional effects of cyclic AMP in cytotoxic lymphocytes. EMBO J 15:510–519

Lauffenburger DA, Horwitz AF (1996) Cell migration: a physically integrated molecular process. Cell 84:359–369

Leberer E, Dignard D, Harcus D, Thomas DY, Whiteway M (1992) The protein kinase homologue Ste20p is required to link the yeast pheromone response G-protein beta gamma subunits to downstream signalling components. EMBO J 11:4815–4824

Leberer E, Thomas DY, Whiteway M (1997) Pheromone signalling and polarized morphogenesis in yeast. Curr Opin Genet Dev 7:59–66

Leeuw T, Fourest-Lieuvin A, Wu C, Chenevert J, Clark K, Whiteway M, Thomas DY, Leberer E (1995) Pheromone response in yeast: association of Bem1p with proteins of the MAP kinase cascade and actin. Science 270:1210–1213

Leeuw T, Wu C, Schrag JD, Whiteway M, Thomas DY, Leberer E (1998) Interaction of a G-protein beta-subunit with a conserved sequence in Ste20/PAK family protein kinases. Nature 391:191–195

Leong JM, Fournier RS, Isberg RR (1990) Identification of the integrin binding domain of the *Yersinia pseudotuberculosis* invasin protein. EMBO J 9:1979–1989

Letsou A, Arora K, Wrana JL, Simin K, Twombly V, Jamal J, Staehling-Hampton K, Hoffmann FM, Gelbart WM, Massague J, et al. (1995) *Drosophila* Dpp signalling is mediated by the *punt* gene product: a dual ligand-binding type II receptor of the TGF beta receptor family. Cell 80:899–908

Leung T, Chen XQ, Manser E, Lim L (1996) The p160 RhoA-binding kinase ROK alpha is a member of a kinase family is involved in the reorganization of the cytoskeleton. Mol Cell Biol 16:5313–5327

Leung T, Chen XQ, Tan I, Manser E, Lim L (1998) Myotonic dystrophy kinase-related Cdc42-binding kinase acts as a Cdc42 effector in promoting cytoskeletal reorganization. Mol Cell Biol 18:130–140

Leung T, Manser E, Tan L, Lim L (1995) A novel serine/threonine kinase binding the Ras-related RhoA GTPase which translocates the kinase to peripheral membranes. J Biol Chem 270:29051–29054

Li R, Zheng Y, Drubin DG (1995) Regulation of cortical actin cytoskeleton assembly during polarized cell growth in budding yeast. J Cell Biol 128:599–615

Lu Y, Settleman J (1999) The *Drosophila* pkn protein kinase is a Rho/Rac effector target required for dorsal closure during embryogenesis. Genes Dev 13:1168–1180

Luttrell LM, Daaka Y, Lefkowitz RJ (1999) Regulation of tyrosine kinase cascades by G-protein-coupled receptors. Curr Opin Cell Biol 11:177–183

Lynch ED, Lee MK, Morrow JE, Welsch PL, Leon PE, King MC (1997) Nonsyndromic deafness DFNA1 associated with mutation of a human homolog of the *Drosophila* gene diaphanous. Science 278:1315–1318

Machesky LM, Insall RH (1998) Scar1 and the related Wiskott-Aldrich syndrome protein WASP regulate the actin cytoskeleton through the Arp2/3 complex. Curr Biol 8:1347–1356

Madaule P, Furuyashiki T, Reid T, Ishizaki T, Watanabe G, Morii N, Narumiya S (1995) A novel partner for the GTP-bound forms of rho and rac. FEBS Lett 377:243–248

Madaule P, Eda M, Watanabe N, Fujisawa K, Matsuoka T, Bito H, Ishizaki T, Narumiya S (1998) Role of citron kinase as a target of the small GTPase Rho in cytokinesis. Nature 394:491–494

Madden K, Snyder M (1998) Cell polarity and morphogenesis in budding yeast. Annu Rev Microbiol 52:687–744

Madhani HD, Fink GR (1998) The control of filamentous differentiation and virulence in fungi. Trends Cell Biol 8:348–353

Mangeat P, Roy C, Martin M (1999) ERM proteins in cell adhesion and membrane dynamics. Trends Cell Biol 9:187–192

Manser E, Leung T, Salihuddin H, Zhao ZS, Lim L (1994) A brain serine/threonine protein kinase activated by Cdc42 and Rac1. Nature 367:40–46

Manser E, Huang HY, Loo TH, Chen XQ, Dong JM, Leung T, Lim L (1997) Expression of constitutively active alpha-PAK reveals effects of the kinase on actin and focal complexes. Mol Cell Biol 17:1129–1143

Manser E, Loo TH, Koh CG, Zhao ZS, Chen XQ, Tan L, Tan I, Leung T, Lim L (1998) PAK kinases are directly coupled to the PIX family of nucleotide exchange factors. Mol Cell 1:183–192

Matsui T, Amano M, Yamamoto T, Chihara K, Nakafuku M, Ito M, Nakano T, Okawa K, Iwamatsu A, Kaibuchi K (1996) Rho-associated kinase a novel serine/threonine kinase as a putative target for small GTP binding protein Rho. EMBO J 15:2208–2216

Matsui T, Maeda M, Doi Y, Yonemura S, Amano M, Kaibuchi K, Tsukita S (1998) Rho-kinase phosphorylates COOH-terminal threonines of ezrin/radixin/moesin (ERM) proteins and regulates their head-to-tail association. J Cell Biol 140:647–657

Mengaud J, Ohayon H, Gounon P, Mege RM, Cossart P (1996) E-cadherin is the receptor for internalin a surface protein required for entry of L. monocytogenes into epithelial cells. Cell 84:923–932

Miki H, Takenawa T (1998) Direct binding of the verprolin-homology domain in N-WASP to actin is essential for cytoskeletal reorganization. Biochem Biophys Res Commun 243:73–78

Miki H, Miura K, Takenawa T (1996) N-WASP a novel actin-depolymerizing protein regulates the cortical cytoskeletal rearrangement in a PIP_2-dependent manner downstream of tyrosine kinases. EMBO J 15:5326–5335

Miki H, Sasaki T, Takai Y, Takenawa T (1998a) Induction of filopodium formation by a WASP-related actin-depolymerizing protein N-WASP. Nature 391:93–96

Miki H, Suetsugu S, Takenawa T (1998b) WAVE a novel WASP-family protein involved in actin reorganization induced by Rac. EMBO J 17:6932–6941

Morize P, Christiansen AE, Costa M, Parks S, Wieschaus E (1998) Hyperactivation of the folded gastrulation pathway induces specific cell shape changes. Development 125:589–597

Mosch HU, Fink GR (1997) Dissection of filamentous growth by transposon mutagenesis in *Saccharomyces cerevisiae*. Genetics 145:671–684

Mosch HU, Roberts RL, Fink GR (1996) Ras2 signals via the Cdc42/Ste20/mitogen-activated protein kinase module to induce filamentous growth in *Saccharomyces cerevisiae*. Proc Natl Acad Sci USA 93:5352–5356

Mukai H, Toshimori M, Shibata H, Takanaga H, Kitagawa M, Miyahara M, Shimakawa M, Ono Y (1997) Interaction of PKN with alpha-actinin. J Biol Chem 272:4740–4746

Mulholland J, Preuss D, Moon A, Wong A, Drubin D, Botstein D (1994) Ultrastructure of the yeast actin cytoskeleton and its association with the plasma membrane. J Cell Biol 125:381–391

Nellen D, Affolter M, Basler K (1994) Receptor serine/threonine kinases implicated in the control of *Drosophila* body pattern by decapentaplegic. Cell 78:225–237

Nern A, Arkowitz RA (1998) A GTP-exchange factor required for cell orientation. Nature 391:195–198

Nern A, Arkowitz RA (1999) A Cdc24p-Far1p-gbetagamma protein complex required for yeast orientation during mating. J Cell Biol 144:1187–1202

Nimnual AS, Yatsula BA, Bar-Sagi D (1998) Coupling of Ras and Rac guanosine triphosphatases through the Ras exchanger Sos. Science 279:560–563

Nobes CD, Hall A (1995) Rho rac and cdc42 GTPases regulate the assembly of multimolecular focal complexes associated with actin stress fibers lamellipodia and filopodia. Cell 81:53–62

Nobes CD, Hall A (1999) Rho GTPases control polarity protrusion and adhesion during cell movement. J Cell Biol 144:1235–1244

Nolan KM, Barrett K, Lu Y, Hu KQ, Vincent S, Settleman J (1998) Myoblast city the *Drosophila* homolog of DOCK180/CED-5 is required in a Rac signalling pathway utilized for multiple developmental processes. Genes Dev 12:3337–3342

Noselli S (1998) JNK signalling and morphogenesis in *Drosophila*. Trends Genet 14:33–38

Obermeier A, Ahmed S, Manser E, Yen SC, Hall C, Lim L (1998) PAK promotes morphological changes by acting upstream of Rac. EMBO J 17:4328–4339

Oehlen LJ, McKinney JD, Cross FR (1996) Ste12 and Mcm1 regulate cell cycle-dependent transcription of FAR1. Mol Cell Biol 16:2830–2837

Pan X, Heitman J (1999) Cyclic AMP-dependent protein kinase regulates pseudohyphal differentiation in *Saccharomyces cerevisiae.* Mol Cell Biol 19:4874–4887

Parks S, Wieschaus E (1991) The *Drosophila* gastrulation gene *concertina* encodes a G alpha-like protein. Cell 64:447–458

Penninger JM, Crabtree GR (1999) The actin cytoskeleton and lymphocyte activation. Cell 96:9–12

Penton A, Chen Y, Staehling-Hampton K, Wrana JL, Attisano L, Szidonya J, Cassill JA, Massague J, Hoffmann FM (1994) Identification of two bone morphogenetic protein type I receptors in *Drosophila* and evidence that Brk25D is a decapentaplegic receptor. Cell 78:239–250

Persson C, Nordfelth R, Holmstrom A, Hakansson S, Rosqvist R, Wolf-Watz H (1995) Cell-surface-bound *Yersinia* translocate the protein tyrosine phosphatase YopH by a polarized mechanism into the target cell. Mol Microbiol 18:135–150

Persson C, Carballeira N, Wolf-Watz H, Fallman M (1997) The PTPase YopH inhibits uptake of *Yersinia* tyrosine phosphorylation of p130Cas and FAK and the associated accumulation of these proteins in peripheral focal adhesions. EMBO J 16:2307–2318

Peter M, Herskowitz I (1994) Direct inhibition of the yeast cyclin-dependent kinase Cdc28-Cln by Far1. Science 265:1228–1231

Peter M, Neiman AM, Park HO, van Lohuizen M, Herskowitz I (1996) Functional analysis of the interaction between the small GTP binding protein Cdc42 and the Ste20 protein kinase in yeast. EMBO J 15:7046–7059

Quilliam LA, Lambert QT, Mickelson-Young LA, Westwick JK, Sparks AB, Kay BK, Jenkins NA, Gilbert DJ, Copeland NG, Der CJ (1996) Isolation of a NCK-associated kinase PRK2 an SH3-binding protein and potential effector of Rho protein signalling. J Biol Chem 271:28772–28776

Ramesh N, Anton IM, Hartwig JH, Geha RS (1997) WIP a protein associated with Wiskott-Aldrich syndrome protein induces actin polymerization and redistribution in lymphoid cells. Proc Natl Acad Sci USA 94:14671–14676

Ramesh N, Anton IM, Martinez-Quiles N, Geha RS (1999) Waltzing with WASP. Trends Cell Biol 9:15–19

Reid T, Furuyashiki T, Ishizaki T, Watanabe G, Watanabe N, Fujisawa K, Morii N, Madaule P, Narumiya S (1996) Rhotekin a new putative target for Rho bearing homology to a serine/threonine kinase PKN and rhophilin in the rho-binding domain. J Biol Chem 271:13556–13560

Ren XD, Bokoch GM, Traynor-Kaplan A, Jenkins GH, Anderson RA, Schwartz MA (1996) Physical association of the small GTPase Rho with a 68-kDa phosphatidylinositol 4-phosphate 5-kinase in Swiss 3T3 cells. Mol Biol Cell 7:435–442

Ricos MG, Harden N, Sem KP, Lim L, Chia W (1999) Dcdc42 acts in TGF-□ signalling during *Drosophila* morphogenesis: distinct roles for the Dracl/JNK and Dcdc42/TGF-□ cascades in cytoskeletal regulation. J Cell Sci 112:1225–1235

Ridley AJ, Hall A (1992) The small GTP-binding protein rho regulates the assembly of focal adhesions and actin stress fibers in response to growth factors. Cell 70:389–399

Ridley AJ, Paterson HF, Johnston CL, Diekmann D, Hall A (1992) The small GTP-binding protein rac regulates growth factor-induced membrane ruffling. Cell 70:401–410

Riesgo-Escovar JR, Hafen E (1997) *Drosophila* Jun kinase regulates expression of decapentaplegic via the ETS-domain protein Aop and the AP-1 transcription factor DJun during dorsal closure. Genes Dev 11:1717–1727

Riesgo-Escovar JR, Jenni M, Fritz A, Hafen E (1996) The *Drosophila* Jun-N-terminal kinase is required for cell morphogenesis but not for DJun-dependent cell fate specification in the eye. Genes Dev 10:2759–2768

Roberts RL, Mosch HU, Fink GR (1997) 14-3-3 proteins are essential for RAS/MAPK cascade signalling during pseudohyphal development in *S. cerevisiae.* Cell 89:1055–1065

Robinson DN, Cooley L (1997) Genetic analysis of the actin cytoskeleton in the *Drosophila* ovary. Annu Rev Cell Dev Biol 13:147–170

Rodriguez-Viciana P, Warne PH, Khwaja A, Marte BM, Pappin D, Das P, Waterfield MD, Ridley A, Downward J (1997) Role of phosphoinositide 3-OH kinase in cell transformation and control of the actin cytoskeleton by Ras. Cell 89:457–467

Rohatgi R, Ma L, Miki H, Lopez M, Kirchhausen T, Takenawa T, Kirschner MW (1999) The interaction between N-WASP and the Arp2/3 complex links Cdc42- dependent signals to actin assembly. Cell 97:221–231

Ruberte E, Marty T, Nellen D, Affolter M, Basler K (1995) An absolute requirement for both the type II type I receptors punt and thick veins for dpp signalling in vivo. Cell 80:889–897

Rudd CE (1999) Adaptors and molecular scaffolds in immune cell signalling. Cell 96:5–8

Rupp S, Summers E, Lo HJ, Madhani H, Fink GR (1999) MAP kinase and cAMP filamentation signalling pathways converge on the unusually large promoter of the yeast FLO11 gene. EMBO J 18:1257–1269

Rushton E, Drysdale R, Abmayr SM, Michelson AM, Bate M (1995) Mutations in a novel gene *myoblast city* provide evidence in support of the founder cell hypothesis for *Drosophila* muscle development. Development 121:1979–1988

Sanders LC, Matsumura F, Bokoch GM, de Lanerolle P (1999) Inhibition of myosin light chain kinase by p21-activated kinase. Science 283:2083–2085

Schmidt A, Hall MN (1998) Signalling to the actin cytoskeleton. Annu Rev Cell Dev Biol 14:305–38

Schoenwaelder SM, Burridge K (1999) Bidirectional signalling between the cytoskeleton and integrins. Curr Opin Cell Biol 11:274–286

Schwartz MA, Baron V (1999) Interactions between mitogenic stimuli or a thousand and one connections. Curr Opin Cell Biol 11:197–202

Sells MA, Chernoff J (1997) Emerging from the Pak: the p21-activated protein kinase family. Trends Cell Biol 7:162–167

Sells MA, Knaus UG, Bagrodia S, Ambrose DM, Bokoch GM, Chernoff J (1997) Human p21-activated kinase (Pak1) regulates actin organization in mammalian cells. Curr Biol 7:202–210

Sells MA, Boyd JT, Chernoff J (1999) p21-Activated Kinase 1 (Pak1) Regulates cell motility in mammalian fibroblasts. J Cell Biol 145:837–849

Snapper SB, Rosen FS, Mizoguchi E, Cohen P, Khan W, Liu CH, Hagemann TL, Kwan SP, Ferrini R, Davidson L, Bhan AK, Alt FW (1998) Wiskott-Aldrich syndrome protein-deficient mice reveal a role for WASP in T but not B cell activation. Immunity 9:81–91

Staehling-Hampton K, Laughon AS, Hoffmann FM (1995) A *Drosophila* protein related to the human zinc finger transcription factor PRDII/MBPI/HIV-EP1 is required for dpp signalling. Development 121:3393–3403

Stowers L, Yelon D, Berg LJ, Chant J (1995) Regulation of the polarization of T cells toward antigen-presenting cells by Ras-related GTPase CDC42. Proc Natl Acad Sci USA 92:5027–5031

Strutt DI, Weber U, Mlodzik M (1997) The role of RhoA in tissue polarity and Frizzled signalling. Nature 387:292–295

Su YC, Treisman JE, Skolnik EY (1998) The *Drosophila* Ste20-related kinase misshapen is required for embryonic dorsal closure and acts through a JNK MAPK module on an evolutionarily conserved signalling pathway. Genes Dev 12:2371–2380

Suetsugu S, Miki H, Takenawa T (1998) The essential role of profilin in the assembly of actin for microspike formation. EMBO J 17:6516–6526

Symons M, Derry JM, Karlak B, Jiang S, Lemahieu V, McCormick F, Francke U, Abo A (1996) Wiskott-Aldrich syndrome protein a novel effector for the GTPase CDC42Hs is implicated in actin polymerization. Cell 84:723–734

Toker A (1998) The synthesis and cellular roles of phosphatidylinositol 4,5-bisphosphate. Curr Opin Cell Biol 10:254–261

Tolias KF, Cantley LC, Carpenter CL (1995) Rho family GTPases bind to phosphoinositide kinases. J Biol Chem 270:17656–17659

Tran Van Nhieu G, Krukonis ES, Reszka AA, Horwitz AF, Isberg RR (1996) Mutations in the cytoplasmic domain of the integrin beta1 chain indicate a role for endocytosis factors in bacterial internalization. J Biol Chem 271:7665–7672

Tran Van Nhieu G, Ben-Ze'ev A, Sansonetti PJ (1997) Modulation of bacterial entry into epithelial cells by association between vinculin and the *Shigella* IpaA invasin. EMBO J 16:2717–2729

Treisman JE, Ito N, Rubin GM (1997) *misshapen* encodes a protein kinase involved in cell shape control in *Drosophila*. Gene 186:119–125

Valtz N, Peter M, Herskowitz I (1995) FAR1 is required for oriented polarization of yeast cells in response to mating pheromones. J Cell Biol 131:863–873

Van Aelst L, D'Souza-Schorey C (1997) Rho GTPases and signalling networks. Genes Dev 11:2295–2322

Van Aelst L, Joneson T, Bar-Sagi D (1996) Identification of a novel Rac1-interacting protein involved in membrane ruffling. EMBO J 15:3778–3786

Vincent S, Settleman J (1997) The PRK2 kinase is a potential effector target of both Rho and Rac GTPases and regulates actin cytoskeletal organization. Mol Cell Biol 17:2247–2256

Virji M (1996) Microbial utilization of human signalling molecules. Microbiology 142:3319–3336

Waddle JA, Karpova TS, Waterston RH, Cooper JA (1996) Movement of cortical actin patches in yeast. J Cell Biol 132:861–870

Wasserman S (1998) FH proteins as cytoskeletal organizers. Trends Cell Biol 8:111–115

Watanabe G, Saito Y, Madaule P, Ishizaki T, Fujisawa K, Morii N, Mukai H, Ono Y, Kakizuka A, Narumiya S (1996) Protein kinase N (PKN) and PKN-related protein rhophilin as targets of small GTPase Rho. Science 271:645–648

Watanabe N, Madaule P, Reid T, Ishizaki T, Watanabe G, Kakizuka A, Saito Y, Nakao K, Jockusch BM, Narumiya S (1997) p140mDia a mammalian homolog of *Drosophila diaphanous* is a target protein for Rho small GTPase and is a ligand for profilin. EMBO J 16:3044–3056

Watarai M, Kamata Y, Kozaki S, Sasakawa C (1997) *rho* a small GTP-binding protein is essential for *Shigella* invasion of epithelial cells. J Exp Med 185:281–292

Westwick JK, Lambert QT, Clark GJ, Symons M, Van Aelst L, Pestell RG, Der CJ (1997) R,ac regulation of transformation gene expression and actin organization by multiple PAK-independent pathways. Mol Cell Biol 17:1324–1335

Whiteway MS, Wu C, Leeuw T, Clark K, Fourest-Lieuvin A, Thomas DY, Leberer E (1995) Association of the yeast pheromone response G protein beta gamma subunits with the MAP kinase scaffold Ste5p. Science 269:1572–1575

Wu C, Lytvyn V, Thomas DY, Leberer E (1997) The phosphorylation site for Ste20p-like protein kinases is essential for the function of myosin-I in yeast. J Biol Chem 272:30623–30626

Xue Y, Batlle M, Hirsch JP (1998) *GPR1* encodes a putative G protein-coupled receptor that associates with the Gpa2p Galpha subunit and functions in a Ras-independent pathway. EMBO J 17:1996–2007

Yorihuzi T, Ohsumi Y (1994) *Saccharomyces cerevisiae* MATa mutant cells defective in pointed projection formation in response to alpha-factor at high concentrations. Yeast 10:579–594

Young PE, Richman AM, Ketchum AS, Kiehart DP (1993) Morphogenesis in *Drosophila* requires nonmuscle myosin heavy chain function. Genes Dev 7:29–41

Zhao ZS, Manser E, Chen XQ, Chong C, Leung T, Lim L (1998) A conserved negative regulatory region in alphaPAK: inhibition of PAK kinases reveals their morphological roles downstream of Cdc42 and Rac1. Mol Cell Biol 18:2153–2163

Zheng Y, Bagrodia S, Cerione RA (1994) Activation of phosphoinositide 3-kinase activity by Cdc42Hs binding to p85. J Biol Chem 269:18727–18730

Zicha D, Allen WE, Brickell PM, Kinnon C, Dunn GA, Jones GE, Thrasher AJ (1998) Chemotaxis of macrophages is abolished in the Wiskott-Aldrich syndrome. Br J Haematol 101:659–665

Zohn IM, Campbell SL, Khosravi-Far R, Rossman KL, Der CJ (1998) Rho family proteins and Ras transformation: the RHOad less travelled gets congested. Oncogene 17:1415–1438

Subject Index

MIX
Papier aus verantwortungsvollen Quellen
Paper from responsible sources
FSC® C105338

If you have any concerns about our products,
you can contact us on
ProductSafety@springernature.com

In case Publisher is established outside the EU,
the EU authorized representative is:
Springer Nature Customer Service Center GmbH
Europaplatz 3, 69115 Heidelberg, Germany

Printed by Libri Plureos GmbH
in Hamburg, Germany